PERSONAL PROTECTIVE EQUIPMENT FOR CHEMICAL, BIOLOGICAL, AND RADIOLOGICAL HAZARDS

PERSONAL PROTECTIVE EQUIPMENT FOR CHEMICAL, BIOLOGICAL, AND RADIOLOGICAL HAZARDS

Design, Evaluation, and Selection

Eva F. Gudgin Dickson, Ph.D.
Kingston, Ontario, Canada

WILEY
A JOHN WILEY & SONS, INC., PUBLICATION

Published by John Wiley & Sons, Inc., Hoboken, New Jersey.
Published simultaneously in Canada.

Library of Congress Cataloging-in-Publication Data:

Gudgin Dickson, Eva F. (Eva Frances), 1957–
Personal protective equipment for chemical, biological, and radiological hazards : design, evaluation, and selection / Eva F. Gudgin Dickson.
p. cm.
Includes bibliographical references and index.
ISBN 978-0-470-16558-4 (cloth)
1. Protective clothing. 2. Hazardous substances–Safety measures. I. Title.
T55.3.P75G82 2012
687′.162–dc23

2012015264

10 9 8 7 6 5 4 3 2 1

MIX
Paper from responsible sources
FSC® C013604
FSC www.fsc.org

CONTENTS

ACKNOWLEDGMENTS

There has been 20 years of explosive growth in the field of CBRN defense since I began a small project on military personal protective equipment in 1991 at the Royal Military College of Canada (RMCC) in Kingston, Ontario. The working environment in Canada and at the college has made it possible for me to be involved in all aspects of CBRN defense, as a result of the combination of teaching and defense environments. I've had the privilege to learn from, work with, and instruct people from around the world and from a huge variety of backgrounds, involved with everything from doctrine and requirements, standards, procurement and life-cycle management, design and evaluation of materials, personal protective equipment, and personal decontamination systems, as well as, of course, the users of the equipment in the field.

Special thanks go to the members of the RMCC CBRN Protection Group and co-authors of various works, including Roy Pottier, Paul Bodurtha, Lisa Prior, Scott Duncan, Michael Evans, Tiny van Houwelingen, Brian Harrison, Tom Beardall, Julie Tremblay-Lutter, Bill Andrews, Bev Kelly, Tes Mengistu and Paul Fedele, and to the numerous organizations that funded the work, including the Department of National Defence, the Centre for Security Science CBRN Research and Technology Initiative, and Defence Research and Development Canada, along with various industrial and international partners. I would also like to thank fellow participants in various standards-setting processes: for ISO, CSA/CGSB, NATO, ASTM, and NFPA. Barret Ruttan was an always willing model for photography.

Finally, special thanks go to RMCC and DRDC for permission to write this book and to my supportive and loving family: Tim, Terry, and Julia, for their patience, as this endeavor took hundreds of valuable hours from our time together.

1 Introduction to CBRN Protection

In this chapter we familiarize the reader with the general concepts that are most important to CBRN protection and personal protective equipment, acting as an introduction to later chapters, where we deal with these topics in more depth.

1.1 WHAT IS CBRN PPE AND WHY IS IT USED?

Personal protective equipment (PPE) is equipment worn to protect the wearer from some external hazard: in this case, chemical, biological, radiological, or nuclear hazards, all of which can be considered to be toxic. The term *CBRN*, an acronym for "chemical, biological, radiological, and nuclear," is used here to describe the particular combination of the hazard environment and the intent of use. The book is focused primarily on protection against deliberate use of CBRN agents in a terrorism or combat environment. The same PPE may be useful in a workplace setting in which CBRN agents are handled; however, as we discuss later, this results in some potentially important distinctions in the concept of use of the equipment.

CBRN PPE almost always has protective or operational requirements in addition to its CBRN protective functions. In most cases, however, the CBRN protection is deemed a primary requirement, with the other requirements superimposed once CBRN protection is provided. CBRN protective equipment may be designed to be worn by:

- Those responding to the use of CBRN agents (e.g., first or later responders)
- Those who are expected to perform their normal functions despite the fact that CBRN agents have been used (e.g., the military)
- Those who are being provided with emergency protection for escape purposes (e.g., civilians located in the vicinity)

In addition, CBRN protective equipment may be worn by those who are performing activities such as remediation, demilitarization, or laboratory investigation, where the environment is more controlled but the possibility of exposure to CBRN agents still exists. Protection against toxic materials has often been treated, conceptually, as an "all or nothing" idea—a person is either protected totally or is not protected at all.

Personal Protective Equipment for Chemical, Biological, and Radiological Hazards: Design, Evaluation, and Selection, First Edition. Eva F. Gudgin Dickson.

As we shall see, this approach is both overly simplistic and counterproductive. The degree of protection required is dependent on many factors, and protection need not be "total" to be effective; however, the protection requirements and expected performance must be well understood, and limitations and use of the equipment must be well defined.

A number of issues need to be considered to understand protection requirements. The first is the nature of the hazard for which protection must be provided.

1.2 WHAT ARE CBRN AGENTS?

CBRN agents consist of any chemical, biological, or radiological/nuclear substance that can be deliberately employed to cause harm to unprotected persons [1,2]. Chemicals may cause damage as a result of specific chemical reactions that happen when the body is exposed to them, disrupting bodily functions. Biological agents are living microorganisms that cause disease. Radiological agents (which may either result from a nuclear explosion or themselves be used) will damage living systems as a result of high-energy radiation interactions. CBRN agents may range from military agents, which have been designed or chosen to be particularly effective when used in a deliberate attack, to toxic industrial chemicals, which may be available more readily or in larger quantity.

There are a number of additional distinctions between C, B, and R/N agents: in terms of how they act on the body, their relative toxicity (Figure 1-1), and how they may be delivered, which is discussed in Chapter 2; nevertheless, it is apparent that they can all be described in general terms as materials that may be hazardous when the body is exposed to them, and there are a number of generic ways in which these hazards can be described, regardless of the class of agent. The most important aspect of these materials in the context of CBRN protection is the idea of deliberate use. Deliberate use implies the features outlined in Table 1-1 compared with those of an accidental release.

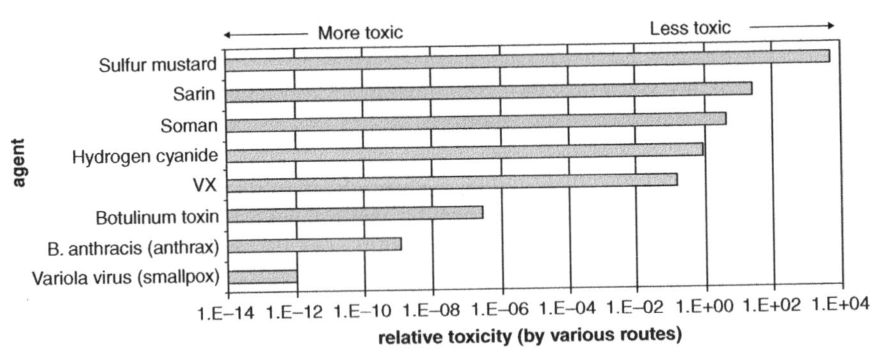

FIGURE 1-1 Approximate relative toxicity (related to mass of agent required to cause effect) of a variety of agents by various routes of entry.

TABLE 1-1 Differences Between an Accidental or Workplace Exposure to, and Deliberate Use of, a Toxic Material

	Accidental or Workplace Exposure	Deliberate Use Against Civilians	Deliberate Use Against Military
Intent	Unintentional; not a criminal event.	Criminal; needs of law enforcement may require alteration of normal response procedures.	Anticipated use will dictate complete change in operational tempo and procedures.
Location and severity of event	Not targeted, and large release events not likely to occur in a highly populated area. Most likely a small area is covered. Hazmat events: often, outdoor release with a relatively small area of effect, as dispersal is natural or passive (e.g., a leak from a tanker after a collision). Infections occur in a normal manner; epidemic events possible. Limited number of civilians involved; large-scale public panic unlikely.	Targeted location and timing, toxic material may be weaponized for efficient delivery, probably used in highly populated areas to cause maximum effect based on targeted population. Delivery designed to cause maximum disruption (e.g., covering a large area with airborne material or targeting many people). More likely to be either indoor release or covering a very large outdoor area, involving active dispersal mechanisms. Larger number of persons and casualties involved; public psychological trauma and panic likely.	Targeted location and timing; likely to cause maximum effect based on selection of optimum agent delivery conditions by a trained opponent. Delivery designed to deny territory, affect morale, change tempo of operations by requiring changes in procedures, burden medical care. Confined to theater of operations, and risk of use is usually identified prior to the event. Limited number of persons involved; training, PPE, and preparedness should limit effects.

(*continued*)

TABLE 1-1 ***(Continued)***

	Accidental or Workplace Exposure	Deliberate Use Against Civilians	Deliberate Use Against Military
Nature of toxic material	Could be low-level routine workplace exposure, chemical accident, or normal biological infection; few military agents likely. Toxicity variable, usually low to moderate; amount of toxic material small except in catastrophic events.	Radiological or nuclear incidents; chemical agents as well as toxic industrial chemicals; atypical biological infections. Toxicity likely to be high; amount may be considerably more than normally encountered.	A well-armed opponent may use conventional or engineered CBRN agents; others could use more easily available industrial or commercial materials. Toxicity is likely to be high; the amount may be considerably more than normally encountered.
Measures in place to reduce risk	For releases involving industrial or commercial materials, specific emergency plan in place; containment contingencies anticipated, specific training implemented. For normal biological infections occurring on a small scale; health care system has predeveloped management strategies.	Planning must be generic; containment strategies ad hoc and broader in scope. Low likelihood of event means that training is difficult to maintain. Resources are rarely in place to deal with catastrophic or pandemic scale of events.	Planning should be effective when based on prior intelligence. Opportunity for appropriate routine user training.
Duration of event	Duration of spill or release generally short; consequence management may require hours to weeks; epidemic biological events may last months to years. May be a single or repeated workplace exposure.	Similar to accidental events, but consequence management may last for months to years. A single (lifetime) exposure.	Military approach may permit shortening of required protection time scale by sacrifice of assets or avoidance of contaminated areas. Multiple exposures possible but not expected.

Ultimately, the worst-case deliberate event is as bad as any accidental event that can be conceived. This does not mean that PPE designed for a deliberate event will then necessarily provide appropriate protection for an accidental event; many factors must be considered, and potentially traded off, to permit the optimum response to the spectrum of events that could occur.

1.3 CONTEXT OF USE AS IT RELATES TO DESIGN, SELECTION, AND PERFORMANCE

To design, select, and use the most appropriate PPE for a job, the context of use must be understood. For each potential toxic substance, user, or exposure scenario, the following questions are important:

- What might the toxic substance be?
  - How toxic is it?
  - Where and how does it enter the body?
- Who may be exposed to the substance?
  - What level of effects resulting from exposure is acceptable for this population?
  - What operations and activities will be performed by them during exposure?
- What might the conditions of exposure be?
  - How long?
  - How often?
  - How large is the potential exposure dose?
  - What is the range of possible environmental conditions?

The three main questions above can be answered once the context of use of the protective equipment is analyzed and understood. The answers to all of these questions together determine the level of protection that is required. Additional questions may affect other important design and selection considerations.

- What other external hazards may exist?
  - Does the wearer, or the equipment, need to be protected against these hazards?
- Under what conditions might the equipment be stored or worn both before and during use?
  - What type of shelf life may be desirable?
  - What type of use life may be desirable?
  - What are its requirements for durability and survivability?
- What other activities must the wearer be able to perform?
- What other requirements may affect use of the equipment?
  - How does it need to integrate with other equipment?

The answers to the questions above may be very different depending on the user; the military, for example, may require that PPE be wearable for several weeks while continuing to protect after multiple exposures or launderings, whereas a first responder may expect to wear equipment once for only an hour or two in a hazardous environment. The military or police may potentially accept a higher level of risk to the wearer to reduce risk from equally potentially lethal hazards compared with an emergency medical worker who may be exposed to more limited or different hazards. These very different contexts of use can have a significant impact on the appropriate design of equipment.

Examples of standards that follow the process as we outline it here are two CBRN PPE standards: Canadian standards for civilian responders [3] and the NATO clothing standards for military users [4], and much of the information given here is consistent with those documents.

1.4 ACQUIRING EQUIPMENT

To actually begin the acquisition of PPE, there is a significant onus on the user to perform a number of activities. Outlined in this section in brief, and throughout the book in more detail, is an approach to acquiring CBRN PPE that significantly increases the likelihood that the equipment that is procured will suit the user's requirements.

1.4.1 How Not to Do It

This is a true story—repeated hundreds, if not thousands, of times over the past decade.

You work for an organization that has been in existence for some time, or even a newly minted user group, and you've just been told that your group must be able to support CBRN operations. You've been given a budget and a requirement to develop an operational capability as quickly as possible to satisfy your superiors, governments, and the public that the issue is being addressed in a timely manner. The strategy is probably to throw a lot of money at the problem up front, with a very short time line for delivery. What's your first step? Of course, you buy equipment, including PPE, for there is nothing like shiny new pieces of equipment to show that money has been spent and action is being taken. But which approach should you take?

1. Browse the Internet and talk to salespeople.
2. Talk to user groups that have already procured equipment.
3. Ask your local expert what to buy.

Unfortunately for you, the answer is most likely (d): none of the above. And, after procurement, you will have spent a lot of money on equipment without ever knowing whether it satisfies your requirements completely (and it almost certainly won't), and the PPE you bought will limit your capability and your safety to the point that you might be putting lives at risk by implementing its use (Figure 1-2).

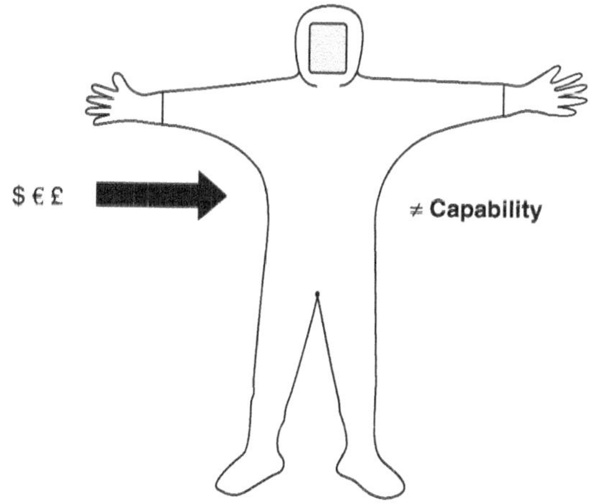

FIGURE 1-2 The usual result of urgency in acquisition.

So, at the end of this exercise, you recognize that this wasn't the best approach, but if none of these people really knew the answer to what to buy, who does? Well, here's the bad news—you (having become your local expert) are the only one that really holds the answer to what you need, and only after considerable work on your part, which will involve the engagement of many people inside and outside your organization.

It's pretty obvious, then, that as much of this work as possible should be done *before* someone arrives on your doorstep with the next parcel of money to be spent on equipment acquisition. It is important to note here that PPE is just one piece—albeit an important one—of the puzzle, and that this exercise must be performed for every type of equipment to be procured to develop an entire CBRN response capability. Nevertheless, since the focus of this book is on PPE, other aspects of the capability development are not discussed further here.

1.4.2 Stage 1: Prior to the Design and Procurement Cycle

Once the decision has been made to procure PPE, it is generally far too late to begin all the work that needs to be done. Therefore, prior to this time, the user should already have worked out a concept of operations that includes CBRN operations. In other words, equipment users should understand fully what they have to be able to do whether or not they are in a CBRN environment; and they should understand that being in a CBRN environment may limit their operational capability, so that the essential must be separated from the desirable operational capabilities to be delivered. The trade-offs that the military commander must consider have been described by NATO [5], which gives "fundamental principles for the guidance of operational level commanders and their staffs in an NBC environment."

First, the organization's non-CBRN concept of operations should be translated into:

- Specific tasks, assessing for each such factor as:
  - Work rate at which it is performed
  - Dexterity and freedom of motion required
  - Situational awareness required
- Normal hazards other than those present in a CBRN environment
- Existing non-CBRN procedures and training
- Ancillary equipment worn or used by the user that may have an impact on PPE performance, or vice versa
- User population characteristics such as:
  - Age, gender, anthropometrics, fitness
  - Education, training, and CBRN operations proficiency level expected
- Minimum and maximum duration and conditions of operations

With all of this information collated, it should be possible to summarize the organization's capabilities when operating in a non-CBRN environment. If there is an existing CBRN response capability, it should be summarized and documented. It may well be that this capability has never been explicitly analyzed despite the presence within the organization of PPE and training. The analysis should include:

- The nature of possible CBRN exposure
- Additional possible hazards other than CBRN in a CBRN environment
- Existing CBRN PPE
- Existing CBRN procedures and training
- How organizational response capabilities are altered in a CBRN environment:
  - Targeted capabilities and tasks
  - Gaps and limitations

With all of the information collated, documented, and updated on a regular basis, the process of acquisition of new PPE can proceed at the optimal pace once the decision is made to proceed.

1.4.3 Stage 2: At the Time of Decision to Procure New PPE

Sometimes when the time has come to procure new PPE, it is prompted by a change in desired operational capabilities; procurement may also occur as part of the normal process of life-cycle renewal of equipment, but in this case a desire for new capabilities will also inevitably result. The following approach will assist:

1. Reassess information from stage 1 for correctness.
2. Perform capability assessment:
   a. List targeted capabilities.

   b. Identify which are existing, which are new, and if any existing capabilities fall outside the target and can be sacrificed.
3. Compare existing standards with targeted capabilities: Is there a response or PPE standard that assists in describing these?
4. Compare available PPE with capability targets.
5. Compare the concept of use of available PPE with user capabilities.
   a. Do not neglect such factors as fitting, sizing, supply, and resupply requirements.
6. Take into account the available level of user participation in the process.
7. Either equipment must meet standards that take into account all relevant user requirements, meaning that less user involvement is required, or
8. Users must prioritize sufficient availability of an appropriate user population (10 to 30 standard users plus one or more user experts) at all stages of the development and selection program in order to address and assess:
   a. Sizing and fitting
   b. Functionality and use
   c. Putting on PPE
   d. Removing PPE
   e. Wear
   f. Range of motion
   g. Situational awareness
   h. Duration of use
   i. Equipment integration
   j. Simulated workplace protection
   k. Training program development
9. Determine time line and budget envelope for acquisition.
10. Decide whether off-the-shelf procurement (stage 3) and/or development (stage 4) is possible or required.

1.4.4 Stage 3: Off-the-Shelf Procurement

Stage 3a: Procurement Against Standards. The equipment must meet specified standards appropriate to the user group and concept of operations. User acceptance is based mainly on:

- Cost and delivery
- Integration requirements
- Interoperability requirements
- Limited operational trials
- Life-cycle management issues
- Additional features that may be provided in excess of standard requirements

Stage 3b: Procurement with a Few Additional Customized Requirements. The user must translate custom requirements into test methods and criteria. In addition to the factors listed in phase 3a, user acceptance of PPE is also based on:

- Ability to meet nonstandard test criteria.

1.4.5 Stage 4: Development Program

A development program is a major undertaking and will be considered only by large organizations and only then when off-the-shelf procurement cannot provide an adequate solution. Depending on existing limitations or capabilities within an organization, certain design options may be more desirable than others. Some considerations are obvious, such as the specific nature and magnitude of CBRN hazards to be protected against, and these will drive the design parameters required to keep the hazard out, as discussed in further detail in later chapters. Some examples of how other types of issues may have an impact on design are given below.

Logistics of PPE Availability and Issue

- Storage
  - Central depot? Carried with user or in vehicle?
  - Space available
  - Environmental conditions
- Size of stockpile? Enough for each person or enough for a subset?
- Time to resupply or recharge? (in theater or in use)
- Weight and bulk when packaged
- Time to respond? To open? To put on? To decontaminate and remove? (Just-in-time or continuous protection?)
- Sizing and fitting strategies (one size fits all, precustomized, presized, etc.)
  - Fitting capabilities: Time of issue? Time of putting on? Both?
- Mechanism of issue
- Disposability or reuse

Duration of Use

- Requirement to change PPE or to recharge air or air-purifying elements
- Weight and bulk of human-portable items
- Hydration
- Physiological burden

Extreme Environments or High Work Rates

- Microclimate control, hydration, fogging
- Durability
- Water, wind, temperature

Other Hazards—Particularly Ranked More Important Than or Incompatible with CBRN Protection

- Blast
- Ballistic
- Fire
- Electrical
- Contaminants
- Oxygen depletion

To lay the groundwork for understanding protection requirements, we focus next on the hazards from CBRN substances.

2 Hazardous Substances

Our intent in this chapter is to familiarize the reader with the many possible CBRN hazard agents that can be encountered and the types of effects that are of concern. The relative significance of the various routes of entry of these substances into the body and how these materials can be disseminated are described.

2.1 GENERAL OVERVIEW OF AGENTS

CBRN agents can be classified in a variety of ways. Much of the discussion that follows describes agents as they fall into various classes rather than as individual agents. The most commonly used descriptors are based on the hazard type of the agent:

Chemical Agents. The chemical (C) agents consist of nonliving chemicals no matter their source; technically, this category also includes the **toxins** (which are often classified with biological agents because of their origin as poisons produced by biological organisms). Chemical agent hazardous effects relate to their direct poisonous or toxic action on the body.

Biological Agents. The biological (B) agents consist of living microorganisms such as bacteria and viruses. They are classified into risk groups based on their hazard, with risk group 1 being reasonably benign organisms and risk group 4 being very high hazard organisms. Biological agents can reproduce inside the body and therefore can be a hazard at much lower quantities than other agents. Biological agents are often lethal as a result of the production of toxins.

Radiological and Nuclear Agents. Radiological (R) agents consist of radiological particles dispersed in the air in some manner by any means other than a nuclear explosion; nuclear (N) agents are produced by nuclear weapons or explosions. In either case, it is their radioactive decay that produces high-energy radiation or particles that are hazardous to the body.

Classified by their original intent, agents can fall into two categories.

Personal Protective Equipment for Chemical, Biological, and Radiological Hazards: Design, Evaluation, and Selection, First Edition. Eva F. Gudgin Dickson.

Classical Military Agents. These include (1) various chemical classes of chemical warfare agents (CWAs), such as the nerve agents (e.g., sarin, soman, VX*), vesicating (blister) agents (e.g., sulfur mustard, lewisite), blood agents (e.g., hydrogen cyanide, cyanogen chloride), choking agents (e.g., phosgene), and others; (2) militarized biological agents such as solid or liquid aerosols (e.g., organisms that cause anthrax or smallpox); and (3) nuclear and radiological agent particulates (e.g., fallout, dirty bomb materials). Not belonging to the classical military agents category, but sometimes considered militarily relevant, are the canister penetrants (e.g., perfluoroisobutylene).

Toxic Industrial Chemicals and Materials (TICs/TIMs). These are chemicals (or other types of toxic or hazard materials) that are produced for industrial and civilian purposes. Some CWAs are, in fact, TICs: for example, hydrogen cyanide. While the list of militarized agents is relatively short, the list of TICs and TIMs is extremely long, and hence comprehensive protection is a difficult issue.

Later, we discuss the hazards posed by these agents in more detail.

2.2 DOSE AND EXPOSURE

All substances are poisons . . . the right dose differentiates a poison and a remedy.
—Paracelsus (1493–1541)

An important fact to recognize about toxicity is that a large enough dose of *any* substance can be toxic. It is therefore clear that the hazard posed by a toxic material is in large part determined by "how poisonous" and "how much," as well as the detailed circumstances of exposure. The same principles apply to infectious or radioactive materials.

Although toxicity or other hazardous properties of a substance cannot be modified, we *can* control the dose, the amount to which exposure occurs. This is the role of PPE—it protects appropriate selected *routes of entry* into the body, with the intent that significant toxic effects will not be observed. Dose can be expressed in different units, depending on the type of material and the route of entry (we elaborate on these later). *Exposure* may be used to refer to the *potential* dose received by an individual (if he or she took no remedial measures to reduce this amount, such as wearing PPE), but the term is also used to include the conditions under which the actual dose was received (including the amount), as in "occupational exposure" to a substance. We discuss how delivered dose relates to effects in Section 3.5.

2.3 ROUTES OF ENTRY

Hazardous substances may enter the body along several different pathways. In some cases, the effects are felt **locally**, at the site of entry; for example, a corrosive material

* The military uses one- or two-letter abbreviations for many militarized CWAs that are more commonly used than the actual chemical names. VX is a persistent nerve agent.

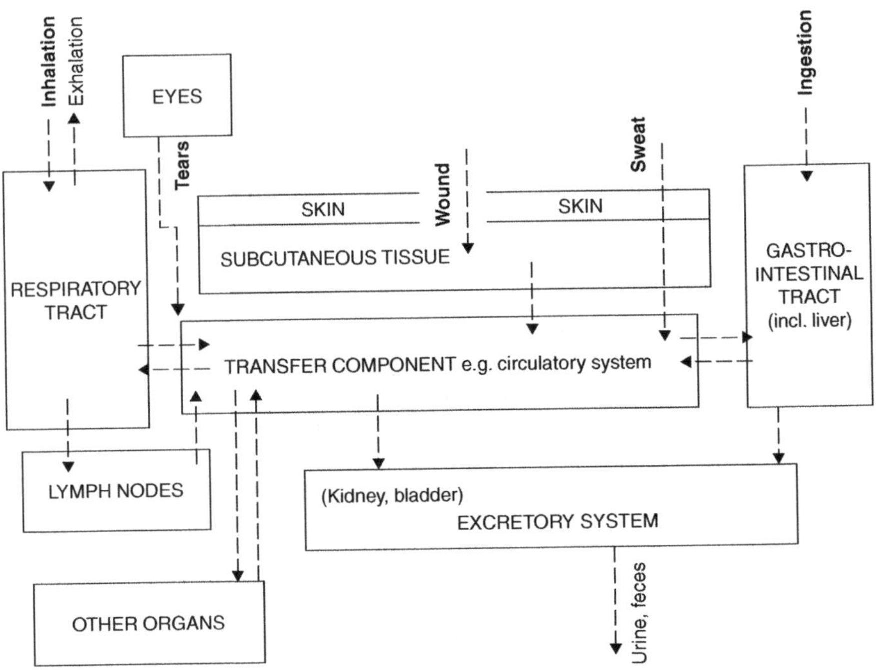

FIGURE 2-1 Main routes of entry and transfer of toxic materials through the body.

affects the body where direct contact is made. Alternatively, in the case of a **systemic** poison, the effects felt by the body are the same regardless of the route of entry. For example, benzene exposures usually result in tachycardia (abnormally rapid heart rate), whether contact results from spilling liquid benzene on the skin, inhaling its vapors, or ingesting benzene-contaminated materials. Radiation can also affect the body at a distance without radiological materials ever entering.

The main routes of entry and transfer for toxic materials throughout the body are shown in Figure 2-1. These also include many of the target organs. Each is discussed in more detail below.

Respiratory Tract (RT). Toxic materials are usually most hazardous when inhaled; this can occur any time a toxic material is airborne, which means that this route of entry is generally the most important to protect. The respiratory system includes the upper airways (nose, mouth, trachea) as well as the lungs (Figure 2-2). Some agents may not reach the deepest part of the lungs, due to various mechanisms of the body that are designed to defend against hazardous substances. The function of the lungs is to exchange gases between the bloodstream and the air in the lungs. As a result, poisons can enter the bloodstream directly and be transported quickly to other sites with minimum interference by the body's defense mechanisms, resulting in systemic toxicity. Local actions also include corrosion and inflammation of the lungs by chemicals, with results that lead to prevention of effective respiration. Similarly,

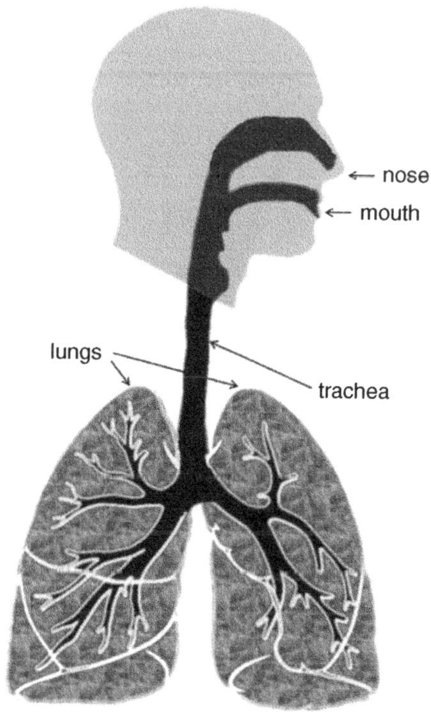

FIGURE 2-2 Respiratory tract.

biological agents can infect the respiratory tract, and inhaled radiological materials can lodge in the lungs, resulting in a significant irradiation mechanism.

Gastrointestinal (GI) Tract. Toxic materials can be ingested (via eating or drinking), following which they proceed through the GI tract (Figure 2-3). Materials are broken down in the upper digestive tract and absorbed primarily in the intestines. The liver and kidneys further process toxic materials and waste products in particular, with

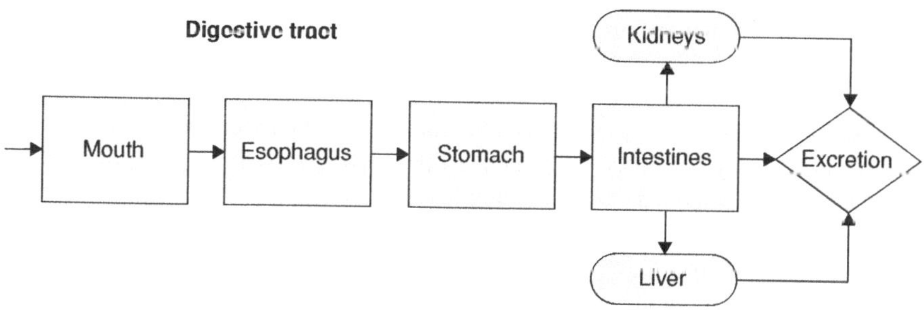

FIGURE 2-3 Schematic of the GI tract and excretory system.

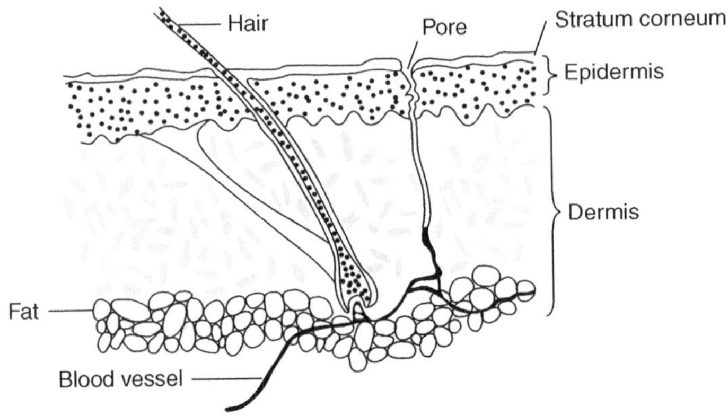

FIGURE 2-4 Skin structure.

the intent of breaking them down and excreting. In terms of symptoms that may result from ingestion of a toxic material, the GI tract may be the organ affected, with toxic symptoms, including irritation of the lining, resulting in vomiting and diarrhea. However, it is common for poisons to be taken into the rest of the body and cause effects elsewhere (often, in the liver and kidneys, which are designed to remove waste or toxic materials from the bloodstream).

Skin. Few substances pass easily through intact skin; for example, biological substances are too large to do so in virtually all cases. Therefore, the most dangerous are generally those military chemical weapons that have been selected precisely *because* they can. The skin (Figure 2-4) is composed of several layers. The outer protective layer (the top part of the epidermis) is the stratum corneum, consisting of dead skin cells. The living layers beneath include the rest of the epidermis, and the dermis, within which lie many other vital structures, such as hair follicles, sweat pores, and nerve endings. Chemical materials that can penetrate intact skin usually dissolve into the stratum corneum and are generally fat soluble. Other chemical substances may succeed in penetrating by other means: for example, through pores. Once the toxic material has penetrated, it may act on the live tissue beneath, causing effects directly, such as irritation or blistering, or alternatively, it can reach the blood vessels and may then enter the circulatory system, where it is distributed throughout the body with potential systemic effects at sites remote from the initial point of contact. Penetration of any type of hazardous agent can occur through a damaged skin barrier reaching the tissue and circulatory system beneath; damage may be preexisting, such as a cut or abrasion, or may be caused by something that deliberately penetrates the skin—a biting insect or a poisoned dart, for example.

Eyes. Toxic materials may often target the eyes, causing irritation or inflammation, damaging the eye's surface, or affecting the muscles of the eyes. In some cases they may enter the body via the tear ducts; infection by biological organisms is quite possible by this route.

2.4 FORMS OF AGENT LEADING TO EXPOSURE

Although the nature and degree of hazard are important in the selection of PPE, as important are the physical form of the agent and the route of entry, since these govern the nature of the protection that must be provided.

2.4.1 Airborne Hazards

Airborne hazards are of several types: vapors and gases (individual molecules of a substance), or somewhat larger aerosols, sprays, and particulates (aggregated molecules of substances, or alternatively, living things). Airborne hazards can move while airborne either by diffusion (normal random motion of small particles) or convection (being carried along by the motion of the surrounding air). The relative size of each type of airborne hazard is shown in Figure 2-5. Loosely defined, **inhalable particles** (which are also those that could be a hazard to eyes or skin) are those that remain suspended in air and lie below 30 to 100 μm or so, while **respirable aerosol particles** are those that reach the deepest part of the lungs and are not subsequently exhaled, lying in the approximate range 0.2 to 5 μm [6]. Below this size range the magnitude of the interaction with the deep regions of the lungs is not necessarily well understood and probably depends in part on the specific molecular nature of the hazard.

Vapors and **gases** consist of individual molecules of an agent in air. Most substances have solid, liquid, and gaseous states. Even when a substance is in its liquid

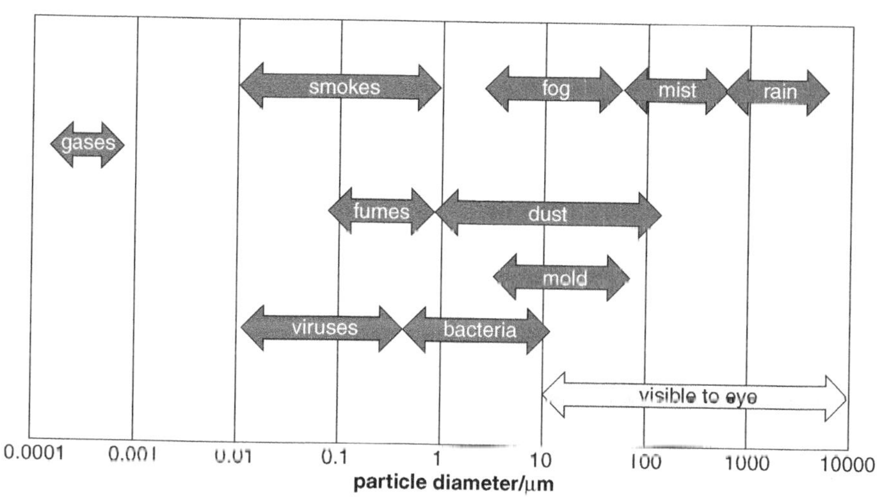

FIGURE 2-5 Size distribution of airborne substances.

or solid state, it is in equilibrium with some finite amount of vapor. The amount of material found in the vapor phase at any given temperature is the *vapor pressure* or the *volatility* of the substance (expressed as *concentration* in air). All liquids have significant vapor pressure as they get near their boiling point and above that temperature are found exclusively as gases. Hence, for a given toxicity of substance, the closer to its boiling point it is at a given temperature, the higher the volatility and the higher the vapor hazard.

> **Vapor pressure** is that *pressure* of the vapor that exists in equilibrium with the liquid. It increases as a function of temperature until the boiling point is reached, at which point the vapor pressure is generally 1 atm. The vapor pressure is directly related to the *volatility*, which is the *mass* of the vapor per unit volume of air under the same conditions. Hence, the magnitude of the volatility increases as the vapor pressure and molecular mass of the substance increase. The volatility is expressed in units of mass concentration of vapor per volume of air (e.g., $mg \cdot m^{-3}$); the concentration of vapor in air cannot exceed the volatility but can be less due to nonequilibrium conditions. Note that the *total* concentration of a substance in air may exceed its volatility, but if it does, it cannot be exclusively vapor (i.e., it must be combined aerosol and vapor).

Vapors will not settle to the ground and deposit on surfaces in the same manner as larger particles. They will be carried by air for large distances, and dilution is a major factor in reducing the hazard. Vapors of high-molecular-mass substances may resist dilution into air, as they have a higher *vapor density* than air, meaning that the vapors will settle into low-lying areas, tending to prevent free mixing. In addition, vapors may absorb into surfaces, or *decompose*, reducing their concentration.

> **Vapor density** is generally expressed as a ratio relative to that of air; vapors denser than air have vapor densities greater than 1. Vapors are denser than air when the molecular mass of the vapor is greater than the average molecular mass of air ($29\ g \cdot mol^{-1}$).
>
> **Decomposition** is the breakdown of a molecule into smaller molecules; in the case of agent vapors, this reaction may occur as a result of contact with oxygen or water in air, or the effects of light or heat.

Particulates occur in a full spectrum of sizes. Aerosols are very small particulates that because of their small size, settle very slowly and, if dry, can easily be reaerosolized by air movement if they do settle. Aerosols at low enough concentrations regardless of particle size are invisible to the eye. Higher concentrations and larger particulates are more easily seen. Aerosols may be generated from solids or liquids: A cloud of water or fog is a liquid aerosol, while a fine dust or pollen is a solid aerosol.

Although aerosols may follow the movement of air in much the same way that vapors do, there are important differences in how aerosols interact with materials, which means that they are removed from an airstream more easily than are vapors. Settling also occurs due to the force of gravity. Submicrometer particles settle very slowly in normal turbulent air (hours to days), whereas very large particles (hundreds of micrometers) settle so quickly that they may travel only a few meters after release in the absence of high winds or explosive forces.

Airborne hazards can pose a problem to any route of entry, depending on the nature of their toxicity and their size. Respiratory hazards are particularly common, and there are virtually no toxic agents that are not effective via this route of entry if delivered properly. Larger particulates, above 1 μm (Figure 2-5), are effectively removed by the upper airways; this localizes the hazard in this region. Smaller particulates and gases reach the lungs, with particles smaller than 0.1 μm tending to deposit in the gas-exchange region of the lungs (the alveoli) because of diffusion [7]. Particles around 0.3 μm in size are least likely to interact effectively with the body, with a large fraction breathed out again, but are also least likely to be removed by common filtration mechanisms in PPE. This effect is discussed in more detail in Section 4.2.2.

It is important to remember that in most cases, the amount of hazardous material per particle is significantly greater the greater the particle diameter: The larger the particle, the more of the hazard agent it contains, and this amount is proportional to the mass of the particle, which is itself roughly proportional to the cube of the particle radius. Thus, the total hazard posed by a particular size of particulate is a complicated function of the route of entry or target of the agent, and thus the likelihood of its removal by various filtration mechanisms, finally taking the total hazard per particle into account.

2.4.2 Contact Hazards

The main difference between a contact hazard and an airborne hazard is that an airborne hazard can travel after release, whereas a contact hazard is encountered when a person moves toward it (or as it is carried away from the hazard area by another object, such as a contaminated victim). This means that contact hazards can be more easily avoided and that the body regions most likely to encounter the hazard are the hands and feet (skin). However, under certain circumstances (e.g., rescuing contaminated victims, moving through contaminated foliage) there may be a significant possibility of contact for the remainder of the body.

The two significant sources of contact hazard are hazardous liquids and contaminated surfaces.

- **Hazardous liquid chemicals** may be neat (undiluted) or consist of solutions such as an acid dissolved in water. Bodily fluids can also be a contact hazard that while rarely used as a primary weapon, can transmit biological agents from person to person. Liquid contamination can be described as varying in size from spray droplets (visible to the naked eye) to puddles. Contaminated or nonsterile drinking water is a contact hazard that primarily targets ingestion.

- **Contaminated surfaces** are also common and more insidious, as the contamination is often not easily observable. Surfaces can be contaminated by CBR airborne hazards that have settled, and by C liquids and vapors that have absorbed into them and will subsequently off-gas even after the surface has apparently been cleaned. Once contamination is transferred to a person after contact, it can subsequently be spread to other routes of entry (e.g., ingestion from contaminated hands or from rubbing eyes).

2.4.3 Radiation Hazards

The last form of hazard is **radiation**, which is formed when an unstable atom's nucleus decays, resulting in the release of a great deal of energy in the form of high-energy particles or electromagnetic radiation. Emitted radiation includes:

- Alpha (α) particles, beta (β) particles (which are helium nuclei and high-energy electrons, respectively), and neutrons
- Gamma (γ)- and x-rays (electromagnetic radiation, whose position in the electromagnetic spectrum is illustrated in Figure 2-6)

These high-energy species pass through air or materials for greater or lesser distances, depending on the type and energy of the radiation. Alpha and beta particles can be stopped more easily than can electromagnetic radiation or neutrons by barriers such as clothing.

The most pertinent hazard from radioactive materials in the context of this work is from particulates that are either themselves radioactive substances (radiological or nuclear aerosols or dusts) or dusts that have radioactive materials deposited onto them, which can then travel through air or deposit on surfaces. They are thus airborne and contact hazards; however, because additionally, radiation travels directly through space and materials, protection against it can be more difficult. As such, radioactive materials can be a hazard to the entire body.

Radiation itself can be a significant challenge for PPE, but it can also be quite easily monitored with standard portable equipment, permitting exposure control by dosimetry on the spot. Further, the hazards are also not always immediate, meaning that decontamination of external body surfaces is also potentially quite effective at reducing the skin and contamination transfer hazards subsequent to exposure. Therapies that can reduce the severity of effects are also available for some types of exposure.

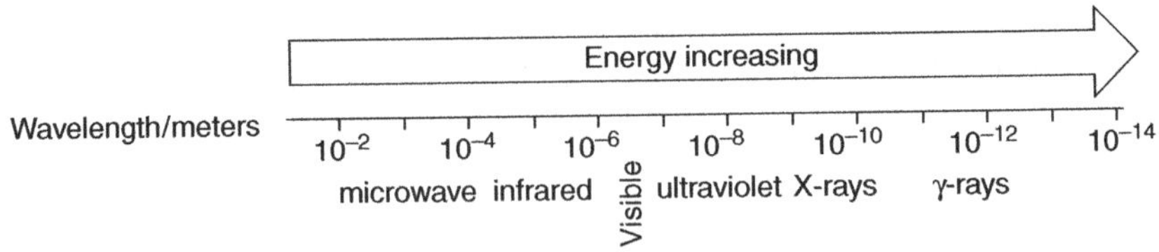

FIGURE 2-6 Electromagnetic spectrum.

2.5 EFFECTS OF HAZARDOUS MATERIALS

The potential effects caused by hazardous materials are as varied as the ways in which the body functions. The disruption of any bodily system can result in illness, and the more essential the system, the more immediately life threatening the effects will be. Many toxic chemicals cause their effects by irritation, inflammation, and/or corrosion of various tissues. This can result in respiratory distress and pneumonia, acute eye irritation or blindness, blistering or ulceration of skin, or various forms of gastrointestinal distress. Other chemicals act specifically on one biochemical system of the body. Targeted systems can include the respiratory pathway, which exchanges oxygen for carbon dioxide (e.g., blood agents); the particular enzyme system that controls neuromuscular function (e.g., nerve agents); and various brain functions (e.g., psychotropic substances). These effects may be immediate or may take hours to weeks to manifest themselves.

Radiation and some chemicals also act by damaging, either selectively or indiscriminately, important cellular functions over the entire body. Radiation sickness results from the widespread cellular or molecular damage caused by the high-energy particles, whereas carcinogenicity results from certain types of damage—specifically to the genetic makeup of the cell. These effects may take days to years to be seen.

Each biological agent acts in a different way: In many cases the injury to the body results from the extreme reaction of the body's immune system as it tries to eliminate the invader. In other cases the toxicity is to specific cell systems through release of a toxin; the symptoms of anthrax, for example, result from release of a toxic protein that triggers cell death. Many of the most toxic chemicals known are toxins of biological origin, such as snake venom, botulinum toxin (produced by the organism that causes botulism), and ricin from the castor bean. Effects from biological organisms require an infection to become established and thus typically take days to be seen. Toxins delivered directly often act very quickly, within minutes of exposure.

2.5.1 Local vs. Systemic Effects

As discussed previously, some toxic agents act locally, directly upon the body system with which they come into contact. Examples of such effects include the blistering caused by blister agents and the corrosion caused by corrosive materials. Cutaneous anthrax lesions are initially a local effect.

Systemic effects, on the other hand, can be felt in body systems that are remote from the route of entry; for example, cardiac failure resulting from inhalation of blood agents or the convulsions and death that can result from nerve agent toxicity by either inhalation or skin absorption. Systemic effects are additive in the sense that no matter what route the agent is absorbed by, once the agent has entered the bloodstream, the dose is cumulative. Therefore, protection must take into account all possible significant routes of entry, although the total dose required to cause effects may be higher than that for a local agent.

Many agents may cause both systemic and local effects: for example, nerve agents may cause local pinpointing of the pupils from eye exposure, and sulfur mustard (a blister agent) may additionally cause systemic effects such as nausea and vomiting.

2.5.2 Acute, Chronic, and Long-Term Effects

The effects that are usually of most concern in CBRN protection are those that result from a single (acute) exposure and will generally occur within seconds to hours of the exposure. However, long-term effects can also occur, such as an increased likelihood of cancer. The probability of effects resulting from chronic (repeated) exposure is of little relevance to CBRN protection, for which case exposure is assumed to be a once-in-a-lifetime event.*

2.6 CHEMICAL HAZARDS

Chemical hazards can result from any nonliving material that can cause damage by chemical reaction with the body. Toxic chemicals can thus include industrial and consumer chemicals, drugs, chemical warfare agents, and toxins (which while chemical in nature, are biological in origin; these are discussed further in Section 2.7). There are literally tens of thousands of candidate chemicals to consider when devising protective systems. Such materials can target any route of entry. For the purposes of PPE development and selection, it is most useful to classify agents by the routes of entry that must be protected based on the form in which they might be encountered.

2.6.1 Chemical Warfare Agents and Their Classification

Chemical warfare agents are the most toxic and effective chemical agents known; hence, they are worthy of individual discussion. These are the agents that the military will generally require protection against, as they have been militarized in the past. In general, chemicals can be classified using the United Nations Globally Harmonized System for Classification of Chemicals [8], which describes how chemicals can be categorized by health hazards, physical hazards, and environmental hazards. The classification of CWAs is focused on specific properties that are relevant to their use by an aggressor. There are three principal ways of classifying or describing CWAs. All of them are useful and each emphasizes some particular property of the agent. Each can be used in conjunction with every other.†

* Although historically, the military could have proposed that soldiers would fight routinely and repeatedly in a contaminated battlefield against an opponent with massive stockpiles of CBRN weapons, the current likelihood of use for most states is asymmetric. In this case an ill-equipped opponent uses a small stockpile of CBRN agents to occasional effect (i.e., it is the threat of use that affects the tempo of operations rather than actual use, and even a single exposure to any given person is unlikely).

† In principle, this classification system can also be used for toxic industrial chemicals.

Physiological Classification of Agents. This classification scheme looks at the target physiological system of the agent and includes the following categories:

- Choking agents attack the lungs, which then fill with fluid (e.g., phosgene).
- Blood agents act by interfering with the handover of oxygen from the blood to the tissues (e.g., hydrogen cyanide).
- Riot control agents cause temporary incapacitation but are unlikely to cause death. Tear gases produce temporary incapacitation by attacking the eyes as well as the respiratory tract (e.g., CS a tear gas). Other chemicals act mainly by respiratory or even skin irritation. Vomiting agents (e.g., adamsite) may also be used as riot control agents.
- Vesicant agents (blister agents) attack the skin, producing extremely painful and slow-healing blisters (e.g., mustard, lewisite).
- Nerve agents owe their toxic action to interference with the nerve pathways between the brain and the voluntary muscles, producing muscular spasms and paralysis [e.g., G (a class of nerve agents) and VX].
- Psychochemical agents disorient the mind, producing hallucinations and irrational behavior (e.g., LSD, a psychochemical agent).

Degree of Effective Damage. Three terms are used to describe agent effects: A *lethal agent* such as nerve gas will kill, an *incapacitating agent* such as tear gas will cause very short-term incapacitating effects, and a *damaging agent* such as sulfur mustard will cause severe illness.

Persistency. Persistency is the ability of an agent to remain in place and effective for a specific period of time after release. This property is described further in Section 2.9.2.

- Nonpersistent agents evaporate, and hence dissipate, rapidly. They are employed to provide a toxic cloud of vapor over or within a target area, lasting for a very short time, meaning that the area is likely to be accessible to the aggressor shortly after use. Vapors are likely to be lighter than air.
- Persistent agents are viscous and evaporate slowly. They are used to produce chemical rain or spray, in order to contaminate personnel, ground, and equipment for a long period. Vapors are likely to be heavier than air and to accumulate in depressions.

A full description of a particular agent could employ all three classifications; for example:

- Phosgene is a lethal, nonpersistent choking agent.
- VX is a lethal, persistent nerve agent.
- Sulfur mustard is a damaging, persistent blister agent.

Select specific agents are discussed later.

2.6.2 Respiratory Hazards

It is a very unusual chemical that would *not* pose a respiratory hazard at some airborne dose if delivered appropriately. Many chemicals target the respiratory tract directly, causing choking effects either immediately or after a few hours. Others simply enter the body via the respiratory tract and cause systemic effects. For a chemical to pose a particular challenge to the ability of a respirator to be able to protect against it in a given incident, it must be at least one of the following:

- Highly toxic
- Present in very high concentrations
- Difficult to remove by conventional air purification means

There are a number of other features that may make the respiratory hazard persist longer after release, such as chemical stability, high vapor density, and/or relatively low volatility. Since the inhalation toxicity values, as well as minimal effect levels suitable for use in occupational exposure, have been estimated for virtually all available chemicals, it is possible to understand and set protection requirements systematically for this route of entry. Although the list of chemicals against which protection is required may seem dauntingly long, by categorizing chemicals using the three properties listed above, it becomes possible to target protection against the most hazardous chemicals first.

Chemicals can be categorized with respect to their toxicity by examining lethal doses by inhalation, as well as doses that may cause incapacitating effects. The list of highly toxic chemicals contains many of the CWAs as well as a large variety of other chemicals; the industrial chemicals phosgene, hydrogen cyanide, and chlorine have all been used as weapons of war.

For a chemical to be present in high concentration, it must either be available in very high quantity (such as that found in a storage tank or facility of some sort, representing probable release over a relatively large area, up to kilometers in size) or if available only in a small quantity, released over a smaller area and probably for a shorter duration (within a building or in an outside area a few hundred meters wide). Chemicals that are difficult to protect against are either difficult to remove from an airstream by conventional means or may penetrate through materials from which facepieces and the like are constructed. These features are discussed later when specific mechanisms of protection are addressed.

2.6.3 Ocular Hazards

Little is given in the literature about the magnitude of doses that cause eye effects independent of respiratory effects. If a chemical is in any way corrosive or irritating to the lungs, it is likely that this effect will also be felt at the eyes at a dose that may be similar to that causing respiratory effects, since in both cases the chemical usually acts by dissolving into tissues or body fluids and then irritating sensitive mucous membranes. Damage or irritation to the eye can be immediately debilitating, resulting in impairment of vision; in some cases, damage is permanent, leading to blindness. A variety of corrosive chemicals may act in this way, and some of the chemical warfare agents (e.g., sulfur mustard) may damage the cornea.

Other noteworthy eye effects from chemicals include contraction or dilation of the pupils, which can significantly affect the ability to compensate for light levels. Notable are the organophosphates and carbamates (nerve agents and insecticides), which cause pupil contraction (miosis) and hence difficulty seeing in low light, as well as loss of peripheral vision, both of which may significantly affect a person's ability to operate in and extricate from a hazardous situation. Dilation of the pupils may similarly cause difficulty seeing in bright light, which in some cases could be equally debilitating. Many riot control agents are lachrymatory agents (tear gases) that cause uncontrollable tearing of the eyes, which lessens over some minutes without generally causing long-lasting effects.

2.6.4 Dermal Hazards

The skin is an excellent barrier to a chemical agent, making the list of chemicals responsible for significant dermal toxicity quite short. Lethality is particularly rare by this route. The relative toxicities of dermal hazards compared to respiratory and eye hazards are usually orders of magnitude less for the same form of agent. However, selected nerve agents may be nearly as lethal by skin contact as by inhalation, and dermal absorption of pesticides may also be lethal.

Aside from absorption through skin, dermal effects include itching and blistering from blister agents as well as corrosive effects caused by acids, oxidizers, and bases. Blister agents such as mustard gas (actually, dispersed as a liquid) can be lethal. If enough of the body is involved, the effects may be similar to those of a severe burn victim, with death from infection possible. Damaging effects may occur at relatively low doses.

Historical note: Blister agents

Mustard gas was first used effectively in World War I by the German army against British soldiers near Ypres in 1917 and later against the French Second Army. It has been used by at least 10 countries in conflicts since that time.

2.6.5 Ingestion Hazards

Ingestion is a significant potential route of entry for poisoning, and most accidental domestic exposures to toxic substances occur by this route. Nevertheless, from the point of view of PPE, further explicit discussion of this route of entry and relevant toxic materials will be limited; prevention of entry into the digestive tract by toxic materials is achieved by the use of effective respiratory protection during potential exposure, and minimization of contamination transfer from contaminated skin after removal of PPE. Other means of preventing this form of hazard include general decontamination, sterilization of drinking water, and protection of food supplies.

Thus, protection of the digestive tract is achieved automatically by a combination of preventing the opportunity for exposure, primary protection of the body, and prevention of any remaining opportunities for secondary contamination by effective decontamination. Protection levels and concepts that will appropriately minimize respiratory hazards will minimize ingestion hazards simultaneously.

In addition to the various forms of systemic toxicity that may be caused by ingestion, local toxicity may include irritation, inflammation, or corrosion of the digestive tract.

2.6.6 Dissemination of Chemicals and Types of Events

Important properties of chemicals that affect their properties and suitability for various types of events are their boiling point and volatility, and their surface tension and viscosity in the liquid phase. How close a liquid is to its boiling point determines how quickly it will evaporate and how much vapor can be in the air. Surface tension and viscosity, which are both related to how a liquid bonds to itself and resists flow, affect both how quickly a liquid evaporates and how it spreads, sticks, and penetrates materials after it is deposited.

Bulk liquid contamination (i.e., puddles of liquid) will occur in the vicinity of storage containers or reservoirs of liquid as a result of breach or spill (or ineffective dissemination by some other means, such as spray). Although initially the hazard will be quite local, it may reach surface water or the water table, in which case contamination may be much more widespread. Liquid chemical agents can be produced from precursors that react together on the spot to form the agent. A munition that contains two precursors, in separate compartments that do not mix until it is fired, is termed a **binary weapon**.

Historical note: Binary agents

The Aum Shinrikyo (the sect that perpetrated the Tokyo subway attack using sarin nerve agent) attempted several releases of hydrogen cyanide gas in the Tokyo subway using separated containers of potassium cyanide and sulfuric acid which when breached should have formed hydrogen cyanide; however, they were unsuccessful in producing a functioning device.

Droplets of liquid may result from some form of airborne release, such as release of bulk liquid from an aircraft or explosive munition, from coarse spray, or from explosive dissemination (which may destroy some chemicals due to the heat involved). Explosive munitions may contain submunitions, spreading the area covered. Thickened agents (formed by the addition of a thickening polymer to the liquid) are more likely to form droplets rather than aerosols.

Liquid aerosol can be formed either by generation of a fine mist of liquid using a pressurized sprayer, from the evaporation of larger droplets as described above, or from the condensation of vapors formed from explosive dissemination. **Solid aerosol** can also result from the condensation of vapors formed from explosive dissemination. In addition, powdered materials can be disseminated by any mechanism that will make them airborne, including compressed gas, explosion, or release into any airstream.

Vapor (or gas-phase) hazards will result from evaporation of another released form of chemical, provided that its volatility is sufficiently high; alternatively, release of pressurized or liquefied gas will result in a large cloud. When a vapor is less dense than air (has a lower molecular mass), it will rise and dissipate from ground level quite quickly. Heavy vapors will tend to hug the ground and remain in low-lying areas. A few agents, such as hydrogen cyanide, exist in the gas phase at warm ambient temperatures.

Historical note: Vapor hazards

In World War I, chlorine gas was released from pressurized cylinders dug into the German front line. Its effectiveness as a weapon was based on the entrenched position of the opposition forces, which could not evade or protect against the release when the heavier-than-air gas filled the trenches.

Contaminated water may result as a by-product of dissemination by other means, or by deliberate addition of a bulk chemical into a water system.

As illustrated in Figure 2-7, in principle, an airborne release of a liquid agent will result in a fallout of liquid contamination downwind of the release and a further airborne downwind hazard both from the primary release and from evaporation of the

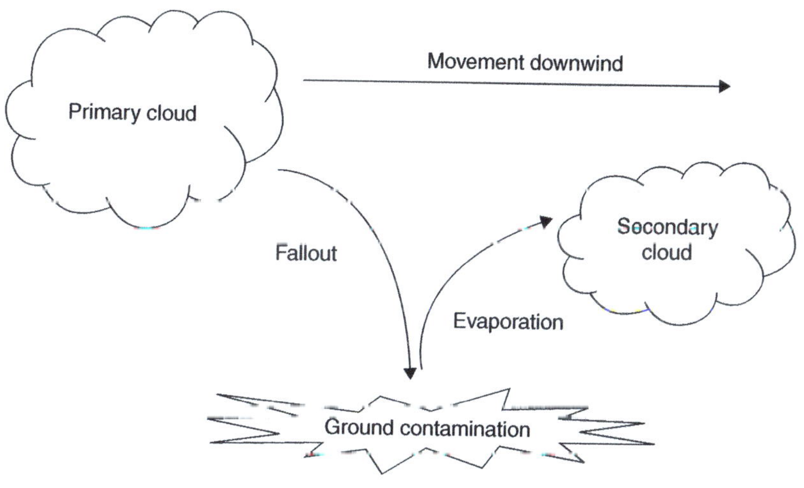

FIGURE 2-7 Downwind hazards from an airborne liquid release.

liquid surface contamination, as shown. Evaporation during or after dissemination will cause a reduction in the area or volume of liquid contamination; a sufficiently volatile liquid that was disseminated as droplets from an aircraft will turn to aerosol or vapor as it falls, while a puddle of volatile liquid will quickly evaporate to vapor. Thus, the physical nature of the hazards may change during or after dissemination, depending on the volatility of the agent.

2.7 BIOLOGICAL HAZARDS

2.7.1 General Background

The primary category of biological agents is living organisms such as the **microorganisms** that can be disseminated intentionally to cause disease. Technically, any living organism such as venomous animals (e.g., snakes or spiders) could be considered a biological weapon if used as such; this approach is unlikely for contemporary use. When an animal is used to transport and disseminate another biological agent, it is termed a *vector*: an example being the rats and fleas that carry plague.

Historical note: Vectors

During World War II, Japan contaminated grain with plague-infected fleas; the grain was then dropped from planes onto China to spread disease and weaken the population [9].

To require the use of PPE, the disease need not necessarily be primarily one of humans; however, if the disease does not in some way infect people, the requirements for the PPE may differ. The focus in this book is primarily on protection against those agents that can infect humans, which requires PPE to prevent infection of the person wearing it. Alternatively, PPE may be worn primarily to prevent people from acting as *fomites*, or passive carriers (e.g., to prevent the spread of hoof-and-mouth disease from one affected farm to another). In this case, the main objective is to provide an outer layer that can easily be removed and decontaminated or destroyed, minimizing the decontamination requirements for the person within. Such PPE may have different design considerations from equipment that is intended to prevent infection of the individual. The types of effects that can be caused by living biological agents are highly varied, ranging from a severe respiratory tract infection through diseases that target the heart, GI tract, circulatory system, immune system, nervous system, or generalized tissues.

2.7.2 What Makes a Potential Biological Agent?

Biological agents need to meet certain requirements for use against personnel, domestic food and draft animals, or plants. Biological organisms are for the most part difficult to produce and keep alive long enough for them to effectively infect any target. It is as much the fear of use as the likelihood of success that tends to produce a significant effect. This means that there is a very short list of agents that actually have a reasonable probability of doing damage, and if these can be protected against, the remainder are likely to be of less concern.

A successful agent would:

- Consistently produce a given disabling or lethal effect
- Be capable of efficient dissemination, by whatever means relevant

and if deliberately disseminated by release (rather than through a living vector):

- Be capable of being manufactured on a large enough scale
- Be relatively stable in production, storage, and transportation
- Be stable after dissemination

It is noteworthy that *very* few agents would actually meet these criteria in practice.

2.7.3 Classification

Biological/toxin agents may be classified in several ways, such as type of agent, objective of attack, severity of effects produced, viability, virulence, communicability, and use. Such classifications are relative and often overlapping.

Type of Agent. This classification includes the microorganisms, the vectors (or disease carriers), and the toxins.* On the basis of structural and behavioral characteristics, microorganisms may be further grouped as follows (in order of decreasing size): fungi, protozoans, bacteria, rickettsia, and viruses.

- **Microorganisms.** Throughout the following discussion, emphasis is placed on certain groups of microorganisms that might be used as potential antipersonnel, or possibly antiplant or antianimal, agents. Although protozoans and other groups, such as algae, commonly occur, they presently have little significance for use as agents. Characteristics and properties mentioned under the general term *microorganism* will thus refer to *bacteria*, *rickettsia*, *fungi*, and *viruses* unless otherwise indicated.
- **Vectors.** These are disease-carrying animals, called vectors of disease, and can include insects and higher animals.
- **Toxins.** Toxins are poisonous products of microorganisms, animals, and plants. Strictly speaking, they are chemical agents but are often grouped with the biological warfare (BW) agents due to their properties and origin.

* Chemical antiplant agents such as plant growth regulators, herbicides, weed killers, defoliants (e.g., Agent Orange), and desiccants are sometimes classified with BW agents, although they are clearly CW agents, as they are chemicals that are not biological in origin. Protection against such agents is not discussed except inasmuch as they may be toxic to humans.

Target of Attack. Agents can be classified as antipersonnel (humans), antianimal (animals used for food, transportation, or biological preparations), or antiplant (plants used for food, clothing, and industrial products). There may be some overlapping between antipersonnel and antianimal agents in that some agents of either group will be effective against both humans and animals (e.g., the organism that causes anthrax, *Bacillus anthracis*).

Severity of Effects. Effects may be either lethal or nonlethal (incapacitating). Lethal or killing agents can produce death in susceptible individuals, but from a practical standpoint death occurs only in a certain percentage of those exposed. The nonlethal pathogenic agents usually do not kill but may produce infection or disease with significant disability among susceptible exposed individuals. **Virulence** is the ability of microorganisms to overwhelm the defensive body mechanism of the prospective host. Virulence depends on a number of factors to do with the particular strain or organism; strain selection for increased virulence among selected microorganisms is a distinct possibility.

Communicability. A communicable, or contagious, disease is one that is transmitted directly or indirectly from one infected host to another by contact, body excretions, coughing, or sneezing. Such diseases as diphtheria, typhoid fever, mumps, and measles are communicable; anthrax, tetanus, and botulism are noncommunicable.

Viability. The viability (ability to live and thrive under given conditions) of microorganisms varies with the species. Since biological agents are living organisms, most of them are affected significantly by environmental conditions. Although most microorganisms and toxins may be killed or inactivated by any of several environmental factors, some of the most delicate agents may survive for prolonged periods if conditions are favorable or if natural reservoirs are established. In general, following dissemination, the spores of fungi and of sporulating bacteria, such as those causing anthrax and botulism, remain viable in the environment for greater periods of time than do nonsporulating organisms.

Transmissibility. The likelihood of effective transmission of a disease can also be used for classification. The viability, route of entry, and communicability of an organism will all significantly affect its transmissibility. Modes of transmission are discussed further in Section 2.7.12.

The **Biosafety risk group** classification of organisms for handling and laboratory safety [10] takes into account the transmissibility and severity of the disease:

- Level 1: never cause disease in humans; low risk (e.g., plant pathogens, some lab strains of *Escherichia coli*, and some animal pathogens).
- Level 2: can cause disease in humans but with limited potential for transmission; moderate risk (e.g., salmonella, measles).
- Level 3: deadly pathogens that can cause severe outbreaks in the community but treatments or vaccine exist; high risk (e.g., viruses that cause anthrax, plague, polio).

- Level 4: deadly pathogens that present a risk to humans worldwide if released; no treatment or vaccine exists; extremely high risk [e.g., Ebola, Lassa fever, (smallpox *)].

2.7.4 The Immune System and Infection

Infection is defined as the invasion of a host animal by microorganisms in such a way as to allow for the continued growth and reproduction of the invading microorganisms (e.g., bacteria, viruses, and parasites).

If successful, infection may lead to disease, and the production of disease is the aim of biological weapons. Following invasion by a microorganism, there is an *incubation period*, the time during which the microorganism is attempting to multiply within the body to a sufficient concentration to cause disease. The incubation period will vary with the microorganism and the dose received, and not all infections will lead to disease; there are many situations in which the host "wins," with the immune system preventing disease. The immune system protects against organisms as they try to penetrate the body at three different levels.

Skin and Mucous Membranes. These are the first level of defense, acting as a physical barrier to infection. The skin forms a tough coat over the exterior of the body and when unbroken is quite resistant to invasion by microorganisms. The mucous membranes are a continuation of the skin and are found lining those body cavities that communicate with the exterior. This membrane is modified to produce a moist sticky substance known as mucus, which tends to trap and hold microorganisms that may enter these cavities. Certain of these membranes (nasal and bronchial) are further adapted in that they contain multiple hairlike projections known as *cilia*. Microorganisms that can be trapped on the ciliated mucous membrane lining of the respiratory tract are moved either to the nostrils, where they can be expelled from the body, or are moved to the back of the mouth, where they are swallowed and thus subjected to the acid conditions in the stomach.

Lymphatic System and White Blood Cells. If microorganisms gain entrance into the body, their protein makeup causes microorganisms to stimulate the body's immune mechanisms. The body responds by sending white blood cells to the site where the microorganisms are located. These cells engulf the invading microorganisms and digest them. If the microorganism eludes the white blood cells, it will gain entry into the lymph channels. The lymphatic system consists of numerous small vessels that

* Although an effective vaccine for smallpox does exist, yielding its worldwide eradication, its extremely high transmissibility and severity of disease, combined with the unknown state of general immunity in the population in the decades since smallpox vaccination ceased and the limited stores of the vaccine available, have resulted in its inclusion as a level 4 agent.

ultimately return fluids (lymph) to the circulatory system. A system of filters that are known as lymph glands or nodes is located along lymph vessels through which the lymph must pass. These filters tend to screen out the majority of foreign materials, including microorganisms, which may be present in the lymphatic fluid.

Liver, Spleen, and Antibodies. If microorganisms can get past the preceding lines of defense, they enter the blood and the circulatory system. The blood must flow through the liver and the spleen during its course through the body. In these organs are additional fixed cells that can digest many of these microorganisms. In addition, the blood-circulatory system contains antibodies, proteins whose function is to combat invading microorganisms. Some antibodies are produced early in a disease state and are nonspecific and less effective, whereas those produced later in the infection specifically target the microorganism.

Immunity is the ability of a host animal to resist infection by microorganisms. There are two major types of immunity: genetic, or innate, and acquired immunity. *Genetic immunity* is conferred by the nature of the species; humans cannot be infected by potato viruses. *Acquired immunity* is due to the response of the body to a previous infection: When the body has previously seen and mounted a defense against a microorganism, it will frequently retain the ability to produce this defense quickly, due to the retention of the ability to rapidly produce the specific antibodies required. Acquired immunity can also be conferred by *immunization*, where a person is given an attenuated or ineffective form of a disease that will cause the immune system to produce specific antibodies, hence preventing subsequent true infection. Immunization is specific to a particular disease or even strain of a disease. Immunization is a form of *prophylaxis*.

> **Prophylaxis** is the process of preventing an effect by some form of intervention prior to exposure or during the incubation period. **Therapy** is treatment after effects are seen. These terms can be applied equally to disease caused by any form or type of agent. For example, prophylaxis against nerve agents consists of taking a drug that prevents nerve agents from manifesting their full effect. The use of PPE is, in fact, a form of prophylaxis, although it is not normally described as such.

2.7.5 General Properties of Microorganisms

Description. Microorganisms are microscopic living organisms that are too small to be seen by the unaided eye, so small that the unit applied in their measurement is the micrometer (μm), which is equivalent to 0.001 millimeter. Viruses, for the most part, are so small that they can be seen only by using an electron microscope. When magnified properly by a microscope, each microorganism is found to be composed of a single cell or a group of associated cells. Each cell, or group of associated cells, is capable of carrying on all the functions of life, including growth and reproduction. Since the microorganism lacks a digestive tract, it acquires food in soluble form through the membrane that surrounds the cell. A microorganism assumes the temperature of its surroundings because it has no heat-regulating system. Many microorganisms resemble plant life and are regarded as members of the plant kingdom while others have characteristics that cause them to be placed in the animal kingdom.

Distribution. Microorganisms are universally distributed in air, water, and soil. Most pathogenic or disease-producing microorganisms of humans, animals, and plants do not survive long or grow well in the absence of a suitable host because favorable environmental conditions are necessary for their survival.

Pathogenicity. Microorganisms capable of producing disease are called *pathogens*. *Pathogenicity* is therefore the ability to cause disease. Since they live within a living host at whose expense they obtain food without benefit to the host, many pathogens are *parasites*. Organisms that multiply in dead rather than in living matter are called *saprophytes*. Examples of some of the harmful saprophytes are the bacteria that cause tetanus and botulism.

Self-Protective Mechanisms. A protective mechanism favorable to the survival of some bacteria is *sporulation*, which is the formation of heavy-walled bodies called *spores* that are conceptually somewhat similar to the seeds of a plant. Bacterial spores are more resistant to injurious or unfavorable influences (e.g., starvation, high and low temperatures, microbicidal chemicals, drying, and oxidation) than are the growing (vegetative) forms. A resistant spore may remain dormant for years without food or water and under extreme ranges of temperature. It may develop into an actively growing vegetative cell when conditions again become favorable.

Next we describe more specifically some types of biological warfare (BW) agents, with a focus on the most relevant pathogenic bacteria and viruses.

2.7.6 The Bacteria

Bacteria are microscopic one-celled plant-like organisms that have no chlorophyll (which is the green coloring matter of plants). They are widely distributed in nature, being found in soil, air, water, the bodies of living animals and plants, and in dead or decaying organic matter. Few are actually pathogenic. Bacteria generally range in size from 0.5 to 10 μm across their greatest dimension, although a few are much larger. Bacteria have historically been classified more or less by their shape (e.g., round, rod-like, corkscrew-shaped). Other ways in which they are classified include their genetic sequence, their route of transmission (e.g., sexually transmitted disease), and by their cellular wall characteristics, which give rise to different staining characteristics under the microscope (gram-negative vs. gram-positive). Bacteria cause many of the common diseases of humans, animals, and plants. Examples of bacteria classified by their shape, and the types of disease they cause, are given in (Table 2-1).

TABLE 2-1 Types of Bacteria, Classified by Shape

Type	Shape	Examples of Disease
Coccal	Round	Staphylococcal: food poisoning; streptococcal: scarlet fever, strep throat; gonococcal: gonorrhea; meningococcal: meningitis
Bacilli	Rod	Tuberculosis, anthrax, typhoid fever, bubonic plague
Vibrio	Comma	Cholera
Spirochete	Corkscrew	Syphilis

The symptoms of anthrax, diphtheria, tetanus, gas gangrene, and botulism are all caused principally by the toxins produced. The bacterium causing tetanus produces its toxin while living in injured and dead tissue of the host; this toxin can then spread throughout the body. The botulism organism produces its toxin in some foods; ingestion of food contaminated with even tiny amounts of botulinum toxin can cause fatal poisoning. Some of the most hazardous bacteria and other microorganisms of concern are discussed in more detail in Section 2.7.13.

2.7.7 The Rickettsia

The rickettsia are intracellular parasitic microorganisms that are intermediate between the bacteria and viruses in size: 0.3 to 0.5 μm in length and about 0.3 μm in diameter. They resemble the bacteria in shape and resemble the viruses in their strict growth requirements for living host cells. Most rickettsia are parasites primarily of lower animals and arthropods (i.e., lice, fleas, mites, ticks) which can then be used as vectors to pass them to a higher animal host. Rickettsia are transmitted to humans and animals by such vectors. They are nonsporulating and are easily killed by heat, dehydration, or disinfectants.

Some of the more severe common human rickettsial diseases include: classic epidemic (human) typhus, an acute louse-borne infectious disease; Rocky Mountain spotted fever, a relatively severe disease transmitted to humans by the bite of a tick; and Q fever, an acute febrile illness that differs from other rickettsial diseases in that it may be transmitted by the bite of an arthropod vector or by ingestion or inhalation of contaminated material. Most are characterized by fever, skin rashes or dark blotches, and central nervous system disturbances.

2.7.8 The Viruses

The viruses are complex organic substances that will live and multiply or increase only in susceptible living host cells. They can reproduce only by "hijacking" the host cell's ability to reproduce itself. They range in size from about 0.01 to 0.27 μm across their greatest dimension. Because of their extremely small size, not all viruses have been observed; however, globular, crystalline, square, rectangular, and spherical shapes have been described. Aside from their form and type of disease caused, viruses are also classified by their genomic structure and sequence, as well as by their replication strategy. The viruses are responsible for many important diseases of humans, animals, and plants. Human diseases caused by viruses include poliomyelitis, rabies, smallpox, yellow fever, encephalitis, mumps, measles, chickenpox, influenza, and the common cold.

2.7.9 The Fungi

The fungi are unicellular or multicellular members of the plant kingdom. Fungi include molds, mildews, smuts, rusts, mushrooms, toadstools, puffballs, and yeasts. The cells of most fungi are larger than bacteria, ranging from 3 to 50 μm in size. They are usually rod shaped and arranged end to end in strands or filaments, although yeast cells are usually oval and may appear singly, in clumps, or in long chains. Relatively few important diseases of humans or animals are attributable to this group of organisms. Fungal diseases in humans are generally less acute than those produced by other organisms and are for the most part low-grade chronic infections such as ringworm and "athlete's foot." Some fungi are capable of producing serious diseases in humans, examples of which are histoplasmosis and coccidioidomycosis.

2.7.10 Toxins

Toxins are relatively unstable, poisonous substances often chemically related to proteins, and may be of microorganism, plant, or animal origin. Recall that they are in truth chemical agents, as they are not living but rather are large molecules that have been isolated from biological sources. It is convenient to classify them with biological agents because not only are they of biological origin but are themselves responsible for many disease symptoms and can often be deactivated by means similar to those for biological agents.

Toxins are isolated from three sources:

- *Toxins of microorganisms* are produced by some bacteria and possibly by some of the other microorganisms, such as rickettsia and viruses. Examples include those causing the symptoms of botulism, tetanus, cholera, and typhoid.
- Those *zootoxins* (toxins of animal origin) that are particularly dangerous to humans are usually found in insects, reptiles, and fishes. *Antivenins* (antibodies developed against venoms) have been prepared against a number of these substances and have been employed with considerable success; these are a form of therapy as they are used postexposure.
- *Phytotoxins* (toxins of plant origin) are rare; many of the poisons produced by plants are not considered to be toxins.

Examples of toxins include:

- Ricin, which is a protein produced by the castor bean plant
- Saxitoxin, which is a neurotoxic alkaloid produced by certain types of algae
- Botulinum toxin, a protein produced by a bacterium, *Clostridium botulinum*

Toxins may have properties in common with either CW or BW agents. Their toxicity lies midway between CW and BW agents. Like a BW agent, they are usually relatively easily deactivated by heat, decontaminants, and ultraviolet light. Also like BW agents, toxins are usually antigenic and induce the production of specific antitoxins (antibodies developed against toxins) in suitable animals. However, since they are CW agents, they are not infectious or transmissible. In their pure form, they are dry powders that can be spread as aerosols or dissolved into food or water; aerosolized toxins are the form of primary concern for use of PPE.

Historical note: Use of toxins

Clostridium botulinum was cultivated by the government of Iraq to produce the botulinum toxin. Ricin has been isolated by a number of people with intent to use in domestic terrorism incidents. Bulgarian dissident Georgi Markov was assassinated in London in 1978 by somone using a ricin-tipped umbrella.

Toxins produce their poisonous actions in a variety of ways; they cause serious or fatal illnesses such as botulism, tetanus, gas gangrene, and diphtheria, producing symptoms of illness such as cardiac or respiratory complications, convulsions, blindness, nausea, and fever. They may be ingested, inhaled, or absorbed through the skin, depending on the toxin. Some toxins do not develop their poisonous effects immediately; they may have a delayed action that varies with the particular toxin (although not as delayed as production of disease by microorganisms). It should be noted that other chemicals isolated from biological sources not technically considered to be toxins could still be used as hazard agents.

2.7.11 Other Important Methods of Control

Biological agents are particularly amenable to other forms of prevention and remediation common in everyday life that may reduce the burden of PPE.

Immunization. A number of diseases have effective vaccines, and many category A and B diseases now have programs for development of vaccines that did not exist previously. Vaccines may not be completely effective when a disease possesses many strains that mutate continuously (e.g., the influenza virus); in this situation, constant maintenance of the vaccine is required to sustain its potency.

Therapy. Many diseases can be controlled by antibiotics (against bacteria) or antivirals. The challenge often comes in recognizing the nature of the infection during the early course of the disease, which is when the organism is most susceptible. Similarly, poisoning by toxins can be countered by the appropriate antitoxin. General supportive care and specific medical interventions can often reduce the likelihood of death or serious disease; however, in an epidemic, medical resources are seriously strained.

Decontamination. Generally speaking, all of the biological agents (including the toxins) are easier to decontaminate than are chemical or radiological agents. The use of enough of virtually any harsh agent (acid, alkali, oxidizer), high temperature, or ultraviolet light will eventually kill virtually all agents, leaving no additional hazardous by-products, although spores may be particularly resistant to temperature. However, because infectious doses may be very low, the decontamination problem may still be significant.

Hygiene Measures. Reducing the likelihood of infection by biological agents can be performed by some basic hygiene measures already commonly practiced, such as protection and sterilization of food and water sources, hand washing, isolation of contagious persons, and sterilization of potential fomites.

2.7.12 Transmission and Dissemination

Biological agents may be of two types: *communicable* or *noncommunicable*. **Communicable agents** infect people with a contagious form of disease, the primary route of transmission being from *person to person* (i.e., via a human vector or, sometimes, via a fomite). This can occur via various bodily secretions that usually target the eyes, digestive tract, or respiratory tract as a route of entry. Deliberate initial dissemination by this means would at some initial point involve a person or persons already infected with the disease; once transmission had occurred into the general population, it would continue to transmit. Such transmission can lead to outbreaks in which a significant fraction of the population becomes infected simultaneously (particularly where general population immunity to the disease is low).

Modes of transmission of disease directly or indirectly from person to person include:

- Sneezing or coughing followed by inhalation of droplets or agent aerosol from the air (e.g., RT illnesses)
- Contact with agent on skin, skin eruptions, or contaminated surfaces which is transferred to eyes, nose, or mouth (e.g., smallpox, many RT and GI illnesses)
- Drinking water or eating food contaminated by microorganisms (often from feces/GI illnesses; e.g., salmonella, hepatitis A)
- Sexual transmission [e.g., HIV (human immunodeficiency virus) various venereal diseases]
- Bloodborne, from open wounds or blood transfusion (e.g., hepatitis B and C)

Noncommunicable agents infect humans only at certain points in their lifecycle or are primarily diseases of animals that may affect humans incidentally; in either case, the disease will have passed through an animal vector before reaching the human and

TABLE 2-2 Routes and Risks of Transmission of Disease

Route of Transmission	Risk	Examples of Disease
Blood, mucus, sexual transmission	Low with simple precautions	Gonorrhea
Casual body contact	Moderate	Ebola
Insect vector	Depends on the vector	West Nile, plague
Fomites, surfaces	Depends on the persistence of the organism	Common cold
Through air	Often high	Influenza, plague, smallpox

is generally noncommunicable from human to human. Examples of diseases caused by these agents include:

- Malaria, West Nile virus: transmitted via the bite of mosquitoes
- Plague and typhus: transmitted via the bite of fleas that have colonized rats
- Anthrax: often transmitted via the carcasses of animals that have died from the disease or soil surrounding the carcass and entering by a variety of routes, including RT, GI, and broken skin
- Avian flu: transmitted from infected animals or carcasses

All of the foregoing modes of transmission that are naturally occurring could, at least in theory, be used intentionally in order to spread disease. Routes and risks of transmission are summarized in Table 2-2.

Most commonly person-to-person transmission through air occurs via larger droplets with a limited range of a meter or two; however, the potentially underrecognized role of aerosol transmission in disease has also recently been discussed [11], and a few organisms can transmit from person to person over significant distances in aerosolized form (common examples are measles and tuberculosis). Many organisms that would not normally be transmitted in aerosolized form could be deliberately disseminated as aerosols.

Historical note: Use of Bacillus anthracis

An attempt by the Aum Shinrikyo cult to release *Bacillus anthracis* did not result in a single infection, emphasizing the technical difficulty of infecting by the airborne route. The anthrax letter attacks were only moderately effective at causing infection, but far more effective at causing societal disruption.

In addition, a disease can be disseminated intentionally by using a cultured organism that is released into the environment in a form that can reach a few or a large number of people. These methods include:

- Aerosolization or droplet formation from a liquid medium using sprayers or cluster bombs, for example
- Aerosolization from a dry powder (a few agents only, either toxins or spores)
- Contamination of food or drinking water (untreated or posttreatment)
- Contamination of projectile weapons (has been used historically in warfare and assassination, particularly for toxins)

Historical note: Salmonella

In Oregon in 1984, the followers of Rajneesh contaminated 10 salad bars with salmonella, infecting 750 people.

Growing, isolating, and keeping an agent alive through the dissemination process is quite difficult for most agents, and hence deliberate wide-area dissemination that does not ultimately involve person-to-person transmission, and does not involve a spore form of an agent, is a particular technical challenge that most aggressors would find difficult.

In summary, there are several ways or types of events in which biological agents may be used that could require the use of PPE:

- Those that involve release of an agent that may cause subsequent disease; during the initial phases of such an event, if it is recognized at all, efforts would be expended on minimizing further spread and assessing who has been exposed, as no actual disease may manifest itself for several days; the organism may or may not be contagious (e.g., plague vs. anthrax); respiratory and eye protection combined with other clothing as a means to minimize spread and ease decontamination are needed.
- Those that involve person-to-person spread of disease; this could be the later stages of a release event that involved a contagious organism or could be started by a human vector; respiratory and eye protection are the highest priority.
- Those that involve transmission via a nonhuman vector, typically a biting insect, which may require skin protection as a priority.

Physical Form of the Hazard. When designing PPE, there are essentially two physical forms of biological hazard that must be considered: contact hazard and airborne aerosols, sprays, and droplets. The nature and scale of the contact hazard may range from isolated organisms deposited on surfaces, through pools of body fluids. For an airborne biological agent, the characteristics of the airborne cloud are as always highly dependent on the mode of generation. Relevant biological agents potentially vary in size from 0.01 μm for the smallest viruses through 10 μm for spores; therefore, the size of a particle containing a biological agent cannot be less than that of the agent itself. Aerosol and spray can be generated by mechanical sources starting from a slurry or liquid solution (easy to produce but difficult to disseminate to generate high concentrations of viable organisms) or dried materials (difficult to produce yielding viable organisms, but once produced, easy to disseminate in high concentrations). Various types or scales of sprayers could conceivably be used (ranging from nebulizers to agricultural sprayers), and generally the total volume of viable organisms that could be spread is proportional to the particle size (i.e., the most potentially effective micrometer-size aerosols are difficult to produce in large amounts, as the rate of production is slow, and shear stress involved in the production is likely to result in a smaller viability of the organism relative to larger droplet sprays).

Aerosol and spray generated by human sources has particular characteristics [12]. Coughing or sneezing generates a substantial quantity of particles, a large number of which are <5 to 10 μm in diameter, as reviewed by Nicas et al. [13]. Particle size changes after emission as a function of evaporation (in dry air) or the tendency to rehydrate (in humid air, such as in the lungs) [14].

Environmental effects on biological agents after dissemination can be significant. The ultraviolet light in sunlight will kill most microorganisms and degrade many toxins. Some microorganisms have a relatively narrow range of temperature and humidity (similar to that of their host) to maintain viability, although cold may also act to preserve the organism. Most microorganisms can remain alive only for a reasonably short time (hours to days) outside their host. Spores that are formed specifically to keep an organism alive under harsh conditions are much more resistant to environmental conditions.

2.7.13 The Agents of Concern

The U.S. Centers for Disease Control and Prevention (CDC) have defined a number of agents of particular concern [15] either because they have previously been militarized or because of their particular virulence and potential for spread by contagion. **Category A diseases and agents** include those that:

- Can easily be disseminated or transmitted from person to person
- Result in high mortality rates and have the potential for major public health impact
- Might cause public panic and social disruption
- Require special action for public health preparedness

The list of **category A agents** consists of:

- Anthrax (*Bacillus anthracis*)
- Botulism toxin (*Clostridium botulinum* toxin)
- Plague (*Yersinia pestis*)
- Smallpox (*Variola major*)
- Tularemia (*Francisella tularensis*)
- Viral hemorrhagic fevers [filoviruses (e.g., Ebola, Marburg) and arenaviruses (e.g., Lassa, Machupo)]

Category B diseases and agents include those that:

- Are moderately easy to disseminate
- Result in moderate morbidity rates and low mortality rates
- Require specific enhancements of current diagnostic capacity and enhanced disease surveillance

The **category B agents** consist of:

- Brucellosis (*Brucella* species)
- Epsilon toxin of *Clostridium perfringens*
- Food safety threats (e.g., *Salmonella* species, *Escherichia coli* O157:H7, *Shigella*)
- Glanders (*Burkholderia mallei*)
- Melioidosis (*Burkholderia pseudomallei*)
- Psittacosis (*Chlamydia psittaci*)
- Q fever (*Coxiella burnetii*)
- Ricin toxin from *Ricinus communis* (castor beans)
- Staphylococcal enterotoxin B
- Typhus fever (*Rickettsia prowazekii*)
- Viral encephalitis [alphaviruses (e.g., Venezuelan equine encephalitis, eastern equine encephalitis, western equine encephalitis)]
- Water safety threats (e.g., *Vibrio cholerae*, *Cryptosporidium parvum*)

Three category A agent diseases—plague, anthrax, and smallpox—pose particularly difficult challenges for protection. **Plague** is caused by *Yersinia pestis*, an organism that has been responsible for three recorded pandemics, occurring in the seventh, fourteenth, and nineteenth centuries. It is now endemic in a number of regions around the world. It is carried by a number of mammals as well as by fleas during their life cycle and can be transmitted directly within and between species by a variety of mechanisms (direct contact with bodily exudates, airborne transmission, ingestion of an infected animal, flea bite).

- *Bubonic plague* is the most common, caused by a flea bite, and causes 30 to 70% deaths if not treated. The bubos are painfully swollen lymph nodes. The blackened dead tissue in the limbs gave it the name "black death."
- In *septicemic plague*, the organism infects the bloodstream without obvious lymph node involvement and the toxins spread throughout the body. It is rapidly, and almost always, fatal.
- *Pneumonic plague* is the most deadly and high transmissible form of disease, and almost impossible to treat. The organism infects through the lungs and spreads via the airborne route. This is the form most relevant to bioterrorism scenarios and PPE requirements.

Anthrax (caused by *Bacillus anthracis*) is in people's consciousness throughout the world, due to its weaponization by Iraq as well as a number of recent domestic incidents in the United States. *B. anthracis* in the form of spores is a dangerous weapon because it is highly lethal, easy to produce in large quantities, easily spread in the air over a large area, easily stored, and highly persistent. Anthrax is not communicable; that is, it cannot be contracted by contact with a live infected person or animal. However, once dead, people and animals are a source of the spores that are the infectious form of the organism. Anthrax is found in the wild as a disease of various ruminant animals, such as cattle and deer, which contract the disease by disturbing spores lying dormant in the ground as they graze.

Anthrax can occur in various disease forms, all a result of poisoning by the toxin produced:

- *Respiratory anthrax*, resulting from inhalation of anthrax spores, is almost always (90%) fatal without treatment.
- *Cutaneous anthrax* results from the entry of the spores through broken skin and is a localized infection of the skin. This form has a lower mortality rate if untreated (up to 30%).
- *Anthrax of the digestive tract* has a relatively high mortality rate, resulting from consuming food contaminated with spores. This form of anthrax is relatively common in developing countries, resulting from eating an animal that has died of anthrax.

There are various historical indications of the viability and lethality of this particular BW agent. It is known that the accidental release of anthrax spores from a germ warfare laboratory in Sverdlovsk resulted in the deaths of 68 people and hundreds of animals downwind of the facility. Viable anthrax spores were found in the King's Cross subway station in London in 1992; these spores came from horsehair embedded in the wall plaster installed in the nineteenth century. Gruinard Island, also-called "Anthrax Island," which was a test site for anthrax biological warfare in the UK in the 1940s, was contaminated for decades afterward until a concerted decontamination program was implemented. Finally, a number of people have died or been infected by anthrax in its cutaneous and respiratory forms, due to the contamination of mail in the United States in 2001.

The causative agent for **smallpox**, the *Variola* virus, is a controversial member of the category A agents. It is to be hoped that it is unlikely to be available for use in bioterrorism, as there are only two known and closely held stores of the agent in the United States and Russia. Smallpox is a disease with acute morbidity and high mortality rates (up to 30%). Its most important symptom is a debilitating rash that covers the entire body in small bumps accompanied by fever. A person is contagious the entire time the rash is present and sometimes before. Although close contact with an infected person or contaminated object is usually necessary for disease transmission, airborne transmission for large distances within buildings has been documented [16], probably resulting from sores that form within the mouth, permitting aerosolization during coughing.

As noted previously, smallpox is classed as a category 4 agent for biosafety risk because although a vaccine to the virus exists, any release of smallpox into the general population would still cause significant infection and mortality until it was possible to produce enough vaccine to totally prevent spread. Although there is limited clinical information on the current vaccine due to the fact that it is impossible to test it in actual use until an outbreak occurs, it is likely that, like its predecessor, the vaccine is effective even after initial infection has occurred provided that no symptoms have yet appeared. Because the current vaccine itself is not without serious side effects, it is reserved for emergency or essential use. There is some recent indication that the immunity to smallpox imparted by the smallpox vaccine previously in use is very lengthy [17].

2.8 RADIOLOGICAL AND NUCLEAR AGENTS

2.8.1 General Hazards

Radiological and nuclear hazards are generally listed independently of each other, because of the difference in the way the hazard is generated. Nuclear hazards resulting from explosion of a nuclear weapon or power source imply a particular combination of hazards likely to be much more severe in magnitude than any radiological source could generate. In either case, the capability of PPE to protect against the effects is limited. Some protection against the particular immediate hazards posed by detonation of nuclear weapons, such as heat and blast/shock waves, is possible, depending on the location of the wearer relative to the source. Longer-term hazards arise from the radioactive materials produced, and protection should address minimization of direct radiation as well as prevention of contact with, or ingestion or inhalation of, the actual radioactive material. Radioactive material will usually be in some solid form (with a few exceptions, such as tritiated water or radioactive iodine vapor). The airborne form (i.e., solid particulates) is of particular concern in designing protection.

The hazards posed by various forms of radiation are summarized below.

- **Neutrons, gamma, and x-ray radiation**, due to their high penetrating power, cannot be protected against by any normal protective equipment while the wearer is in the vicinity of the radiation. Only high-atomic weight materials (e.g., lead) of a significant thickness can provide any protection against high energy radiation, although the lower-energy neutrons are stopped sufficiently easily that effects within the body will be nonuniform (with higher effects facing the radiation source).
  - Dosimetry to monitor exposure, distance from the source of radiation, and reduction of exposure time are the best defense against these forms of radiation.
  - PPE can contribute to protection by preventing contact with, or internalization of, emitting materials, thus easing subsequent decontamination.

  - Neutrons may have varying energy and penetrating power; a nuclear explosion will produce high-energy neutrons. Neutron radiation will only be found immediately after a nuclear weapons or reactor incident.
- **Alpha and beta radiation** in general have low penetrating power through materials; only very high-energy beta radiation has high penetrating power.
  - Low-energy alpha and beta particles are protected against relatively easily using clothing but are still dangerous when radioactive materials are internalized (by inhalation or ingestion).

To prevent internalization of radiation, the initial focus is on protection of the respiratory and digestive tracts. Secondary focus is on protection of skin against aerosol deposition that may add to overall body dose and be a reservoir for contamination transfer. The *half-life* of the agent (the length of time it takes to decay to half its original strength) and the energy of the radiation determine the overall radiation hazard from a given isotope. Shorter-half-life agents emit radiation at a faster rate but decay away more quickly, and certain energies of radiation interact with the body more effectively than others do, causing damage.

2.8.2 Examples of Hazardous Radiological and Nuclear Agents

Table 2-3 outlines some of the most hazardous R and N agents of concern. The R agents may be found in various types of mostly commercial devices that produce high-energy radiation for power generation, sterilization, or radiotherapy. From the long list of known radioactive isotopes, only a few stand out as being particularly suitable for radiological terror. In addition to those in Table 2-3, ^{90}Y, a daughter product of ^{90}Sr, ^{192}Ir, ^{226}Ra, ^{241}Am, and ^{252}Cf, may be a possible terror agent [2,18,19]. Other possible hazards include stolen nuclear wastes, which may contain additional isotopes, such as ^{233}U, ^{237}Np, ^{243}Am, and ^{99}Tc [2].

2.8.3 Types of Short- and Long-Term Effects

Acute radiation exposure results in the symptoms of radiation poisoning: skin lesions, gastrointestinal illness, hair loss, sterility, and immune suppression. In general,

TABLE 2-3 Radiological and Nuclear Agents of Particular Concern

Isotope	Origin of Source	Radiation Type
^{90}Sr [from $SrTiO_3$ or $Sr(NO_3)_2$]	Radioisotope thermoelectric generator	β
^{137}Cs (from CsCl]	Industrial irradiator	γ
^{60}Co (from Co metal)	Industrial irradiator	γ
^{239}Pu (from Pu metal)	Nuclear weapon	α
^{238}Pu (from PuO_2)	Radioisotope thermoelectric generator	α

actively proliferating (rapidly growing and dividing) cells are most sensitive to radiation. These include cells in the digestive tract, blood, skin, and bone marrow, and their selective death results in many of the classic symptoms of radiation poisoning.

Radiation often induces programmed cell death, usually by fragmentation. Nonlethal changes in cellular function can occur as a result of lower radiation doses. Dose-dependent inhibition of mitosis (cell division for the purposes of replication) is particularly common in actively proliferating cell systems. Cell growth may also be retarded. The motility of a cell may be decreased following irradiation. Local tissue damage after external irradiation includes the resulting swelling, blistering, and ulceration.

Radiation can also result in chromosomal and cellular repair abnormalities, which can be responsible for susceptibility to illness and carcinogenesis. Death after a large whole-body exposure may result in a few days from neurovascular effects, or weeks after exposure from sepsis from immune system failure. Permissible exposure levels for workplace exposure are determined by the likelihood long-term effects that occur at much lower dose levels; these effects include life shortening, carcinogenesis, cataract formation, chronic radiodermatitis, decreased fertility, and genetic mutations. It is noteworthy that many radioisotopes are also chemically toxic; for example, heavy metals such as uranium are toxic to the liver and kidneys even in the absence of radiation, and therefore ingestion must be prevented for this reason also.

2.8.4 Route of Entry

Inhalation is the most likely route of internalization for which PPE is relevant. Some radionuclides exist in the form of gases or vapors (e.g., radioactive iodine and tritiated water) and will enter the body easily by this route. The relative amount of deposition of particles in the various parts of the airways is governed by the particle size, as outlined in Section 2.4.1. The International Commission on Radiological Protection (ICRP) respiratory tract model [20] and improvements that include the nasopharyngeal region [21] describe how dose is distributed among the various parts of the airways. Insoluble particles will remain and irradiate respiratory tissue for years, resulting in scarring and inflammation. Particles that are cleared in the mucus may subsequently be swallowed and end up in the digestive tract, from where some soluble materials may be absorbed (e.g., cesium) and some will be insoluble and will clear in 1 to 5 days (e.g., plutonium, radium, and strontium). The only radionuclide that can easily pass through intact skin is tritium in radioactive water.

2.8.5 Dissemination

As noted previously, the primary difference between radiological and nuclear agents is their source. Radiological agents are essentially radioactive material that is intended for peacetime uses, whereas nuclear agents are disseminated exclusively as a result of a nuclear reaction (either because of an accident in a nuclear reactor or an explosion of nuclear weaponry). These differences result in different isotopes and physical forms and particle sizes; however, in principle, these forms of radioactive hazard are no more different than toxic industrial chemicals are from chemical warfare agents (i.e., they are all radioisotopes).

Historical note: Polonium poisoning

The death of Alexander Litvinenko in 2006 resulted from exposure to and ingestion of microscopic quantities of ^{210}Po. Although an assassination was presumed by Litvenenko himself as well as the authorities, it has not been proven.

Radiological materials can be used to poison people through contamination of food, water, and surfaces. Radiological agents would be disseminated most effectively to cause the widest hazard as an airborne powder. The "dirty bomb" or *radiological dispersal device* (RDD) concept exemplifies this, that is, a small-yield explosive intended to spread a fine radiological powder over a large area, resulting in significant local surface contamination and an airborne cloud [19]. The explosion itself is not necessarily intended to create the particulate. If the source is not already powdered, the result will probably be larger chunks of contaminated material. Regardless, the radiological hazard that results is that of the original isotope released and any decay products over time. RDDs would be particularly effective at yielding contaminated wounds.

On the other hand, a nuclear explosion generates a large variety of radiological hazards, much less well defined than for an RDD. The nuclear explosion generates radiation and radiological particulates and vapors, along with the intense light flash, ultrahigh-energy blast, and thermal effects [22]. Subsequent to the release, radiological materials also adhere to environmental particulates as well as those formed in the explosion in the atmosphere, coming to ground eventually as fallout downwind. Additionally, neutrons released in the explosion "activate" other materials, such as soil, rendering them radioactive.

Fallout from a fission weapon can continue over a period of years after such an explosion [2], with the radioactive composition changing over time and depending on the altitude of the explosion, with a high atmospheric explosion resulting only in the longer-lived isotopes falling out:

- Short-term fallout in the first days to weeks contains such isotopes as ^{131}I and ^{140}Ba, as well as unspent ^{233}U, ^{235}U, or ^{239}Pu.
- After a few months, somewhat longer-lived isotopes, such as ^{141}Ce and ^{89}Sr, dominate.
- Eventually, over a period of years, isotopes such as ^{144}Ce, ^{106}Ru, ^{147}Pm, and finally, ^{137}Cs, ^{90}Sr, and ^{239}Pu remain.

The factors that determine the extent of the anticipated fallout hazard relate to the specific design of the weapon and whether it is fission or fusion, the altitude of the burst, the surface composition if it is released near the ground, meteorological conditions, and time after the explosion.

Release from a nuclear reactor is another scenario that has occurred all too frequently. The releases from Chernobyl in 1986 and Fukushima in 2011, although not deliberate, demonstrate the short- and long-term hazard that could result from an attack on a nuclear facility. NATO ranks the level of hazard from such releases as ranging between source term categories 1 and 5. Fission products of concern for a containment or evacuation operation include isotopes of barium, cesium, iodine, krypton, lead, rhenium, rubidium, ruthenium, strontium, technetium, tellurium, xenon, and yttrium. Internal doses to organs are dominated by ^{90}Sr, ^{131}I, ^{137}Cs and ^{239}Pu. Further information on types of hazards is given in International Atomic Energy Association documentation [23].

2.9 SUMMARY OF DISSEMINATION OF CBRN AGENTS

2.9.1 Methods of Dissemination

Most of the methods summarized in Table 2-4 have already been mentioned where they are relevant to each type of CBRN agent but are summarized here for ease of reference. Only those that are pertinent to the selection and use of PPE have been included here; for example, contamination of food and water has little relevance in PPE selection, as it is dealt with by other means. The choice of dissemination method could depend on the agent and its physical form, the way in which it is available or stored, the availability of a means to disseminate, and/or the nature of the target. Biological agents are usually in some form of aqueous solution or bodily fluid, although dry aerosols are also possible for spores. To permit dissemination; most radiological agents are in some form of powder; a very few are gases. Chemicals may take many forms.

TABLE 2-4 Methods of Dissemination

Method of Dissemination or Spread	Class(es) of Agent for Which It Could Be Used	Physical Form of Agent
Explosive dispersal (weapon, explosives)	C, B spores, R, N	Any
High-pressure release from/rupture of container or storage vessel	C, N	Any
Sprays and mists	C, B, R	Aerosol evolving into vapor
Release into air circulation systems	C, B, R	Aerosol/vapor
Small packages and envelopes (passive spread of powder)	C, B, R	Aerosol
Fomites, spread via contaminated objects or people	C, B, R	Contact hazard
Communicable diseases	B	Aerosol, spray and contact hazard
Biological vectors	B	

2.9.2 Persistency and Environmental Effects

Persistency is the tendency of an agent to remain a hazard over time; the less persistent an agent is, the shorter the time protection will be required. A variety of factors will have an impact on persistency, relating to the agent, the method of dissemination, and the surroundings.

Physical State. The first aspect of persistency is related to the tendency of an agent to stay in one physical form rather than evolving into another. Solids and liquids with low volatility (i.e., low tendency to evaporate) are likely to remain unchanged in physical state for a long time after dissemination. Liquids with high volatility will tend to evaporate quickly into gases. As the temperature increases, the rate of evaporation also increases; once the boiling point of a liquid is reached, it exists only as a gas. Very low temperatures may cause liquids to freeze, making them potentially less of a hazard. Bulk liquid may be a very high local hazard, which also evolves a spreading, lower-level hazard, due to evaporation. This aspect of persistency is very important, as it determines what physical form of agent must be protected against.

Surface Contamination. Surface contamination may persist long after other forms of agents have ceased to be a hazard. Aerosols will tend to stick to surfaces, reducing the hazard in the air over time but increasing the contact hazard and the likelihood of spread by transfer from contaminated surfaces. Many surfaces will absorb chemical agents, which will then off-gas slowly for a significant period, causing a low-level but persistent hazard. Liquid chemical agents that have a low volatility and a high viscosity (gooey or oily) are a particular hazard in this regard. Radiological materials can also penetrate porous surfaces. In general, the result of these surface effects is that the contact hazard (often relatively low level) will persist much longer than a hazard from airborne agents or from bulk contamination, which can be seen and removed relatively easily.

Particle Size Distribution. Another issue is the tendency of aerosol particles to change size after dissemination, for a variety of reasons. The size of the particulate has a significant impact on the ease with which it remains suspended in air (generally speaking, 0.1- to 30-μm particles can remain suspended in air for some time, with those outside this size range being referred to as ultrafine and supercoarse) and the efficiency with which the particular type and size of particulate can be removed by filtration systems. High-volatility aerosols will evaporate, decreasing in size and eventually transforming completely into vapors; low-volatility aerosols will tend to agglomerate, increasing in size (particularly biological and radiological, as they are usually hygroscopic). Additional factors affecting the rate of agglomeration include the surface structure and charge of the particulate. Aqueous aerosols will generally evaporate more quickly in dry weather, but their change in size is less easy to predict in humid weather. Larger particulates will fall out of the air or adhere to surfaces, while slightly smaller ones will remain suspended longer. As a result, the particle size distribution of an aerosol will be continually evolving.

TABLE 2-5 Form of Agent Relative to Release Point

	Toxic Material and Its Form			
	C Vapor Including TIC	C, B, R Airborne (Droplet, Aerosol)	C Liquid, Including Thickened	N Explosion
Location and nature of hazard resulting from initial release	Vapor close to release point, downwind, or within building; progressing to far from release point, hundreds of meters to kilometers downwind	Close to release point, downwind outdoors or within building; vapors or fine persistent aerosols may be carried far downwind; off-gassing of vapors, emission of radiation; in the vicinity of infected persons (sneezing, coughing)	Falling, spray (not thickened), or explosive liquid droplets or bulk—close to release point	Fire, blast overpressure, intense heat, and extreme radiation exposure up to kilometers from explosion
Hazard resulting from contamination transfer to another location (at any time after release) or from persistent agents at later times (e.g., recovery phase)	Off-gassing from contaminated environment or surfaces	Reaerosolization, off-gassing, radiation, contamination transfer from contaminated environment or surfaces	Off-gassing or liquid contamination transfer from contaminated environment or surfaces	Reaerosolization, transfer, and radiation from contaminated environment or surfaces

Degradation and Dilution. Environmental factors have a significant impact on the stability of chemicals, the viability of organisms, and the tendency of a hazard to spread, all of which affect the concentration of agent present in a particular location. Ultraviolet light in sunlight degrades most chemicals and kills most biological agents. Many organisms require moist air to remain viable, and an aerosol of biological organisms suspended in water will evaporate more slowly under humid conditions. Rainfall will dilute any agent, washing it out of the air and off surfaces; water also degrades some chemical agents. Wind and air movement dilute vapors and aerosol suspensions while spreading a hazard over larger areas. High temperature may also degrade chemicals and kill certain organisms, although spores tend to be resistant to decay. Radiological agents decay naturally over time, depending on their half-life.

2.9.3 Summary of CBRN Hazards and Their Location

Table 2-5 summarizes where and when the various forms of CBRN hazard may exist relative to their initial release. The initial hazard covers the time period shortly after the release, including the response and intervention phases of a terrorism event. The hazard remote from the release, or later, may persist or be transferable to other locations.

3 Setting High-Level Requirements

In this chapter we help the reader understand how to set the stage for detailed requirements and specifications development by understanding how the equipment will be used and what is required to maintain optimum human performance.

To select or design protective equipment, the conditions of exposure and use, as well as the level of acceptable effects, must be understood. In this chapter we discuss establishing high-level requirements: What type or level of protection does the user need based on his or her overall operations? What types of factors might influence the ability of the equipment to perform? What is the objective of providing protection in terms of the capabilities that are being assured? Trying to provide the same "one size fits all" PPE for every possible user within an organization, in every possible concept of use, is an impossible task. A compromise must be found in which appropriate equipment is provided to the largest number of users while not sacrificing essential performance characteristics, and initially, it will not necessarily be obvious how to achieve these compromises. Hence, at the early stages of the requirements-setting process, the most possible information should be retained without simplification. Any differences in high-level requirements that are identified between the operational needs of various user groups, or that are imposed by different response or event types, should be documented. Later, during the design or procurement process, further decisions may need to be made to achieve the optimum selection of PPE. The process of setting specifications (i.e., detailed performance values that must be achieved in specific tests) are discussed in Chapters 6 and 7.

3.1 DEFINING CONCEPTS OF OPERATIONS

The process of equipment design or procurement must have a clear focus on what the equipment will be used for, including understanding the intended users, and how and when the equipment will be used. The first step in the process, then, is to identify all possible user groups of interest that may have specialized requirements. Differences in these requirements arise from the nature of a group, the types of tasks that the group performs, their duration, and their location. This step in the process should be documented by discussion with each possible user group. Often, the user group may have a fairly poorly defined concept of operations for a CBRN environment, since

Personal Protective Equipment for Chemical, Biological, and Radiological Hazards: Design, Evaluation, and Selection, First Edition. Eva F. Gudgin Dickson.

it has rarely been needed, due to the fortunate infrequency with which such events occur. Hence, before the process can proceed further, this concept of operations must be sufficiently refined.

- Tasks to be performed and types of hazard may differ depending on the type of CBRN event that has occurred, and hence a full review of all such possible events must be performed; this process is discussed in Section 3.4.
- All relevant tasks and duties within a CBRN response must be identified and divided up among all the general or specialized user groups. Even if an organization performing this task is responsible for only part of the response, they must be assured that assignment of all tasks is agreed upon with other cooperating organizations; examples are given in Sections 3.2 and 3.3.
- Specific operational issues may exist for each user group, based on the roles that they fill, which must be documented.
- Requirements may also ultimately differ between groups based on such outside factors as regulations or labor agreements that put limits on working conditions or exposures; exposure limits are discussed in Section 3.5.

The concept of operations also includes such issues as whether the PPE is carried by the person and when various available levels of protection might be implemented [24].

In the sections that follow we outline some of the relevant user groups and operations that have been identified within the military and civilian contexts: that is, support of wartime and deployed operations vs. domestic terrorism events. These lists are not intended to be exhaustive, but contain numerous examples of the types of operational issues that will be important in setting high-level requirements. These two different contexts of operations are governed by different rules and requirements with regard to the nature of the response and the level of risk that may be expected to be borne by the organization and by each person within it. The resources available to military vs. civilian organizations may be quite different as well. This can depend on the importance that is given to the capability of each group to be able to support CBRN operations, the rapidity with which they need to be deployed, the availability of surge capacity from other organizations, and the location of the operation.

3.2 MILITARY OPERATIONS

3.2.1 Concept of Operations

Although the primary focus for many military groups within organizations such as the North Atlantic Treaty Organization (NATO) will be deployed operations overseas, domestic response support is increasingly becoming part of their mandate. These two different theaters of operation may imply significantly different expectations. Operational objectives that have a direct impact on the use of PPE, the types of events, and the roles to be played within them may be quite different.

Domestically, response to and containment of the CBRN event is the priority; the event is more likely to be isolated rather than a part of sustained operations; and interoperability with civilian responders is required. The information in Section 3.3 should be reviewed in this context. The priority in the military theater is strategic, involving mission success and sustainment of operations. Countering the CBRN threat is not, in itself, a primary goal but rather a part of operating within a battle space in which many other hazards exist. The use or threatened use of CBRN agents can severely hamper the ability of a force to maintain its tempo of operations, and protecting the individual can contribute to this issue in either a positive or a negative way. Both individual and *collective protection** [25] must be provided.

In overseas operations in the current political climate, use of CBRN weapons is most likely to be a tactical occurrence of limited frequency that must be managed within a larger operation [26]. Sustainment of operations involving many thousands of individuals must be performed, even though the cold-war scenario of prolonged fighting in a "dirty" environment is no longer considered likely. This means taking into account both the necessity to continue to perform normal operations as well as the ability to sustain CBRN defense capability; all CBRN PPE has a limited lifetime in use. Interoperability with other country's forces is likely to be necessary, and differences in protective capabilities need to be understood and planned for.

In general, individual users are likely to be issued their own equipment, either upon deployment or once a threat situation has been identified, making issues of sizing and fitting manageable through use of a sufficient stockpile of PPE. Specialized equipment must be pre-deployed, as the time taken to resupply could be lengthy. A high-level CBRN requirements concept document for NATO is available at [27].

3.2.2 Military User Groups

Military organizations tend to be extremely large and multifaceted, with a variety of specialized tasks that all have the potential to be required in CBRN operations. Many individuals (e.g., infantry) may have need for general-purpose PPE that can be worn under many different circumstances, whereas other groups will have particular needs that must be addressed individually. A representative sample of various military user groups and some of the pertinent operational issues are outlined here, based on information prepared for NATO protective clothing standards.

Combat Soldier, Dismounted. Dismounted combat soldiers potentially face the widest possible variety of exposure environments while having high physiological stress levels. They are asked to perform a large variety of tasks at high activity levels and use many different types of equipment, and may spend a great deal of their time requiring individual protection rather than in any form of collective protection. PPE design and integration that do not significantly degrade performance of individuals

* Collective protection is protection for groups of people who will not be wearing complete PPE; examples include shelters, buildings, and various means of transport, such as vehicles, ships, tanks, and aircraft.

and their associated equipment, while remaining functional and protective in many environments, are a particular challenge.

Combat Soldier, Mounted. Armor, infantry, artillery, and logistics personnel perform their duties within the environment of a small moving ground vehicle. Military vehicles are subjected to severe fire, flame, and explosion hazards, while their occupants suffer greater heat stress and space constraints. In enclosed vehicles, air filtration and cooling systems may be built in. In open vehicles, an increased risk of CBRN challenges at elevated wind speeds exists, due to the movement of the vehicle.

Special Operations. Special forces personnel need high mobility and low signature protective concepts, and may be exposed to a large variety of environmental conditions. They may need PPE to integrate with specialized equipment; their decision to wear PPE (what type, and when) may also be based on a very specific mission requirement that is understood in advance.

Fast Jet Pilot and Other Closed Airframe Personnel. The interior of an airframe is similar to that of an enclosed vehicle, described above. Respiratory protection is often provided by powered air or breathing gas systems, providing compatibility with provision of oxygen. Ditching over both land and water is a design consideration for PPE components.

Helicopter Pilot and Other Open Airframe Personnel. Most of the issues here are similar to those for closed airframe personnel, but with a potentially increased risk of exposure to vapor and aerosol challenges and high airflows during operational flights. As well, tactical helicopter personnel may have a number of ground roles; the respiratory protective system must be mobile and not tied exclusively to an air or oxygen supply within the airframe.

Aircraft Rear Crew. Aircraft rear crew experience much the same environment as mounted ground personnel, but with the potential for higher wind speeds.

Air Force Ground Crew. These crew members face a relatively known environment and can work under cover. However, their exposure is high to various contaminants, such as jet fuel, that can degrade the protective performance of materials.

Amphibious and Maritime Personnel. Naval personnel experience most, if not all, of the same considerations for PPE as those of mounted personnel in open vehicles. In addition, PPE should be buoyant and water resistant. All maritime personnel perform damage control duties, including firefighting, so their clothing must meet minimum standards of heat protection and flame retardancy.

CBRN Operational Specialist Personnel. These specialists may perform a variety of roles, such as reconnaissance [28], sampling, detection and identification [29], and decontamination. One particularly challenging role for these specialists with regard to compatibility with PPE is equipment decontamination. These personnel may have to function at a high work rate for a prolonged period while being exposed to challenges such as various decontaminants under pressure as well as the hazard agents they are decontaminating. Water- and agent-impermeable systems provide the highest level of protection but also impose the highest levels of physiological burden, making sustainment of effort over several hours a challenge.

Medical Personnel. Medical personnel will be active on and off the actual battlefield [30]. They will require various levels of protection, up to the same level of protection as dismounted infantry but may also need to be protected in various types of transport vehicles. They may have additional considerations, such as protection from blood-borne hazards and the need to maintain excellent visual acuity and fine motor control, to carry out treatment (often incompatible with CBRN protective respirators and gloves).

Explosive Ordnance Disposal and Related Tasks. Specialized equipment for these tasks must protect primarily against blast (overpressure and shrapnel); this type of PPE confers considerable protection against CBRN agents within explosive ordnance as well, being released potentially at high velocity. CBRN undergarments worn beneath the bomb disposal overgarment are usual for the application.

Additionally, many roles that may be fulfilled by civilians in war or in peace could require provision of PPE for emergency use, because agent use will not necessarily be confined to the battlefield.

3.3 DOMESTIC RESPONSE

3.3.1 Concept of Operations

Response to a domestic CBRN incident may involve many different emergency response organizations, either present at the scene of the event, or providing support at a distance. Coordination within these organizations to assure that each person understands the portion of the response for which he or she is responsible, and what that response will entail, is critical.

Overall, responders to a domestic CBRN event have the following responsibilities:

- To protect themselves from both immediate and delayed consequences of exposure while maintaining response capability
- To perform rescues
- To identify hazards and prevent further damage, dispersal of agent, and public injury by controlling the scene
- To assist existing and potential casualties ("triage, treat, and transport"), provide prophylaxis, and evacuate or shelter in place
- To perform criminal investigations
- To deal with the identification and disposition of remains
- To remediate contaminated areas

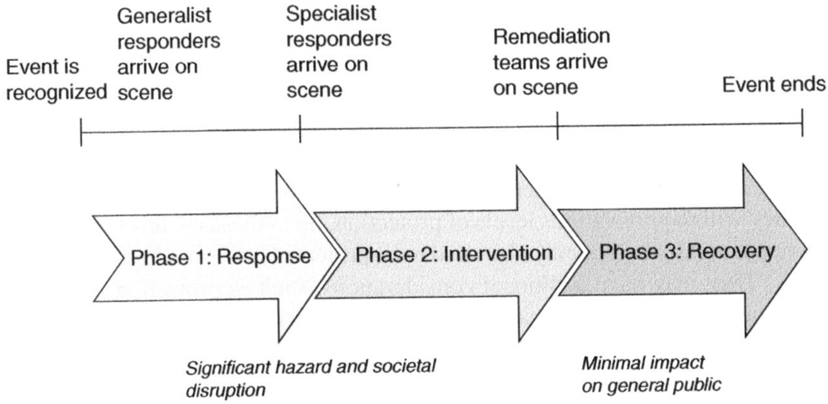

FIGURE 3-1 Phases of a CBRN release event.

It is noteworthy that the pace with which each of these functions is performed may differ greatly, depending on when and where they are performed. An event can be described as having three phases, and the differing nature of these phases has a direct impact on the PPE requirements. For example, the Canadian first responder standard [3] outlines conceptually the phases of a domestic CBRN release event* and many of the activities that would be occurring (Figure 3-1). In most events, the hazard would decrease significantly as time passed, with a few exceptions involving highly persistent agents, and a contagious outbreak where the severity of the event would increase initially through phase 2.

Phase 1: Response. Immediately upon recognition that an event has occurred, generalist responders will be the first involved. This recognition may be in advance of the event (if there is warning) or after the response has started, in which case there will probably be poor identification of the nature and magnitude of the event. Initial management of the event will be confined to attempts to contain the scene and perform critical medical management and evacuation. This phase will last for minutes to hours (or days if it is a contagious outbreak event involving human-to-human transmission).

Phase 2: Intervention. Specialists having specialist CBRN PPE and CBRN response training now participate directly in management of the event. Identification of the hazard will proceed, yielding better information with which to select PPE. More effective mitigation of the event will become possible, due to a combination of better information and tools and the presence of more highly trained responders. Criminal investigation will be performed. Responders located away from the initial site where the event was recognized will become more involved as medical management and

* The names for these phases vary by jurisdiction, but the general principles are the same.

criminal investigation proceed. The duration of this portion of an event will be hours to days for a release event (weeks to months for a contagious outbreak event).

Phase 3: Recovery. In this phase, the event is essentially contained such that there is minimal risk to the public; the focus is on remediation and return to normal operations (i.e., the emergency no longer exists). The response is measured in tempo, and the hazards are generally understood. The level of risk that is expected to be assumed by the worker in this environment should exposure to agent occur is low compared with phases 1 and 2, as time is no longer a major factor and the public is not at risk; normal workplace safety standards will be applied. This phase is likely to include mostly responders from non-front-line and emergency response organizations, representing both government and industry.

These phases have also been divided into crisis or emergency phases and consequence management phases, with generalist responders having awareness-level training and specialist responders having operational-level training [31,32]. These divisions are somewhat arbitrary, but from the point of view of PPE selection there is an advantage in using the three-phase system outlined here, as the way that selection is performed can be tailored to the information available and personnel responsible in each phase. In all phases, appropriate PPE must be selected for the task, location, and known or unknown hazards present. Different PPE selection options are based on knowledge of the event and the capabilities and training of the responders as well as the acceptable level of risk appropriate to the phase of use [3]. In general, specialist personnel will be more likely to be issued their own equipment, at least in part. Many sizing, fitting, and training issues will be addressed thoroughly in advance, with refresher training being relatively frequent; some selection may be performed on the spot from various options based on the hazard assessment.

Other personnel involved may be expected to use equipment with specific, limited capabilities; each person will need to understand and respect those limitations as they pertain to that person's expected role in the event. Some or all PPE is likely to be issued from a pool; each organization must address the question of how to issue properly fitting equipment on short notice and ensure that each person is protected. Although it is possible to presize a person, few will remember their size. Some equipment may require individual fitting in addition, and regular training will be required to ensure that appropriate procedures are followed. It is particularly important for these user groups that all of these issues be taken into consideration *at the time of equipment selection* rather than attempting to deal with them after the equipment has been chosen.

3.3.2 Domestic User Groups

The user groups that fulfill these roles often have normal emergency response duties that may be similar in nature to those performed in a CBRN event; many would be expected to have at a minimum an awareness level of CBRN training. In addition, any of these roles may be filled by operational CBRN specialist persons or teams that have enhanced or dedicated CBRN training.

Hazmat Teams. These teams are already trained to respond to a release of hazardous materials, although their normal role is usually in a noncriminal event. There is a strong overlap in their normal training and equipment with that required in a CBRN event, although the particular nature of a CBRN event may be somewhat outside their usual experience in that the scale and location of the release as well as the type of agent may be unusual. They may be capable of dealing with many facets of a CBRN response, including identification, rescue, containment, and decontamination. Domestically deployed **specialized military response teams** may also be available to fulfill many of these functions in a domestic incident, due to the strong overlap with military capabilities and training for these particular roles.

Firefighters. Firefighters may be the first on a scene and in the initial response phases may be dealing with evacuation, rescue, aid, and emergency washdown of exposed persons as well as dealing with containment of a fire or explosion.

Health Care Workers. This group includes paramedics, emergency medical workers, hospital workers, coroners, and public health personnel. Although health care worker functions will be similar to those performed in any mass casualty or contagious disease event, such functions will need to be performed in appropriate PPE. Since some types of CBRN event could continue for days to months, managing the use of PPE is particularly important for this group. They are most likely to perform their functions away from high-hazard areas, and the most significant hazards they face are likely to be brought to them on the exposed persons, particularly in an event involving a contagious disease.

Tactical and Emergency Response Police. The situation may require intervention by these teams into high-hazard areas if there is a possibility that perpetrators are still on the scene. Low-signature, high-mobility equipment is required with additional ballistic protection.

Explosive Ordnance Disposal. Requirements will be similar to those stated previously for the military.

General-Duty Police. Aside from the very initial portion of the response where they may be on-scene without specialist backup, general-duty police are unlikely to have any significant function in high-hazard areas. They may perform duties such as evacuation, perimeter control, and crowd control, resulting in possible exposure to contamination transfer on persons within the hazard area who are attempting to leave, and to airborne hazards.

Forensic Identification Teams. Their roles in characterizing the crime scene and examining evidence would be essentially identical to that in any potential criminal event, but appropriate PPE must be worn.

Heavy Urban Search and Rescue. Their role in a CBRN event would be similar to that in a non-CBRN event; however, appropriate PPE must be worn and the risks to be managed will be different.

Remediation Teams. These specially trained teams would come into play in the recovery phase of an event, their role being to assess and clean up contaminated areas. The environment would be well characterized at this stage, compared with earlier phases of the response, and PPE selection and use would probably be performed after the tempo of activities slowed.

Coroners, Medical Examiners, and Pathologists. Responsible for determining the cause of death, their activities could be near the scene in portable facilities, or in a hospital or morgue off-site where the level of hazard is more easily controlled. For selected disease outbreak events, specialists in animal disease could also have a role here.

People Involved in Removal and Disposal of Remains. Their role and timing in the response to the event will depend on the urgency of the matter and the nature and severity of the hazard.

3.4 HAZARD ASSESSMENT

3.4.1 General Comments

The process of hazard assessment involves examining all of the possible ways in which a wearer of equipment might experience a hazardous environment. Hazard assessment for design or procurement of CBRN equipment must be tailored to the possible context of use and should be formalized so that the equipment is not ultimately used outside the design parameters. Little information is available on standard procedures for hazard assessment in the CBRN environment; however, information can be derived from a variety of existing standards and national occupational health guidelines. One example of the hazard assessment process for RPD selection in general is given in the Canadian Standards Association (CSA) standard for selection and use of respirators [33], and the National Fire Protection Association (NFPA) has standards that pertain to response procedures and competencies for personnel involved in hazardous materials and CBRN events [31,32].

First, all possible scenarios in which PPE might be used are examined. This requires input from the user community (who may have limits on the types of scenarios in which they would be involved based on their roles in an event), and from the technical and intelligence communities, which will have an understanding of the types of agents that might be involved and the manner of their release. Hazards other than CBRN must be taken into account; these will often be unique to the role of the user.

It is beneficial to repeat the hazard assessment process closer in time to an anticipated or actual use (i.e., prior to deploying or wearing the equipment), since additional information may be available that will permit refining an understanding of the actual hazards. For the military, this reassessment is generally performed over a period of time, with some part of it usually happening prior to deployment to a particular location. Intelligence and monitoring will be constant when a threat of use is deemed to be present, and the user will therefore have preselected the type of equipment to be used in many cases. In contemporary warfare where the battlefield is mobile, it may often be true that the use of PPE by the wearer will be ended by leaving the contaminated area rather than by the hazard having been mitigated.

In a civilian environment, there may be little or no warning and considerably more uncertainty about the possible scope of the hazards. Although there is no substitute for comprehensive preplanning, there will always be a component of on-scene hazard assessment that may permit selection among PPE items with different capabilities. In addition, it is expected that the event will evolve over time with different roles and tasks to be performed, and with the hazard lessening over time as the event is contained and characterized. Nevertheless, the overall length of time that PPE may be required for the various participants in the response is likely to be long, as the contamination of the scene may endure for some time, and scene investigation and remediation will be necessary while wearing PPE.

3.4.2 Scenarios and Modeling

The intent of the use of CBRN agents is to cause the maximum possible disruption of society or of military operations. Although large numbers of casualties could ensue, the psychological effect and the subsequent burden caused by the requirement to operate as though future use is possible is a significant problem. Despite the fact that events that cause few serious casualties are more likely, when designing and selecting PPE it is necessary to plan for reasonable worst-case exposure conditions. A number of organizations have discussed possible CBRN terrorism scenarios (see, e.g., [34] and [35]); warfare scenarios generally involve the use of purpose-built munitions systems, whose details are generally classified.

The considerations in determining worst-case exposure conditions include:

- The type, amount, and physical state of the agent
- How and where release would occur
- The environmental conditions of the release, which can result in differing degrees of reduction of the hazard over time
- Where and when the user would operate relative to the release or other sources of contamination

The first important factors in determining the exposure conditions relate to the physical properties of the agent and the way it disperses in the environment (agent dissemination and dispersion were discussed in Chapter 2).

Second, any limits on the total amount of agent and the rate at which it can be released are important. Some highly toxic agents may be difficult to obtain in large quantities. There may be a relationship between the agent, the method and location of dissemination, and the amount and rate at which it can be released. An attack on a fixed storage installation of chemical that causes a release over a few minutes can be modeled specifically as an example of a worst case with relatively known conditions. Alternatively, release of an agent in a targeted location means that it must be transported to that location and some specific release device used, placing limits on the amount or speed of release. A contagious outbreak event may involve a certain number of infected people transmitting an agent by coughing or sneezing at a particular frequency.

Third, both typical and worst-case environmental conditions should be understood. These will take into account such factors as:

- The way in which air may move to disperse or dilute the agent, and any spatial confinement that prevents dissipation of the agent; such issues as wind or ventilation, indoor vs. outdoor releases, and the profile and type of the terrain will be important.
- Temperature, which can affect air movement, evaporation of liquids, and viability of biological organisms.
- The presence of precipitation or humidity, which may decrease or increase the persistency of the agent as a result of hydrolysis, dilution, or evaporation of water.
- The presence of ultraviolet light, which can reduce the viability of biological agents as well as causing degradation of some chemicals—relating to time of day, cloud cover, and interior vs. exterior release.
- The presence of various surfaces that may cause agents to adhere or adsorb, changing the nature of the hazard from airborne to surface contamination.

Although there is no substitute for actual experimental data obtained during real or simulated events using agents or agent simulants, it is also true that the very large number of possible events means that computer simulation must be used to supplement the understanding of possible exposures. A combination of simple and sophisticated modeling is often useful, where the simple modeling can rapidly yield trends and order-of-magnitude data for a wide variety of release types useful for setting requirements, while the sophisticated modeling can validate the limits of use of the simple models.

The ultimate output of the process ideally is a full picture of the concentration–time profile of the agent involved, in all of its possible physical forms, over the entire area of the event, for a large variety of types of events. Even when the process is simplified by breaking agents into classes with representative physical properties such as viscosity, volatility, particle size, and persistency, and when other factors, such as release locations, mechanisms and environmental conditions are treated generically, this remains a huge undertaking in which many assumptions will need to be made. For example, estimating the persistence of any particular physical form of an agent requires modeling or measuring the evolution over time of evaporation of liquid or liquid aerosols into vapors, or the agglomeration, settling, and surface deposition of aerosols, under many different environmental and release conditions; degradation of agents as a result of environmental conditions is often poorly understood, as is persistence on contaminated surfaces.

Generally speaking, in a CBRN scenario, types of events can be described as outdoor or indoor release (unknown, C, R/N, or B), or contagious outbreak (involving person-to-person transfer of a biological contagion from an infectious source) [3]. Military and civilian scenarios may differ based on the type and sophistication of the dissemination device or the magnitude of the release, but the same categories apply. These categories are useful, as they establish the parameters for generically modeling the concentrations and durations of exposure to establish the agent challenge conditions that PPE must meet. It is important that the models and experiments explore a spectrum of agents whose properties differ significantly from each other, and that those that are particularly difficult to protect against by various available styles of PPE, as well as those that are most likely (due to availability), are included in the process.

Examples of some of the models that have been used for outdoor and indoor release scenarios to estimate hazards of chemicals include ALOHA [36], HPAC [37], VLSTRACK [38], RAP2000 [39], SCIPUFF [40], and COMIS [41] and, for indoors, various convection [42] or computational fluid dynamic approaches. Radiological releases have been modeled using models and systems such as ARAC [43], ARGOS [44], ATSTEP [45], Gaussian plume [46], NARAC [47], RIMPUFF [48], and URD [19], with shortcomings of existing modeling (as applied to accidental nuclear releases) outlined by Gering et al. [49] and actual RDD simulation [50]; the combination of radiological agent dispersion and human exposure likelihood based on patterns of movement has been modeled [51].

Outdoor biological releases have been modeled using LODI with the capability to simulate complex parameters such as particle size distributions, wet deposition, gravitational settling, dry deposition, and deactivation by ultraviolet radiation [52]. Indoor release of *Bacillus anthracis* and circulation through buildings and subways has been modeled and measured extensively [53–59]. Outdoor biological release has been modeled for *B. anthracis* [60] and for bacteria in general based on cropspraying [61].

Probabilistic modeling of the hazard from human-generated aerosols of contagious organisms has also been performed [3,62–64]. A general review of the many approaches and models for dispersion modeling of particulates has been provided by Holmes and Morawska [65].

Simplified models for describing an event are often particularly useful in setting requirements. Figure 3-2, a conceptual illustration of the zones that may occur in a release event, is based on information in the North American *Emergency Response*

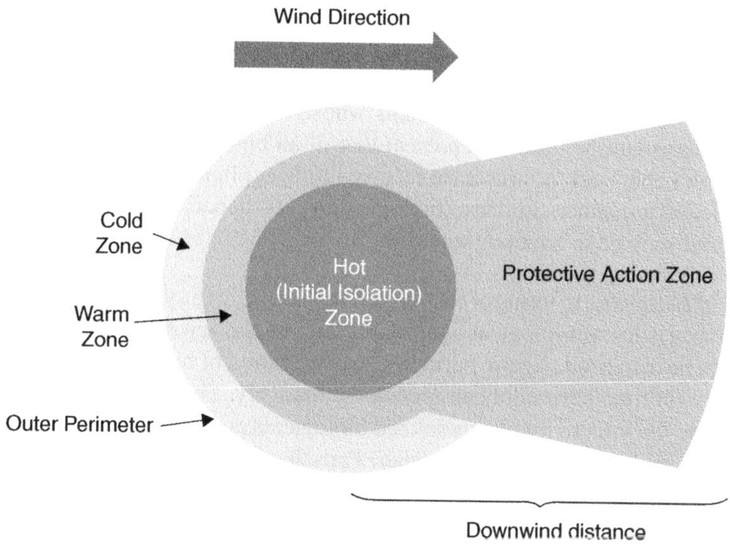

FIGURE 3-2 Response zones in a release event (based on [3]).

Guidebook (ERG) and U.S. and Canadian standards [3,66,67]. The zone shading (or color when used) is intended to indicate the relative hazard in each zone, with the cold zone being relatively free of hazards. The hot or isolation zone is the area around the initial release that requires high levels of protection to enter; the protective action zone is the area downwind of the release in which everyone should be evacuated or protected by some means; and the warm zone is the area upwind of the release that transitions to the cold zone and in which operations such as decontamination will occur. Using this concept, the ERG reports the results of simple modeling to estimate the size of the initial isolation and protective action zones for a large number of hazardous materials. The Canadian standard [3] uses these zones to assist responders in determining the type of PPE that should be selected for a given type and magnitude of release event.

Chemical Hazards. Airborne agent dispersion has been modeled as described above, and such models are becoming more able to predict concentration, taking into account details of air movement based on meteorology and the morphology of the environment. Prediction remains challenging for factors such as aerosol particle size distribution, which depends on the method of generation and will evolve over time as an agent evaporates. There are few, if any, theoretical models in the open literature that predict liquid drop size formation and evolution as a function of the method of dispersion, nature of the chemical, and time after release. Droplet evaporation in sprays [68] has been modeled numerically. Liquid evaporation has been modeled to determine the remaining potential hazard on undecontaminated persons exposed to liquids within the first few minutes or hours after an event [69], including numerical modeling of evaporation from surfaces [70].

Generic chemical characteristics such as surface tension, volatility, and polarity may be used to group together chemical hazards to develop quantitative structure–activity relationships (using a multivariate data analysis approach) [71] to develop predictive models for chemical simulants whose behavior and interaction with protective systems can be used to represent that of an entire subset of similar hazards. Solid aerosol chemical hazards have received little explicit consideration; toxins can be modeled in a manner similar to biological agents, depending on their release as a powder or dissolved in aqueous solution.

Biological Hazards. In terms of interaction with protective systems, the most important characteristics of biological agents are those of the aerosols or liquids in which they might be dispersed. Again, particle size distribution of biological aerosols and its evolution with time have not been modeled effectively. This is a particularly important issue, as the infectivity of aerosolized biologicals is definitely dependent on their particle size and not in a simple manner. Particle size determines both the number of organisms per aerosol drop and whether that drop can infect the person or will be removed by various defensive mechanisms at the routes of entry, as demonstrated for *Bacillus anthracis* and *Yersinia pestis* as well as aerosolized ricin [72].

Biological agents are unlikely to be dispersed deliberately in bulk liquid form, so it is their presence in human body fluids that is likely to be the hazard of most concern in terms of protection against liquids. The relative magnitude of the hazard from contamination transfer by such fluids is poorly understood in general.

Radiological Hazards. There has been extensive modeling of airborne dispersion of radiological agents as a result of nuclear accident [23,73] as well as of the nature of the isotopes involved; some modeling of radiological releases has also been carried out. Again, information on particle size distributions is largely empirical, based on measurement during simulated and actual releases; modeling to demonstrate how it might change as a function of time would be of some benefit. The characteristics of the particles depend significantly on the initial source of the release for nuclear events [74]. Deposition onto surfaces has been measured and modeled extensively (e.g., onto indoor surfaces [75], showing a linear relationship between deposition velocity and size for particles within the mass median aerodynamic diameter range 0.5 to 5.5 μm).

3.4.3 Conditions of Use

The range of possible conditions under which the exposure might occur must be understood, as these affect both equipment and user performance:

- The maximum duration of each type of possible CBRN exposure must be defined.
- The amount and type of use the equipment might have had before CBRN exposure may affect subsequent performance.
- The full range of environmental conditions of use needs to be specified.

Duration of Use. It is important to understand and take into account the distinction between the duration of use of an item of PPE and the protective duration both during and after exposure. These parameters will define the test regime to be used to validate the equipment performance and should be articulated clearly in the statement of requirements for the user and in any performance standards for the PPE. It is to be expected that the equipment will be worn for some time before exposure. This period may be very short—for example, if the PPE is specialty equipment that is put on immediately before entry into a contaminated environment—or very long—military ground forces may wear PPE for days or weeks in a threat environment, and may expect it to be laundered and reused many times.

The protective duration required may range from minutes to hours and must include both the period of exposure and the length of time before the equipment can be removed. Once the equipment has been contaminated, in many cases it will hold out the agent for only a specified time after that, even if the equipment has been removed from the contaminated environment and/or decontaminated. Examples of typical protective durations follow.

- A firefighter wearing CBRN protective turnout gear or an explosive ordnance disposal technician may require this protection only as long as it takes to escape the hazard area and decontaminate and remove the equipment, a relatively short period both during and after exposure.
- Hazmat personnel may intend to enter a contaminated scene a number of times over the course of a shift, and the equipment must therefore both protect for the course of the exposure period and keep hazards from working through the equipment over the length of use; for example, a chemical may dissolve into and subsequently permeate through materials over the course of several hours after the initial exposure.
- Military personnel may have a mission that requires "fighting dirty," such as continuing to pursue an objective for hours or possibly even days after an exposure, which may be brief or continual during that time, depending on the agent and the form of attack.

Environmental Conditions. The full range of applicable environmental conditions is required to take into consideration:

- The relative hazard from various agents under those conditions
- Environmental test conditions for PPE, usually based on various extremes of temperature, humidity, and airflow
- The impact on a person's thermal status and how equipment design may have an impact on maintaining appropriate core temperatures under extreme conditions

All possible locations and conditions in which the equipment might be worn should be considered. NATO describes the possible environmental conditions for various theaters of operation worldwide [76], taking into account all possible extremes of temperature and relative humidity. Information on typical and extreme weather conditions in particular locations can also be gleaned from meteorological services online. Indoor vs. outdoor exposure conditions should also be considered. It is noteworthy, however, that extreme temperature conditions may have an impact on the concentration of agent present; for example, low temperatures may cause liquid agents to freeze and decrease the volatility significantly, resulting in a lowered airborne and contact hazard in general. Therefore, extreme environmental conditions should be considered hand in hand with agent challenge concentrations. In addition, temperature, relative humidity, and wind can affect equipment performance.

3.4.4 Limits to Operations

Once the nature of the pertinent hazards for the given user group has been established (type, duration, location), it is necessary next to establish what limits may exist with regard to the location and duration of operation of the user group relative to the hazard. Ultimately, this is likely to be an iterative process. Many groups would like to operate freely with as few restrictions as possible; in practice, this may be incompatible with providing a person with protection of sufficient scope and duration. Hence, some a priori assumptions of limitations may be made based on experience with the capabilities of personnel and existing PPE. For example, in the case of a release, select user groups may enter the immediate area of the release while others will operate at a distance. The duration of operations of a given person may be limited by factors such as physiological burden when wearing equipment and the ability of equipment to provide protection over long periods (although limited availability of personnel may demand longer shifts). Difficulty in decontamination of equipment may prevent repeated use or use in the presence of particular types of hazard.

3.5 EXPOSURE LIMITS

3.5.1 General

Recall that exposure in general relates to the concentration and length of time of exposure; PPE is then used to reduce the concentration to acceptable values, while limits to operation can be used to reduce the duration of exposure. Therefore, to permit the development of quantitative protective performance requirements, the estimated magnitude and nature of hazard in a particular location for the duration of operations must be combined with the allowable exposure values to the person by each of the possible routes of exposure. In this section we discuss how allowable exposure limits for a person may be chosen. In some cases, the level of exposure of the equipment itself to a hazard is equally important, as the survivability of the equipment is necessary to protect the individual. An example of this is exposure to nuclear flash, where the PPE must survive intense heat to protect the wearer [77].

The area of setting allowable exposure limits for a worker in an emergency situation may be somewhat ill defined depending on the jurisdiction, and will be even less well defined when it comes to specific CBRN agents whose toxicological effects may not be well established. Documentation available for many chemical and radiological and nuclear agents permits setting some exposure limits; there is remarkably little for biological agents, and no meaningful exposure limits exist.

Hazard Materials for Consideration. When considering broad-based protection against CBRN hazard materials, one of the primary considerations is determining the *most hazardous* materials to which exposure could occur under the conditions of exposure and PPE use. The magnitude of the hazard to an unprotected person in the absence of protection results from the relative toxicity of the material by the various potential routes of entry, as well as the amount and form in which it is delivered. When selecting and designing PPE, it is also important to consider the ability of certain materials to more easily defeat various modes of protection. The most hazardous of those agents that are also difficult to protect against often become the performance-limiting materials against which protective capabilities will be gauged.

Route of Entry. Exposure limits are different for each route of entry, and while some hazards may have established exposure limits for inhalation exposure, these limits may not exist for other routes of entry, such as eye or skin. Furthermore, such limits are often developed from some form of aggregate data on human exposure in the absence of protection; when different routes of entry are protected to a different degree, it may, as a result, be difficult to judge how much protection is needed for each individual route based on existing human data.

Emergency Response vs. the Workplace. In the area of CBRN protection, normal workplace exposure limits have particular applicability for those workers in the recovery phase of the response. Here, workers will be performing duties such as cleanup that may go on for weeks under relatively controlled conditions. However, the applicability of these values to emergency response varies. Some jurisdictions or organizations having authority will permit a higher one-time (single, acute) exposure for a person performing an emergency function to enhance public safety compared with a normal workplace procedure (where exposure may be prolonged or repeated). Each organization must therefore investigate the legislative framework within which it operates and perform the necessary due diligence to support the limits used.

3.5.2 Chemical Exposure

Because of the context of use we are considering, in theory every toxic chemical that is potentially available for use should be considered, and to develop effective protective systems, the toxicity of each of these should be known by every important route of entry. The type of effect may range from lethality, through serious incapacitating effects such as difficulty breathing or vomiting, to mild, reversible effects such as irritation of the eyes or skin.

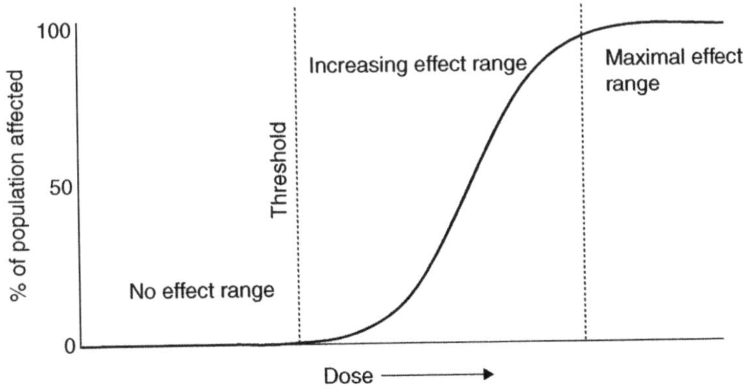

FIGURE 3-3 Classic dose–response curve of likelihood of effects in the population vs. total dose delivered.

The concept of dose–response [78] is fundamental to setting exposure limits and can be expressed simply with reference to the classic dose–response curve (Figure 3-3). At low doses, the body has defense mechanisms that can deal with the toxic material to prevent any effect from occurring. As the dose increases, a certain threshold value is reached [79] at which the more sensitive members of the population are affected. Increasing the dose further results in more people being affected, until at a high enough dose, everyone is affected.

Population variation in effect dosages can be quite significant. Thus, the effective dose for a given effect may vary significantly depending on the population under investigation: Does it include the very young or very old, or the chronically ill? When deriving protection requirements for CBRN protection, it may be valid to use values derived for the healthy adult population as representative of a fit responder or military workforce.* It is also worthy of note that knowing the value that has a 50% probability of causing effects, which is often used to characterize toxic materials relative to each other, does not yield any information about the effect, dosages for other population percentages. The shape and width of the dose–response curve is unique to each toxic material and route of entry. Finally, in using these values, it is important to recall that the *smaller* the effective dose value, the more toxic the agent.

As the dose delivered increases, additional dose–response curves may be required to describe different effects resulting from the same chemical (Figure 3-4); for example, a chemical could cause irritation, vomiting, or other effects at lower doses, and illness or death at higher dose values. There is often overlap between the dose ranges that cause various effects. Some of the terms that are in common use in describing toxic effect levels, such as ECt_{50} and LCt_{50} in Figure 3-4, are explained next; although many are inappropriate as permissible exposure levels, they are often used for ranking the toxicity of chemicals and determining those that are the greatest hazard.

* Some of the values derived may, however, have bias against females, as the methods by which these were derived are often based on the historical workforce or military composition, heavily weighted towards, males.

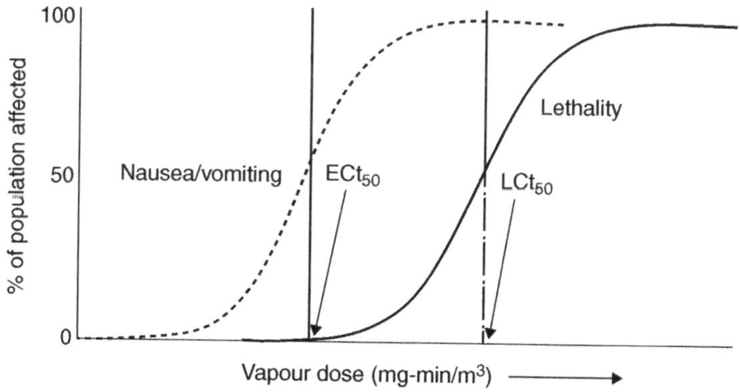

FIGURE 3-4 Dose–response curves for vapor exposure for two different effects arising from the same chemical.

Acute Toxicity. This is the ability of a substance to cause systemic damage within a short time of a single exposure. This is the most relevant index of toxicity when setting requirements for the response and intervention phases, as other forms of toxicity usually require repeat exposures. LD_{50}, LC_{50}, and LCt_{50} are the most common ways in which acute toxicity is expressed: These are the midpoints on a dose–response curve, where there is 50% probability of lethality in the population. Dose in the previous terms is expressed in units related to mass, concentration, or concentration $\times$ time, respectively. Generally, the LD_{50} value is used to rank toxicity by ingestion, injection, or skin exposure to solids and liquids; the others refer to airborne routes of exposure. Lethality is not used in general as an allowable effect for setting exposure limits.

Toxicity values for a given route of entry are expressed in the following general form of $\boldsymbol{ED_x}$ **(route)**, where:

- *E* stands for the type of effects being specified (e.g., *L* for lethal, *I* for incapacitating); or alternatively, *E* can be used where the particular effect is additionally specified.
- *D* refers to the dose, usually of liquid or solution, delivered by the route specified within the parentheses: orally, intravenously, or by skin absorption, usually expressed as mass per unit body weight.
- *D* may be replaced by *C*, in which case it refers to airborne concentration and is generally used for effects by inhalation, or by *Ct*, the vapor dosage (i.e., concentration multiplied by time of exposure).
- The subscript *x* refers to the fraction of the population exhibiting the given effect at the dose specified (e.g., 50 for half the population).

Therefore, LCt_{50} (inhalation) is the lethal vapor dosage by inhalation to 50% of the population, and ED_{10} (oral) for vomiting is the mass per unit body weight delivered orally that would cause vomiting to 10% of the population.

Note that technically, exposure by every route has a related time dependency; in other words, exposure for a longer time is invariably more serious than exposure for a shorter time, and if the hazard is removed, the likelihood of effects is lowered. This distinction is often neglected for ingestion and injection, where the dose is likely to be delivered all at once. However, for other routes of entry, *LD* and *LC* values always have an implied time dependency that may be difficult to extract from the information given. For example, if you are breathing a toxic chemical, every breath you take will deliver a certain amount of agent to your body through your lungs: The more breaths you take, the more will be delivered. If your breathing rate does not change, the dose you receive will be directly related to how long you have breathed the contaminated air. Hence, if using EC_x values to set requirements, the time of exposure used to derive them must also be known. The *ECt* values have expressly taken into account the time dependency by assuming that *Haber's rule* is valid.

Haber's rule assumes that it is the total mass of material taken up by the body that results in the likelihood of effect, not the time over which it is delivered; and that as long as the material is present, the body will continue to take it up at a fixed rate. Further, for the most toxic agents, the body can't detoxify the agents over exposure times of minutes to hours. This means that for a vapor exposure, the product of concentration and time, *Ct*, determines the effective dose delivered.

ED_x values for skin contact with liquids are also difficult to interpret, as the actual dose delivered to the body depends on the length of the exposure, the amount of liquid that evaporated during the exposure, the size of the area of skin contacted, and the associated vapor dosage that the skin saw during contact with the liquid. The dose that will cause effects is also ultimately dependent on a person's total weight. This factor is rarely taken into account when examining effects resulting from airborne agents, because the exposed surface of lung or skin is proportional to the body mass to some degree; thus, an increased mass corresponds to an increased dose taken up by the body, roughly canceling out the requirement for a greater total dose to cause effect. In the case of oral delivery, injection, or liquid skin contact, the mass of the dose delivered is independent of overall body mass, and thus the effective dose must be described in units of amount per unit weight (e.g., $mg{\cdot}kg^{-1}$).

In many cases, and for obvious reasons, reliable values for the human population are not known. While respiratory toxicity values are relatively well known for a large number of chemicals, toxicity by other relevant routes is generally poorly quantified. The large majority of toxicological values are based on animal studies that have been consolidated to provide human estimates, and these are most often studies that determine the dose that is lethal to 50% of the population. Limited data on humans may also be available for mild effects, which may have been observed in controlled studies or in the workplace.

Setting Exposure Limits. Chemical exposure limits are often regulated for the more common chemicals. Nevertheless, it should be recognized by regulators that the most likely exposure type in the response and intervention phases can be considered to be a single, acute exposure. In other words, the exposure limits set for repeat exposures, relevant perhaps to the recovery phase, may be overly conservative for a single exposure. Exposure limits generally refer to airborne dosages, as it has been assumed that any other form of exposure can be avoided. Much of the discussion here revolves around the more challenging exercise of setting exposure limits for the response and intervention phases. The recovery phase can be managed such that conservative workplace exposure limits are more applicable.

Exposure limits will not be set using values that will result in a 50% likelihood of lethality. In theory, exposure limits are set by picking a point on the curve corresponding to a low probability of non-life-threatening or reversible effects; depending on the quality of the data, additional safety margins may be added. If the curve is not well defined by the data available, the center point of the curve may be used and the curve's width estimated. Values may also be derived from animal data, and some form of safety factor will then be added to take into account differences in human response. There is some evidence that the dose of aerosol required to cause an effect is equivalent to the dose of vapor; for example, similar doses of HCl vapor and HCl aerosol caused comparable effect levels [80].

Occupational Exposure Limits: Threshold Limit Values (TLVs). These exposure limits are published by the American Conference of Governmental Industrial Hygienists (ACGIH) and are used commonly in North America. They represent scientific opinion based on a review of existing peer-reviewed scientific literature by a committee of multidisciplinary experts. Most legal limits for a worker exposure are derived from threshold limit values. Threshold limit values refer to the airborne concentration of a chemical substance to which *workers* can be exposed on a regular basis without ill effect; note that the general population may include those significantly more sensitive than the worker population (e.g., the very young, the infirm, and the elderly).

TLV-STEL (short-term exposure limit) is the 15-minute time-weighted average (TWA) exposure that must not be exceeded at any time during the work day, while the TLV ceiling is the concentration that must not be exceeded for any time period. Other TWA values are available for longer exposures, usually an 8-hour day.

Occupational exposure limits may be different in different jurisdictions and may include different substances, but in most cases are expressed in a manner similar to that described above. In some jurisdictions, values for some substances are legally binding, while others, while regulated, are deemed "administrative" or informative. In general, it is important to be aware of the obligations with regard to minimizing such exposure within a given jurisdiction. However, it is also important for regulators to recognize when these limits may have limited applicability, such as in emergency response, since the premise around which the values were derived may imply a different approach to risk management.

Some of the additional guidance values that are available may imply a higher level of risk, more appropriate to emergency response and crisis management. Some of these values are described next.

Emergency Response Planning Guidelines (ERPGs). Developed by the American Industrial Hygiene Association (AIHA), these are intended as a planning tool for predicting health effects at certain concentrations. These values have no built-in safety factor and are applicable to 1-hour exposure periods. They are not protective of hypersensitive persons.

Acute Exposure Guideline Levels (AEGLs). Developed by the U.S. National Advisory Committee for AEGLs, these are emergency response guidelines for the general public, including susceptible individuals ("nearly all people"), for rare events. They include three levels of effects and five exposure durations (10 min, 30 min, 1 h, 4 h, 8 h). In setting exposure limits, the most relevant is AEGL-1, which is the airborne concentration above which it is predicted that the general population could experience notable discomfort, irritation, or certain asymptomatic effects. However, the effects are not disabling and are transient and reversible when exposure ends.

Other values that have been used include **emergency exposure guideline levels** (EEGLs), which are applicable to military personnel (young, healthy adults) for exposures of 1 to 24 hours in nonwartime situations. They permit temporary effects that are nondisabling, but avoid severe acute effects and long-term or chronic injury. Similarly, **military exposure guidelines** (MEGs), for use by U.S. military during deployments, assume a healthy and fit population and take mission requirements into account.

Concentrations Immediately Dangerous to Life or Health (IDLHs). The purpose of defining this concentration level is to ensure that a worker can escape from a given contaminated environment in the event of failure of the respiratory protection equipment, with 30 minutes considered the maximum time for escape to be completed. Based on this premise, various types of toxicological data have been used to derive the values, all based on the 30-minute exposure window, and taking into account such issues as explosive concentrations and capability for severe respiratory irritation.

These values are recognized by the U.S. National Institute of Occupational Safety and Health (NIOSH) and are defined in numerous standards. According to NIOSH [81], an IDLH condition "poses a threat of exposure to airborne contaminants when that exposure is likely to cause death or immediate or delayed permanent adverse health effects or prevent escape from such an environment." The CSA definition [33] is similar but defines an IDLH atmosphere as one that "... poses an immediate threat to life, would cause irreversible adverse health effects, or would impair an individual's ability to escape." The method of derivation of the IDLH value has particular merit for consideration as a regulatory limit when applied to PPE selection for CBRN response; other values, such as AEGLs and EEGLs, may also be applicable.

Finally, it is worthy of emphasis that there are no exposure limits regulated for skin; at best, a "skin notation" may be available to indicate that a chemical may be systemically harmful by this route and that some form of dermal protection is advisable, or a chemical may be noted as corrosive where local effects are possible. Such notations rarely distinguish between the many possible types or severities of harmful effects, and toxicological data are sparse. Whether airborne exposure is sufficient to cause an effect is rarely known; it is generally presumed that avoidance of liquid contact with the body is the primary requirement for protection. Typically, only the CWAs have significant available dermal exposure dose data. It is not usually

TABLE 3-1 Negligible Military Impact for Dermal Vapor Exposure to CWAs

Agent[a]	Permissible Dosage ($mg \cdot min \cdot m^{-3}$)
HD	50
GD	350
VX	30

[a]HD, distilled sulfur mustard; GD, soman.

necessary to give consideration to long-term or chronic effects such as carcinogenicity in setting chemical exposure limits for the purposes of emergency response, except for the recovery phase, as with few exceptions such effects are unlikely to result from an acute exposure at sufficiently low levels (i.e., when protected by PPE).

Exposure Limits for Chemical Warfare Agents. Limits for exposure to these agents are now becoming more commonly available, usually expressed in terms of airborne (vapor) exposure (here most relevant for PPE applications). Values for sulfur mustard [82] and VX [83] have been developed for dermal vapor exposure as a function of body region and for various CWAs with no body region variability incorporated [84], as outlined in Table 3-1. U.S. values for some CWAs for unprotected individuals are given in Table 3-2.

3.5.3 Radiation Exposure

The strength of a radioactive source is determined by how many nuclei decay each second. The modern unit of source strength is the becquerel; 1 Bq is equal to 1 disintegration per second. The older and more convenient unit is the curie; 1 Ci is equal to 3.7×10^{10} disintegrations per second. A 1-Ci source is considered large; a 100-Ci source is extremely dangerous.

TABLE 3-2 Acute Exposure Limits ($\mu g \cdot m^{-3}$) for Selected Chemical Warfare Agents

	General Population Limit	Worker Population Limit	Short Term Exposure Limit	IDLH
Exposure duration	24 h (12 h for sulfur mustard)	8 h	15 min	30 min or less
Agent				
Tabun, sarin	0.001	0.03	0.1	100
VX	0.0006	0.001	0.01	3
HD	0.02	0.4	3	700

Source: [85–87].

The radioactivity emitted by an isotope is inversely proportional to its **half-life**, which is the amount of time it takes for 50% of the atoms in a sample to decay. The shorter the half-life, the more intense the radiation, as more atoms are decaying per unit time. The **specific activity** is expressed in terms of the strength per gram; therefore, a slowly decaying heavy atom has a low specific activity. Unlike chemical and biological agents, the most relevant concentration unit relates to the time-integrated activity per unit volume rather than the time-integrated mass per unit volume, as it is not the isotope itself but the radiation emitted that is the primary hazard.

The **absorbed dose** (in units of grays, or $J \cdot kg^{-1}$) indicates the amount of energy deposited in a given mass of tissue. Although the absorbed dose is a useful unit, it does not take into account the difference in magnitude of effects of different types of radiation on tissue. The **equivalent dose** concept (in units of sieverts) normalizes these different effect types by multiplying the absorbed dose by a weighting factor for each type of particle. The weighting factor is high for alpha particles, which cause a much greater effect internally compared with gamma and beta particles.

The **effective dose** concept (in units of sieverts) further takes into account that different tissues are more or less sensitive to radiation damage; to obtain the effective dose for a given tissue, the equivalent dose is multiplied by a tissue weighting factor.

Finally, biological effect levels must take into account the fact that radioisotopes taken up by the body continue to be hazardous until they are excreted or have decayed away. The **committed effective dose** (in units of sieverts) is the total dose estimated to be received by various tissues over 50 years following exposure, resulting from the amount of the radioisotope incorporated into the various tissues of the body. This dose is related to the type, form, and amount of radioisotope taken up by the body, the type of radiation, and the organ uptake. The various weighting factors have been published by the ICRP [88].

The **relative biological effectiveness** (RBE) is the ratio of the dose of radiation under consideration to that of a standard amount of a particular energy of radiation (such as 250-keV x-rays) required to produce a specified biological endpoint, such as skin erythema. The RBE is derived from epidemiological data and is more accurate than the weighting factor described above, which is somewhat more arbitrary and generic.

All jurisdictions and organizations have defined acceptable radiation exposure levels. Civilian exposure limits in each jurisdiction will be based on those of national scientific and governmental bodies, often based on ICRP recommendations, and will differ for workplace exposure, the general public, and accidental release. Exposure limits may be managed differently in an emergency [89] than in routine civilian or workplace exposure (i.e., a higher exposure may be permissible for a rescuer to save a life than in the normal workplace*). NATO works with **radiation exposure states** (RESs) which are graded depending on the risk of effects, and compared with the overall risk or urgency of the operation and the previous exposure to radiation [90].

*For example, according to the Institute of Medicine [90], "Dose limits apply only to practices. For interventions—where the primary purpose is to accomplish the emergency action—dose limits are not used."

TABLE 3-3 Radiation Exposure States

RES	Dose Range (centigray)	Probable Effects (Long-Term Risk of Cancer and Other)[a]	Probable Effects (Short Term)
0	General public exposure level permitted	20% lifetime risk of fatal cancer	None
1	<75		
1A	0.05–0.5	Up to 0.04% increased risk of cancer	None
1B	0.5–5	0.04–0.4% increased risk of cancer	None
1C	5–10	0.4–0.8% increased risk of cancer	None
1D	10–25	0.8–2% increased risk of cancer, increased morbidity from other injuries, risk of temporary male sterility	None
1E	25–75	2–6% increased risk of cancer	None
2	75–150	Risk of immune system defects	Transient mild nausea; vomiting in 5–30% of personnel
3	>150	Risk of cataract formation	<300, transient mild to moderate nausea and vomiting in 20–70% of personnel, mild to moderate fatigue and weakness in 25–60% of personnel, risk of permanent male sterility; higher doses begin to result in increase in loss of bone marrow cells, likelihood of infection, and death

[a]Injuries observed at lower RESs also occur at higher RESs.

NATO provides guidance for exposure of military groups to nuclear weapons [91] and individuals to other low-level radiological sources [92] (Table 3-3). The ICRP has published a number of guidelines intended to assist regulators, including their 1990 Summary Recommendations [93], which suggests that RES 1E is appropriate for emergency responders [89], an exposure state that may increase the lifetime risk of cancer by up to 5%.

Exposure limits are often focused on the prevention of long-term effects; much higher levels of radiation exposure are required to cause short-term debilitating effects, and onset of effects is not usually particularly rapid until near-lethal doses on the order of 1 Gy are reached. Since radiation damage is often highly nonspecific (i.e., many different body systems are damaged in different ways), radiation effects may not be as easily reversible or recoverable as in the case of chemical exposure. Any effect manifested may be symptomatic of a serious exposure with potential significant or long-term health consequences. (In contrast, for most chemicals, low-level reversible irritation to eyes, the respiratory tract, or skin is the first effect seen, with no long-term consequences predicted. Chemicals known to be carcinogens are the exception.)

In the case of radiation, conforming with permissible exposure requirements is made much easier by the availability of various simple forms of radiation dose monitoring that do not require much prior knowledge of the nature of the hazard (other than that it might exist). The challenge, then, in applying radiation dose limits is more in understanding, for preplanning purposes, how PPE may assist in reducing the doses that reach the individual. Current PPE in general has little effect in shielding from highly penetrating radiation. On the other hand, some radiological agents are a hazard only when internalized, whereas others may deliver a dose of radiation from outside the body whose magnitude may change because of intervening barriers such as PPE that may reduce the amount of radiation or radiological particulate reaching the skin. PPE may also further aid in dose reduction by easing decontamination and reducing the likelihood of contaminant carried on the skin being subsequently inhaled or ingested. Dose in most cases is additive, as radiation is delivered by the various routes of entry, which must be taken into account when trying to estimate in advance how PPE might reduce exposure.

3.5.4 Biological Agent Exposure

For the toxins, as they are nonliving, the dose terms used for chemical agents apply. Botulinum toxin and ricin have LD_{50} values of about 0.001 and 4 $\mu g \cdot kg^{-1}$, respectively [94]. However, for living microorganisms, the relevant dose is the *infectious dose*, the dose that can cause infection. Reported infectious doses are usually based on sparse epidemiological data, although classified sources of human infection data exist. In general, these values are best estimates and are difficult to use for exposure guidance; the estimated ranges can vary by several orders of magnitude. The reason for the uncertainty is twofold: first, data are absent in most cases, due to the absence of good models (either theoretical or animal) for infectivity of humans, and there is a (happy) paucity of human exposure data that are suitably interpretable; and second, dose–response for humans to exposure to microorganisms is extremely variable, due to how the organism enters the body (its route, particle size), its viability, and a large variation in innate human defense mechanisms based on age, general health, and the genetics of both the organism strain and the host. Furthermore, many biological agents could be disseminated as aerosols that are not the normal mode of infection for humans and may well have different infectivity.

TABLE 3-4 Infectious Doses

Disease	ID_{50} (Inhalation) (Number of Organisms or Spores)
Anthrax	<10,000 spores for illness; usually leading to death if untreated [95]
Pneumonic plague	<100 [94]; usually leading to death if untreated
Tularemia	1 for infection [96]; increasing doses cause death
Q-fever	10 for fever [94]; increasing doses cause more severe illness

There have never been regulated biological agent exposure limits. For those agents that are particularly likely to infect by the airborne route, the infectious doses can be so low that any exposure can potentially cause effects; some values are given in Table 3-4. These doses are expressed in terms of total number of organisms or spores required to cause infection in a single (acute) exposure; it is poorly understood what the duration of a single exposure should be defined as; for example, is a single exposure of a hospital worker an 8-hour shift or is exposure cumulative over days?

3.6 HUMAN FACTORS AND TASK REQUIREMENTS

Additional human factor issues must be taken into consideration when setting requirements for equipment (both PPE and the equipment that must integrate with it). The user groups and general operations having already been identified, it is necessary to examine in more detail what particular tasks may need to be performed or non-CBRN hazards may be encountered that would affect particular PPE design or selection requirements. It is sometimes difficult to recognize in advance what these will be without having considerable experience in attempting to perform tasks while wearing PPE and an understanding of what particular factors can compromise PPE performance, which are further dependent on the specific PPE and its construction.

Some of the more obvious issues or requirements that must be considered as a result of the tasks to be performed are:

- The durability of PPE items: requirement depends on tasks and hazards, duration of use, and environmental conditions.
- The integration of PPE: particular tasks or ancillary (non-CBRN) PPE may cause integration failures due to interference with or dislodging of equipment.
- Interference with physical and sensory capabilities: certain tasks require unimpeded use of hands or various senses.
- The physiological burden: high-work-rate tasks and extreme environments impose a much higher physiological burden, which will significantly reduce the time on task when wearing PPE.

It is probably more important to manage physiological stress to the wearer than it is to manage protection against other hazards. For example, the primary cause of both nonfatal injuries and death among firefighters, who face life-threatening situations on a daily basis, is physiological stress [97]. The environmental conditions in which the PPE will be used and the workload of the wearer combine to be responsible for much of the physiological stress, resulting in potential overexertion, anxiety and heat exhaustion. Such stress needs to be managed through a combination of appropriate PPE design and monitoring of the individual and his or her environment.

In addition to physiological stress, wearing PPE can cause many other problems, resulting in physical and psychological impairment; these can include:

- Vision may be distorted and field of view limited.
- Hearing is affected by the layers of material and, additionally, the sounds the PPE itself may produce (e.g., material crinkling, breathing, blowers).
- Communication with others is also limited by voice distortion and the inability to identify others when wearing PPE.
- Touch and tactility are limited by the gloves.
- Freedom of movement is in general limited by restrictive PPE items.
- Carrying extra weight may cause a further shortening of a work cycle over and above that caused by thermal stress; weight may also be poorly distributed, causing a loss of balance or strain.
- PPE restricts the ability to perform many bodily functions.
- Discomfort results due to overtightening, chafing, and sweat accumulation inside a respirator.

The implications of these impairments can be serious in that they can affect a person's survival in a hostile CBRN environment, as well as the likelihood of mission success. Therefore, given that some impairment is inevitable, the minimal acceptable amount of impairment should be understood. It is also important, however, not to set unrealistic requirements that cannot be met, and to accept that some alteration in operations will be required, as already noted in Section 3.4.4, to overcome the limitations imposed.

A few examples, not intended to be all-inclusive of task requirements and their implications on PPE selection are given here. More detail is given in subsequent sections on the fundamentals of many of these performance effects.

Law Enforcement, Military Ground Forces, and Special Operations. The requirement to be able to fire many types of weapons has a particular impact on the design of a respirator: poorly located eyepieces or filters may interfere with sighting or aiming, and weapon recoil may cause dislodging of the respirator seal. Stealth functions may require a very low visual and auditory signature from the PPE as well as, potentially, a low-infrared signature. Ballistic protection may improve the protection provided by PPE but will impose an additional physiological burden.

Emergency Medical or Field Medics. The requirement to perform even simple triage and medical intervention generally means that medical personnel must be able to use certain types of diagnostic and treatment methods and equipment; good hearing, vision, and tactility are important.

Firefighters or Shipboard Damage Control. It is possible to add some level of CBRN protection to equipment that provides fire or thermal protection, or vice versa, but it must be done without compromising the dual nature of the protection required. For example, if it is to be reused, PPE that has been subjected to high heat environments or firefighting solutions must still protect subsequently against CBRN hazards. Thermal burden in the closed state is usually worsened by the addition of the CBRN protective layer.

Specialist Decontamination Teams. Decontamination solutions may make gloves slippery and hard to use. High-pressure and high-volume decontamination solutions may penetrate air-permeable material or insufficiently tight closures. Designs that may provide sufficient resistance to decontaminating solutions may also impose significant physiological burden, incompatible with the high work rate required for the task.

3.7 EXAMPLES OF HIGH-LEVEL REQUIREMENTS DEVELOPMENT DISCUSSIONS

3.7.1 Coroners, Medical Examiners, and Pathologists

The process of beginning to identify high-level requirements for PPE is best performed by bringing together a large group of relevant experts in a workshop or focus group format. In this section, the type of discussion that this process can result in is illustrated using the example of coroners, medical examiners, and pathologists [98]. This group of people is responsible for determining the cause and nature of unexpected death; they are assisted by teams that may recover, decontaminate, and transport the body for examination off-scene, as well as forensic examination teams, each of which will have different requirements.

The first step in the process is to identify the nature of the population to be protected. Depending on the jurisdiction within which they operate, coroners, medical examiners, and pathologists may have varying educational backgrounds, coming from medical, legal/investigative, or social science arenas, and may have different detailed responsibilities. Coroners having only legal training are responsible for determining cause of death, but do not perform autopsies; pathologists may perform autopsies and other examination of recovered tissue; medical examiners can perform the entire gamut of functions. People filling these roles will be of varying age and physical fitness.

The next step involves understanding what tasks might need to be performed, where, and when. These tasks can be generally described as:

- Examining the body and the scene, performed at the location of the CBRN incident or death, in phase 1 or 2 of the event
- Autopsy, performed at a morgue, in phase 2 or 3 of the event

Because these two tasks are potentially very different in location and timing, they are discussed separately. At this stage it is important to remember that the best solution to operating in a CBRN environment may involve redistribution of tasks so that the minimum number of people is exposed to the high hazards, and therefore fewer people require substantial training and fitness in order to wear PPE safely while operating.

The on-scene examination, if performed exactly as it would be in a non-CBRN incident, would involve the coroner personally entering the high-hazard area as soon as possible after the death, to examine the evidence surrounding the death in situ.* The location could be anywhere, indoors or outdoors, with various environmental issues such as weather and temperature, or unsafe terrain or structures. As in any hazardous environment, the coroner would examine the body only if the scene had been declared safe by other responders responsible for scene safety. If the hazards decay with time, then because this function can reasonably be delayed for several days to protect the individual, the PPE need not be designed for the highest level of hazard. However, if the hazard were persistent, the coroner would only be able to enter the scene while wearing PPE. It is possible that the nature of the agent would have been identified and its concentration assessed by the time the entry was performed; however, neither the body nor the scene would have been decontaminated. The discussion that follows illustrates how PPE selection can be optimized by analyzing the specific roles and tasks and potentially modifying them to result in the best outcome.

There are a number of issues that would argue against having coroners enter a CBRN hazard zone. Coroners perform a specialized function that is often shared with other duties and do not normally have a physical fitness requirement. Wearing PPE may be problematic as a result of requirements for corrective eyewear, physiological burden imposed by PPE, and little opportunity for PPE use or refresher training. It could be argued that a specialized group of coroners meeting the fitness and training requirements could provide this function within a given jurisdiction. This approach could be effective where the jurisdiction for a given coroner's office is large enough for there to be a large enough pool of people available. On the other hand, as an example, in England and Wales coroners have a very narrow geographical jurisdiction and therefore many different people would have to be supplied with PPE and given training, with a very low likelihood of use.

An alternative approach would be to use other groups of responders, those already having on-scene functions, as the eyes and ears of the coroner. These people would require appropriate forensic training, and audio–video transmission and recording devices so that information could be fed back to the coroner, located in the cold zone nearby, who could then provide guidance on what information should be acquired. In this case, no PPE is required for the coroner in this phase of the response.

The second category of tasks, in phase 2/3, involves medical examination of the decedent. These procedures would be performed in a designated morgue, whether

* The location of death could also vary, depending on whether the decedent had escaped from the hazard area before succumbing.

TABLE 3-5 Comparison of the Activities Performed by Coroners, Medical Examiners, and Pathologists in Determining the Cause and Nature of Death

Phase 1 or 2 On-Scene	Phase 1 or 2 Off-Scene; Phase 3
Work may be performed in hazard area	Work is performed in controlled surroundings
Hazards are higher and often undetermined	Hazards are lower, may have been quantified or determined
Danger to the public as well as potential other hazards to the worker	No danger to the public from the decedent or the work environment; no additional hazards to the worker
Climate uncontrolled	Climate controlled
Potentially limited power, limited breathing gas supplies	Available power and breathing gas supplies

temporary, set up near the scene, or permanent, within a hospital or other structure. The distinction between these two cases could be relevant to PPE selection; a temporary morgue could have more limited climate and ventilation control. Temporary morgues may be considered as the more desirable option since the permanent morgue will be required for other uses throughout the event. Keeping the hospital environment free of contamination during and after a CBRN autopsy may be a nontrivial exercise, particularly if the agent causing death is not already known and it is therefore impossible to monitor for it.

One noteworthy difference between activities performed in a morgue and examination of the body on-scene is the fact that in a morgue, tasks are confined to a relatively small area that has the potential to be well supplied with power and breathing gas and cooling air. This opens up options for the type of PPE to be selected and the resulting comfort of the wearer, allowing less-fit individuals to function for protracted periods within PPE. In either case, the activities may be performed at a relatively high work rate, if only for short periods that may be separated by breaks. There is not necessarily any time constraint on completion of the tasks, although other factors may come into play. Fine motor skills may be required, although where the environment or the body itself is sufficiently hazardous, very limited procedures may be performed, depending on the limitations imposed by operating in PPE.

The issues discussed above are summarized in Table 3-5, highlighting some of the differences that would suggest that different PPE be selected for the two applications, and in fact different personnel might be appropriate for the two phases.

3.7.2 Law Enforcement

As mentioned previously, collection of the necessary information to develop high-level requirements is best performed through workshops and user focus groups that contain experts from many different relevant fields. Another published example of this process was that performed in the United States to develop requirements for CBRN protection for law enforcement personnel responding to a (potential or actual) release

TABLE 3-6 Tasks Performed by Various Law Enforcement Roles

Task	First Responder	Perimeter Control	Tactical Operations	Criminal Investigation
Weapons proficiency	√	√	√	
Operate equipment	√	√	√	√
Close-quarter battle (tactical)	√	√	√	
Hand to hand	√	√	√	
Fire and movement	√	√	√	
Engage moving targets	√	√	√	
Weapons transition	√	√	√	
Night/low-light engagement	√	√	√	
Self-defense	√	√	√	
Suspect and victim control	√	√	√	
Weapon retention	√	√	√	
Traffic direction and crowd control	√	√		
Evacuation	√	√	√	
Site security	√	√	√	√
Assistance to other responders	√	√	√	√
Rescue	√	√	√	√
CBRNE sampling, monitoring		√		√
CBRNE evidence collection			√	√
CB perimeter characterization		√		
Vehicle operations	√	√	√	√
Unassisted equipment donning	√	√	√	√
Decontamination (personnel—any)				
Decontamination (equipment)				
Radio communications	√	√	√	√
Face-to-face communications	√	√	√	√

event, such protection to be contained in the uniform [99,100]. Those requirements that could be defined by the user, relating primarily to human factors and integration, were developed.

This process was begun by identifying tasks to be performed within each role (Table 3-6). These tasks were then related to duration of role (Table 3-7), specific physical activities to be performed, and potential resulting human factors and integration issues.

TABLE 3-7 Duration of Law Enforcement Roles On-Scene at Release Event

Role	Duration (h)
First responder, perimeter control	8–12
Tactical operations	4–6
Criminal investigation	8–12

The list of activities derived from the task list included the following:

- Running
- Crawling
- Kneeling
- Twisting
- Jumping
- Climbing
- Standing
- Lifting
- Lying prone
- Pushing
- Pulling
- Sitting
- Walking
- Drinking
- Talking or responding
- Sighting a weapon
- Facial gesturing
- Listening
- Looking
- Writing
- Using keyboard or keypad
- Other manual-dexterity tests

Many human factors issues were identified as important in achieving the desired outcome of a protective uniform, with some being more easily achievable than others. The highest-ranking issues that the focus group felt required attention included physical sustainability of officers over the requisite number of hours (heat stress in particular), communications, and standardization of the uniform (by role). Other difficult issues identified included dexterity, putting on and removing PPE, visual acuity, equipment integration, including inability to sight weapons, bulk from plate armor, and restricted equipment access. All of this information was incorporated into subsequent projects to develop relevant PPE for law enforcement.

4 Designing for Appropriate Protection and Performance

In this chapter we discuss all the factors that must be considered to design effective CBRN personal protective equipment, covering the impact of the type of hazard, how materials and systems protect, and the human factors that must be considered. The characteristics of protective materials and systems that can be chosen and optimized are described, and the use of modeling to improve the design process is discussed.

4.1 THE HAZARD

Protection mechanisms are generally strongly associated with the physical state of the hazard. PPE must protect against CBRN agents in three physical states: gas or vapor, liquid, or aerosol (a solid particulate dust or liquid mist). Gases, vapors, and aerosols are airborne, whereas liquids are short-range contact hazards only, as described in Section 2.4.* In addition, reactive protection mechanisms depend on the (bio)chemical nature of the hazard. If we consider the hazards associated with each of these forms of agent, the following observations affecting requirements for protection can be made.

4.1.1 Airborne Hazards

Airborne hazards can be a greater or lesser problem for protection than that for contact hazards. Airborne agents can travel to the wearer, spreading over large areas, and therefore are the most likely hazard to be encountered, although concentrations decrease significantly as the agent spreads. Transport Canada's *Emergency Response Guidebook* (2012, updated regularly) [66] outlines the areas over which hazardous concentrations of an accidental release of a chemical, biological, or radiological material may exist in the vicinity of a release and downwind (Figure 3.2), based on the amount released and the air stability. For example, the guidebook indicates that under stable (nighttime) wind conditions, the area within which unprotected

* Nonparticulate solids may also occasionally be hazards, but protection against them is generally provided by designs that protect against the other forms.

Personal Protective Equipment for Chemical, Biological, and Radiological Hazards: Design, Evaluation, and Selection, First Edition. Eva F. Gudgin Dickson.

people are at risk could be up to 4.5 km downwind for sarin released as a weapon, or 8 km for a large chlorine spill. Respiratory and eye effects can happen at very low concentrations for the more toxic agents.

As already noted, airborne hazards can be in the form of vapors or aerosols. Liquid aerosols, liquids, and some solid aerosols will also evaporate or sublime, producing a simultaneous vapor hazard and changing the aerosol characteristics. Solid particulates may persist in solid form indefinitely. The unpredictability of the challenge requires design of systems capable of protecting against various evolving physical forms.

Airborne hazards will be ubiquitous after any release but in a few ways may be less of a protection problem than contact hazards may be because:

- They will tend to be nonpersistent unless trapped by enclosed spaces indoors, or terrain features that prevent dissipation such as gullies.
- Few airborne hazards are sufficiently dermally active, persistent, and capable of dissemination at sufficiently high concentrations to pose a direct dermal hazard.

Protection of the respiratory tract and eyes is paramount against a large variety of airborne hazards in both the vapor and aerosol states. The skin should be protected more strategically and not more than necessary, recognizing the following considerations:

- Protection against dermally active agents is required if they could be encountered at a sufficient dose.
- Both the PPE and the skin may act as a reservoir for deposition of hazardous materials that may off-gas or reaerosolize with time.
- Given current technology, covering the body in protective clothing that completely prevents ingress of outside air guarantees that human performance limitations will result associated with limited air supply, thermal burden, or impairment of dexterity.

When designing for protection against the full spectrum of possible hazards, those few airborne agents of potential dermal concern will drive body protection levels higher. Therefore, being able to quantify the nature of the hazard is highly beneficial in permitting a downgrade of dermal (and possibly respiratory) protection.

4.1.2 Contact Hazards

For most people, liquid hazards would be less likely to be encountered than airborne hazards, but when they are encountered they present a very large potential contact hazard, due to the hazard being more concentrated. The considerations for liquid protection are:

- The lower the boiling point of the liquid, the higher the vapor pressure and the shorter the duration of the contact hazard (both as a source of hazard and on the surface of PPE), whereas high-boiling liquids or solids can be very persistent.

- The larger the drop size of the liquid, the greater the hazard to the wearer of the PPE, both because of the larger total amount of hazardous material, and because larger amounts are more difficult to hold out of PPE effectively.
- Surfaces may also be contaminated with deposited materials that will off-gas, reaerosolize, or transfer on contact. The persistence of this type of hazard can be very long, and the presence of such contamination can be difficult to identify.
- Liquid chemical hazards can absorb or adsorb into surfaces such that even after decontamination of the surface, off-gassing can persist for months.

4.1.3 Protection Priorities and Issues Related to the Hazard

Important characteristics of an agent that must be taken into consideration when designing or selecting protection are:

- Hazardous dose by various routes of entry
- Physical form (liquid, solid, vapor, particulate, adsorbed into surfaces)
- Vapor pressure and volatility, which affect the likelihood that it will remain as a condensed form or aerosol vs. vapor
- Concentration
- Particle or drop size
- Solubility in various types of materials
- Stability to chemical reactions that might be used in self-decontaminating materials

Because airborne agents are always the most likely hazard encountered, and the eyes and respiratory tract are particularly sensitive routes of entry, respirators, in most cases full-face, are the highest protection priority. The remaining requirements for protection against potential contact and skin hazards should be considered in the context of:

- The likelihood of encountering such hazards
- Their toxicity and amount
- The potential efficacy of decontamination procedures in removing material that is not a dermal hazard from clothing and skin (to prevent entry by other routes after PPE is removed)

Hazards become difficult to protect against when they have particular characteristics, such as being highly persistent or being capable of penetrating conventional protective equipment; certain agents have been weaponized specifically for these reasons. A few examples follow:

- Sulfur mustard dissolves into and permeates a great many materials.
- Fluorinated chemicals, and low-molecular-mass chemicals in general, pose a particular challenge to activated carbon.

- Corrosive materials such as acids and bases can cause degradation of protective barriers and affect clarity of components such as visors and eyepieces.
- *Bacillus anthracis* is highly persistent and difficult to decontaminate.

It may also be true that particular physical forms of such agents pose a greater challenge to protective systems than other forms do, depending on the design of the equipment. The most highly toxic agents are a challenge simply because they place a high requirement for maintaining system integrity to prevent penetration; examples are nerve agents and biological agents with very low infectious doses, such as smallpox. In addition, the method by which the hazard is disseminated may add additional requirements, depending on the role of the wearer. For example, explosive dissemination may require additional protection against blast overpressure or thermal effects for certain user roles.

4.2 MECHANISMS OF PROTECTION

There are several mechanisms by which equipment may protect against the ingress of hazardous materials. These include:

- Providing an impermeable barrier between the agent and the person
- Removing hazardous material by filtering, adsorbing, or reacting with it
- Using overpressure or directional airflow to keep contaminated air away from routes of entry

These mechanisms are discussed in general terms in this section, while more specific examples of material and design options that take advantage of these mechanisms are given later in the chapter.

4.2.1 Barrier Materials and Hardening

In some cases it is possible to design hardened equipment through which air or liquid cannot penetrate or permeate, by providing an impermeable barrier that cannot be penetrated because of its intrinsic chemical and physical structure. Figure 4-1 illustrates several processes that can result in defeat of hardening. **Penetration** is the process by which an agent (C, B, or R, starting in any airborne or condensed form) works its way into or bypasses a material barrier, due, for example, to defects or holes in materials or imperfect seals. If it emerges on the other side, it will be in essentially the same physical form as that in which it started—that is, a penetrating liquid comes through as liquid. **Permeation** is the process by which a chemical (starting in any airborne or condensed form) dissolves into a material and migrates through it. Given enough time, the chemical eventually breaks through to the other side and is emitted as a vapor. Materials such as metals and glass are intrinsically quite impermeable, owing to their structure. However, polymeric and flexible materials are rarely totally impermeable (see [101]).

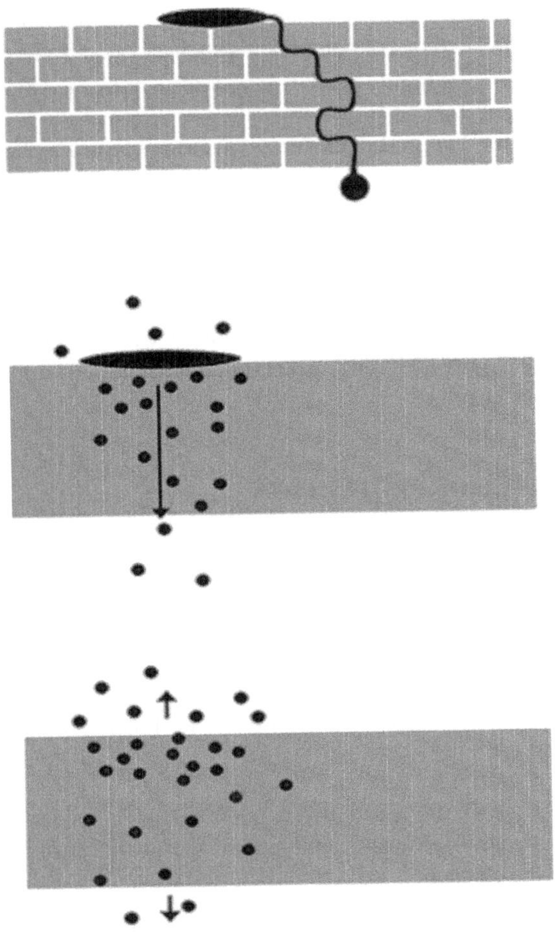

FIGURE 4-1 Penetration (top), permeation (middle), and reemission (bottom).

After the challenge agent has been removed, if chemical agent has dissolved into the material as a result of permeation, **reemission** or off-gassing of the agent as vapors will occur for some time, from either the inside or the outside of the equipment. If high enough, reemission can cause difficulties in safe removal and subsequent handling of equipment for the purposes of disposal, decontamination, or reuse.

Impermeable barriers can be created by layering multiple materials together, each of which is resistant to different agents. Nevertheless, it should be recognized that for many such materials, *impermeable* is a relative term, for although they may apparently be impermeable for some time, eventually breakthrough will occur. Barrier materials that are not sufficiently impermeable themselves can also be hardened by coating with another more resistant material or by treating them to change their surface properties. Some materials are *selectively permeable* (as described in Section 4.5.4) and it may be possible to use this to advantage to allow desirable materials through while keeping out undesirable materials. Currently, this is applied in allowing moisture vapor transmission while resisting permeation of harmful chemicals.

Hardening of materials or items of equipment should remove cracks and crevices that can sequester agent and should choose barrier materials that are unlikely to allow permeation, as well as preventing reactions that might weaken them. The process of material hardening focuses primarily on chemical agent hardening due to the fact that C agents can be a much more difficult problem than most B and R agents; a material that is properly hardened against chemicals is generally hardened against other types of agents by default. Removal of R agents is based strictly on thorough surface washing, whereas for B agents other procedures that may kill the agent can be added to the washing procedure. B and R agents become inaccessible to this form of treatment only through working into cracks, although charging of radiological particles may increase attractive forces, making them more difficult to remove from particular surfaces. Some chemical agents, on the other hand, permeate and penetrate equipment by combining the solvent properties of paint strippers with the spreading characteristics of penetrating and releasing oils; this makes their complete removal from contaminated surfaces particularly challenging. Self-decontamination of surfaces by evaporation and reemission of persistent C agents may take hours to weeks, even for surfaces into which no appreciable penetration has occurred. Hardening also increases the survivability of equipment and materials in general, as penetration and permeation can cause loss of structural integrity.

4.2.2 Air Purification Processes

If air is to be allowed into a PPE system for the purposes of breathing, drying, or cooling, it must be cleaned of all relevant hazardous materials to a safe level. Those portions of the PPE that perform the breathing air purification, some examples of which may individually be called canisters, cartridges, or filters, are generically called **air-purifying elements** (APEs). All air purification processes are based on the principle that the hazardous materials have properties different from those of the air, based on size, mass, or charge (for particulate materials) or size and chemical structure (for vapors). These properties cause them to stick to or react with the surface of the air-purifying element materials by processes such as *adsorption* and *chemisorption*, which act to remove molecules from the airstream, as well as *filtration*, which acts to remove particulates; these mechanisms are described in more detail in the following sections.

A major property of the material that will determine its utility for various forms of air purification is its porosity. Several factors are of importance in determining the suitability of its porosity for the air purification application.

- **Surface area.** The higher the surface area, the greater the removal capacity in general.
- **Density.** If a material contains too few pores, it will be of high mass density, having little surface area. If a material contains too many pores, it may have poor

structural integrity and be of low mass density, requiring too large a volume of material to be practical in use.

- **Pore size distribution.** The pore size distribution is particularly relevant for removal of molecules rather than particles and determines the number of pores in each size (diameter) category:
  - **Ultramicroporous:** <0.5 nm
  - **Microporous:** 0.5 to 2 nm
  - **Mesoporous:** 2 to 50 nm
  - **Macroporous:** >50 nm
- A narrow pore size distribution will permit size-selective removal; a broader pore size distribution results in a more generically applicable material.
- **Network interconnectedness.** The most broadly effective air purification media are highly porous, having a thoroughly interconnected pore structure that allows air to pass through them with the lowest flow resistance possible while maintaining good removal capability.

A high degree of porosity will mean a higher filtering capacity; that is, the pores will take longer to become saturated (filled) by the species being removed from the air. Also, the longer the path that the air must travel to traverse the APE, the more likely removal is, although this longer path also increases the resistance to flow. Macroporous materials are suitable for particulate filtration and have less resistance to airflow. Pores of the atomic and small-molecule scale are required for gas separation and removal from an airstream. These are found in micro- and mesoporous materials such as zeolites, silica gel, and active carbon. For chemical removal by adsorbents, too many macropores will result in a low surface area relative to the density, meaning that a large volume of material might be required. Too many micropores will result in high airflow resistance. Ultramicropores remove only the smallest molecules. The more highly porous a material, the less dense and durable it will be as a rule, so using a high-strength molecular structure as a basis is usually important. Aside from its polarity, the chemical nature of the surface structure determines its preference for adsorption of particular molecular structures; the presence of oxygen or nitrogen groups will probably enhance the bonding of polar molecules, for example.

Filtration. Strictly speaking, filtration applies to the process of air purification by passing through a macroporous material that selectively removes particulates. At the molecular level, the removal process is called *adsorption*, which is discussed later. The pores in the material may be intrinsic to the structure, but for air filtration purposes, most often the material consists of a fibrous mat of glass or polymeric material, with the "pores" generated by the gaps between the fibers. An electron micrograph of such a fibrous material is shown in Figure 4-2. Most of the discussion of filtration mechanisms that follows focuses on fibrous filtration media as the major example.

For very large particulates, filtration occurs by a sieving process wherein the particles are larger than the pores. As the particles become smaller, the forces that

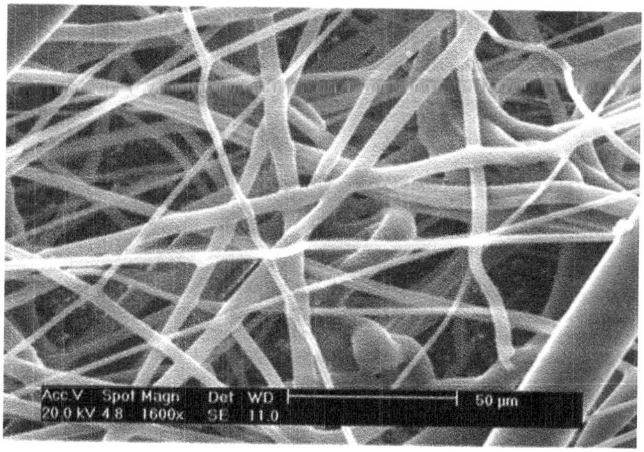

FIGURE 4-2 Electron micrograph of a fibrous filter material.

result in removal are dominated by adhesion of particles to the surfaces of the fibers, due to various surface-attractive forces that remove the particle from the airstream, as illustrated in Figure 4-3 and described further in Table 4-1.

Sieving and *gravitational sedimentation* apply only for very large particles, such as nuclear fallout or other hazardous materials that have adhered to the surface of atmospheric dust, or large particles produced by explosion of a radiological source. For smaller CBR aerosols, these mechanisms have negligible roles in filtration [102];

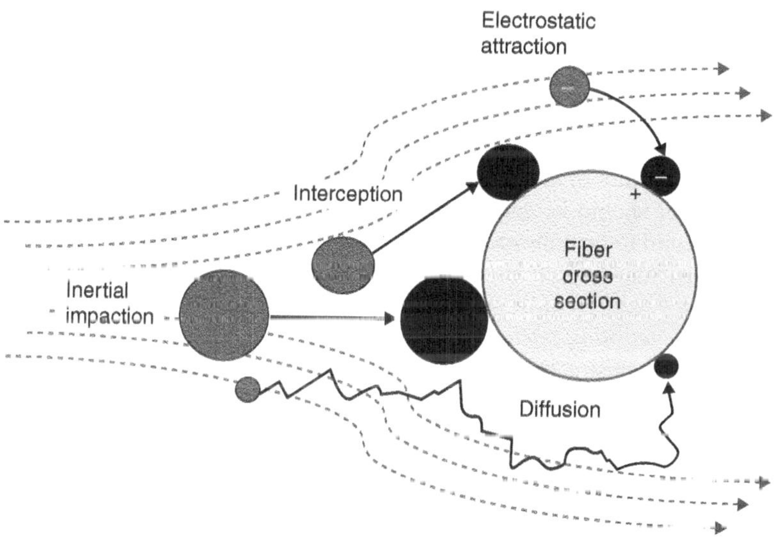

FIGURE 4-3 Various mechanisms responsible for capture of particulates by single fibers as an airstream flows by the fiber.

TABLE 4-1 Aerosol Capture Mechanisms

Mechanism	Description	Contribution to Total Filtration
Sieving	Capture when particles are too large to pass through the pores in the filter	Small contribution, only for very large particles
Gravitational sedimentation	Deposition of particle onto a fiber as a result of gravity	Small contribution, increases with particle size, becoming important over 10 μm
Interception	Capture when particle impacts a fiber due to its normal trajectory when following the fluid flow	Increases with particle size, becoming important over 0.1 μm; tends to dominate in the range 0.1–1 μm
Inertial Impaction	Capture when particle impacts a fiber due to inability of heavy particles to follow fluid flow streamlines around fibers	Increases with particle size, becoming important over 0.3–1 μm, depending on filter structure and flow velocity
Diffusion	Capture via contact of particles with fibers resulting from Brownian motion; dominates for small particles	Significant effect only on particles below 0.1 μm
Electrostatics	Capture resulting from acceleration of particles toward fibers due to charge attraction	Most significant when fiber and particle are oppositely charged, but some contribution when only one has a net charge

thus, *diffusion*, *interception*, and *inertial impaction* are likely to be dominant, with *electrostatics* playing a role when the filtration medium is charged electrostatically.

The magnitude and relative contribution of each of these mechanisms can vary significantly, depending on the airflow, the charge, the density and size of the particle, the fiber diameter, the packing density, the structure and pore size of the filter, and the chemical structure of and charges on the filtration medium. Hence, designing or selecting a filter should be accompanied by an understanding of the nature of the particulate hazard to be removed. In addition, some filter structures or compositions may have particular limitations that should also be recognized: for example, their tendency to clog, restrict airflow, or lose charge under certain exposure conditions.

The *most penetrating particle size* (MPPS), at which the combination of all of the removal mechanisms is least effective, has often been taken as a fixed value of 0.3 μm independent of the design of the filtration medium. In fact, the exact particle size for this minimum can vary with the material and due to a variety of other factors, and therefore the implications of this should be considered when setting requirements and designing test methods (see Table 4-1 [103]). The efficiency of filtration in most cases is very high for particles either below 10 nm or greater than 10 μm in diameter, with the MPPS generally between 0.05 and 0.3 μm, depending on the nature of the filter.

Recent investigations into the filtration of *nanoparticles* have been performed due to concern that previous methodologies did not necessarily capture filtration performance at small particle sizes. Current studies have shown that no deviation from single-fiber filtration theory occurs down to particles as small as 4 nm; in other words, filtration is efficient by diffusion, as predicted for the range >4 nm up to the MPPS [209]. There is some evidence that at sizes below 2 nm, filtration is less effective than predicted, indicating that rebound of very small particles at higher velocities reduces filtration efficiency [104].

It is noteworthy that gas separation can also be performed using size-based filtration [105], when a microporous air-purifying element with a narrow pore size distribution behaves as a molecular sieve. For example, *zeolites* are a class of aluminosilicate minerals of repeating regular porous structures, with pore and channel sizes that can be varied in the range 0.3 to 1 nm, depending on the detailed zeolite structure. Other microporous materials remove gases and vapors by adsorption rather than filtration.

Adsorption and Reaction. *Adsorbents* are materials that remove gases from an airstream by collecting them on their surface in an adsorbed layer. To carry out this process effectively, an adsorbent must have a large accessible surface area. The most common types of adsorbent adsorb those molecules that have a strong attraction to the surface because their particular shape and chemistry match those of the adsorbent's pores. The nature of the adsorption forces that capture the vapor molecules within the pores can be twofold: due to physical forces alone and due to chemisorption. The physical forces behave much like those weak forces that can cause a liquid to condense from a vapor and are called *van der Waals forces*. These forces cause the vapor agent to stick to the surface of the pores in the adsorbent in a process called *physisorption*. The physisorption process is almost irreversible for very large molecules under conditions of normal use when the adsorbent structure is suitable, but the forces become less strong as the hazard agent becomes smaller. The sites that are the most effective match to the structure of the hazard molecule are filled first and most irreversibly; as the surface area fills, less favorable sites are filled, and the binding is less permanent. Pores and cavities that are smaller than a given target molecule will not be able to hold it effectively, while those that are too large may not attract the target molecule as strongly (Figure 4-4).

The important characteristics of an adsorbent from the point of view of its removal capacity are the pore size distribution, pore volume, and surface area; these can be determined by techniques such as nitrogen adsorption isotherms and mercury porosimetry [105,106]. Low-boiling toxic agents such as hydrogen cyanide may not physisorb to a surface irreversibly; however, after physisorption, they can be permanently bound or decomposed if they react further with *impregnants* in the adsorbent in a process called *chemisorption*.

Impregnants are reactive atoms and groups: for example, metal species such as particular oxidation states of silver, manganese, molybdenum, and zinc, or groups such as chlorine or amines; these are added to the surface of the adsorbent. The chemisorption reaction is irreversible, meaning that once the agent has reacted, at

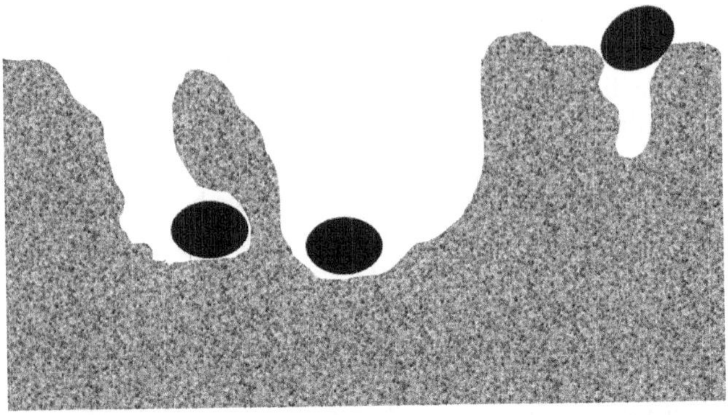

FIGURE 4-4 The pore on the left matches well the hazard molecule to be adsorbed (the dark ovals), whereas the pores in the center and on the right will not adsorb as strongly, due to size mismatch.

least part of it remains trapped and the reactive site is consumed by having encountered the agent. As illustrated in Figure 4-5, molecules adsorb from the air onto the surface of the adsorbent. The molecule in the pore on the left is physisorbed, meaning that it can be removed intact with the addition of energy or an air or solvent wash, while the molecule on the right has chemisorbed to the surface as a result of reaction with an impregnant, and cannot be removed intact.

Each impregnant's reactions tend to be relatively specific, and thus many different impregnants are generally required if a broad spectrum of agents is to be removed. Some hazard agents may break down into smaller products as they chemisorb; if these smaller products are hazardous and do not themselves chemisorb, a new hazard has

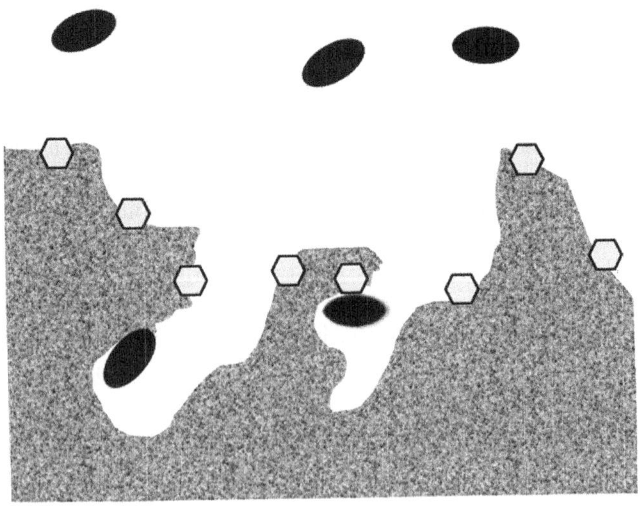

FIGURE 4-5 Physisorption (left) and chemisorption (right) of hazard molecules onto the surface of an adsorbent that has reactive surface impregnants (the hexagons).

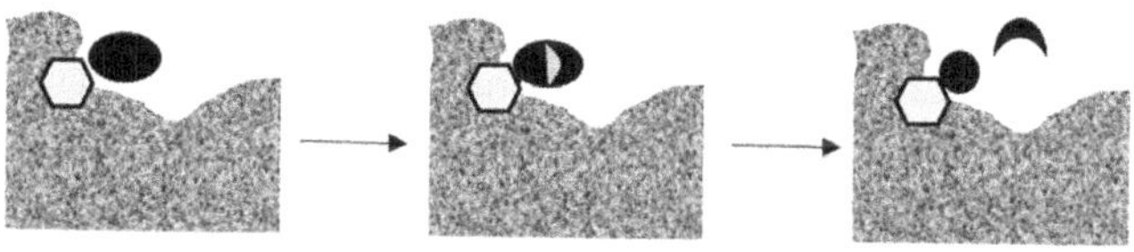

FIGURE 4-6 A molecule physisorbs (left) immediately followed by chemisorption (center); the chemisorption reaction results in the release of a smaller high-volatility breakdown product (right).

been created (Figure 4-6). Therefore, a thorough understanding of the nature of the possible reactions involved is required to assess for possibly hazardous by-products from a particular hazard agent. Further, it should not be assumed that because a low-molecular-mass agent has been removed from the airstream it has effectively chemisorbed; some physisorption can still occur, and subsequent desorption of the agent into the airstream is likely, particularly if conditions such as temperature, humidity, or airflow change.

Reactive species can also be used to surface-treat materials that do not have a strong adsorption component, such as polymeric clothing materials or the surfaces of filter fibers. In this case, reactive species may be targeted toward either chemical or biological agents that have come into contact with the surface, with the intent of achieving *self-decontamination*.

4.2.3 Airflow and Overpressure

All systems that use air filtering or overpressure for CBRN protection are based on directional airflow. Contaminated air can be forced through a filter, or forced away from the area to be protected and out through valves or leaky closures. For example, in a protective mask, contaminated air must flow in through the purification system, or alternatively, clean air must be provided from an air supply; afterward, air must be exhaled through a suitable exhale valve that does not allow backflow of outside contaminated air. Supplied and powered air systems take additional advantage of the fact that overpressure is created within the facepiece of the respirator, which helps to prevent backflow of contaminants. The same principle applies in overpressure suit systems. Note that it is quite difficult to assure that a protective system that is under positive pressure will remain so at all times without resorting to total encapsulation. For example, positive pressure in respirators can potentially be defeated by deep breathing, whereas in suits, bellowing effects resulting from motion can draw air in through closures.

4.3 HUMAN FACTORS

4.3.1 General

The requirement to design equipment that will work with a human being means that any number of factors relating to the person must be understood. Such factors may be different for different user groups and tasks, which in turn will affect the selection of particular material or protective system concepts. Many of these factors are discussed in BS EN 13921 [107]. Generally speaking, two main areas should be addressed to improve the performance of the wearer: maintaining the person's health and comfort and maintaining the necessary situational awareness, safety, and functionality of the person so that tasks can be performed. Additionally, it should not be forgotten that human wearers have human foibles and will be more likely to select and wear equipment that looks fashionable or "cool." In general, the absence of comfort may cause a wearer to reject the equipment [108].

In the category of maintaining health and comfort, the following factors should be addressed:

- Air quality and air supply
- Work and effort: breathing resistance, weight and weight distribution, resistance to movement
- Thermophysiological balance: provision of heat and cooling
- Moisture balance and water supply
- Nutrition
- Management of body wastes
- Pain and irritation

Maintaining situational awareness, safety, and ability to perform tasks should be addressed through the following areas:

- Maintaining cognition
- Maintaining or enhancing sensory and communication interfaces: visual, auditory, verbal, and tactile
- Sizing and range of movement (appropriate to user group anthropometry)
- Maintaining ability to interact effectively with other equipment and the environment
- Reducing the signature: camouflage, heat signature, and noise

Obviously, there is some overlap: Maintaining health and comfort will improve the likelihood of mission or task success, and improving the chance of mission success is likely to reduce the likelihood of harm coming to a person. In this section we discuss some of the more important background information that is available to assist designers in understanding the important human factors issues and to set criteria as discussed in Section 6.4.

4.3.2 Thermophysiology

Thermal stress results from the additional heat burden imposed by wearing extra clothing, combined with the potentially higher work rates required in emergency situations. All of the items of PPE can add thermal burden by impeding body cooling. This discomfort may significantly affect how well people comply with proper PPE use without regular effective education and training in PPE. Further, increasing the core body temperature results in lower endurance. As a general rule for reasonably fit adults, when the core temperature exceeds 39.2°C, or when the heart rate exceeds 180 beats/min for any extended period, morbidity due to heat stress can be expected, causing personal injury and decline in operational capability. A person's age and physical fitness will also affect the likelihood of heat stress. Environmental factors that may have an influence on thermophysiology are discussed further in Section 4.4.

The energy consumed in performing any activity is ultimately converted partially to heat, proportional to the energy required for each activity. The amount of heat that must be dissipated by the body over its surface area ranges from about 65 $W \cdot m^{-2}$ when resting to more than 500 $W \cdot m^{-2}$ at the maximal work rate [109,110,113]. If heat cannot be dissipated from the body, the body temperature will rise, resulting in overheating and thermal stress.

Cooling can occur by several mechanisms [111], as illustrated in Figure 4-7. **Radiation** consists of heat energy moving through space as light and can occur under any circumstances. It is normally a minor contributor to cooling. **Convection and conduction** as cooling mechanisms while wearing PPE require that a cooler medium (such as air or a cooling fluid) be passed over or near the body's surface so that heat can be exchanged to it and removed. It is possible for these to be significant cooling mechanisms when the exchange rate of the cooling medium is high; convection becomes particularly important when combined with evaporative cooling.

Evaporative cooling results from the evaporation of water from skin or through respiration; the act of evaporation requires heat, which is removed from the body in this manner. For this mechanism to be effective for skin, dry air must be near the

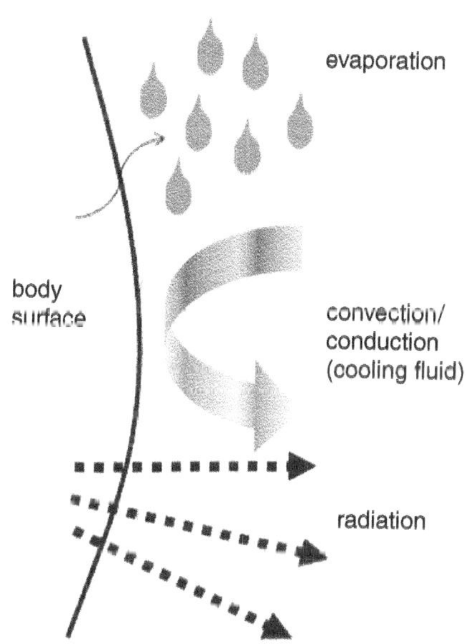

FIGURE 4-7 Various cooling mechanisms.

skin so that the air can take up the excess moisture. The air next to the skin will become rapidly saturated with moisture, so this effect will be negated by wearing sealed systems with air-impermeable barriers, as well as in conditions of high environmental heat and humidity. However, respiratory cooling can occur in PPE as long as the breathing air is not overly hot and humid, because moist exhaled air must eventually leave the PPE environment.

Convection and radiation together may account for less than 10% of total heat loss under normal circumstances, with sweating and water loss through respiration accounting for the remainder. Wearing an air-impermeable respirator and gloves, which essentially eliminate the cooling effect due to sweating on the hands and face, prevents convective and evaporative heat loss from these body regions; however, this represents a relatively small body surface area overall, and the impact is reduced as a result.

Just how comfortable a person feels when wearing PPE depends not only on how much heat is removed but also on the balance of body cooling by various mechanisms. If the normal balance is adjusted, discomfort will probably result, even when the total heat loss is the same (e.g., breathing hot humid air even though the body is relatively cool). There is some correlation between heat stress and cognition, and mental state (e.g., anxiety) may also affect heat production and storage. The correlation between perceived increase in thermal stress and wearing of respirator facepieces, regardless of actual core body temperature, has been discussed [112].

A significant component of the stress resulting from wearing PPE is the increased rate of water loss as the body attempts to cool itself by sweating; the less effective this mechanism is, the more the body will sweat. Water loss while exercising in PPE can be substantial, and water must be replenished regularly.

4.3.3 Breathing and Respiratory Physiology

Additional physiological and psychological stress can result from issues related to breathing when wearing a respirator. If either inhaling or exhaling is not performed at around 1 atm pressure, breathing becomes difficult; this is particularly true for exhalation, which is normally a passive process of relaxing the muscles. As the breathing rate increases at higher work rates, stress will result, resulting in a person either having to work harder to breathe or having to restrict his or her work rate.

Work is a thermodynamic term that describes how much energy is consumed in performing an action.

Work rate describes how fast energy is consumed to maintain a certain activity level. Oxygen is required to use stored energy in a process called *metabolism*. The energy consumed in work is eventually converted to heat that must be dissipated.

The **work of breathing** describes how much energy a person uses simply to breathe. Hence, as a person works harder, requiring more energy and oxygen, the work of breathing also increases; the presence of a higher breathing resistance while wearing a respirator also increases the work of breathing.

The International Organization for Standardization (ISO) standard 16976-1 on metabolic rates and respiratory flow rates [110] outlines the relation between work rates, the generation of heat, and various metabolic demands relevant to task performance when wearing PPE. The type and intensity of *work*, along with the person's general fitness, combined with the weight and resistance to movement of the PPE, affect the metabolic rate (energy expenditure) of the wearer. The work of breathing, which depends primarily on the breath frequency and depth, is added to by the presence of breathing resistance from wearing a respirator. ISO standard 16976-4, *Work of Breathing and Breathing Resistance*, is under preparation to describe parameters associated with the work of breathing, such as inhalation and exhalation resistance and the breathing resistance caused by restrictive harnesses.

The metabolic rate is directly correlated with oxygen consumption, which determines the respiratory demands and flow rates. The relationship between oxygen consumption and metabolic rate is described in ISO 8996 [109] and is summarized in ISO 16976-1 in terms of five average work rates that can be maintained for full work shifts (class 1 being at rest and class 5 being very heavy work). ISO 16976-1 adds classes 6 to 8 as work rates that can only be maintained for finite durations of less than 2 hours for class 6 down to 5 minutes for class 8. Examples of various work activities that can give rise to these work rates are also outlined.

Importantly, the standard also discusses how these work rates are related to breathing parameters such as minute volume (total volume of air exchanged in 1 minute) and peak inspiratory flow (highest flow rate achieved in a single breath during inhalation); it also discusses how speech affects peak inspiratory flow. Unrestricted peak inspiratory flow requirements can be extremely high, up to 300 to 400 L·min^{-1} [109,113]. Figure 4-8 illustrates a generic sample breathing waveform (see [113,114] for some specific examples). It is, of course, true that every person breathes differently, and that the shape, frequency, and amplitude of breathing waveforms depend on various factors, including resistance to breathing and the activities being performed by the person; there is evidence, for example, that breathing against a resistance changes the shape of the breathing curve, resulting in more rapid changes of flow rate [115]. More work is merited in the area of assessing realistic breathing waveforms in the presence of respiratory protective devices (RPDs) of varying breathing resistance and air supply style.

Human life is strongly dependent on an adequate supply of oxygen to support the metabolic processes that produce energy; less than 19.5% oxygen in air is considered

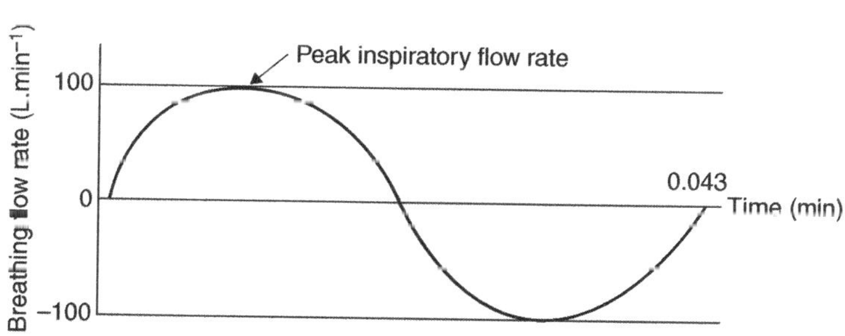

FIGURE 4-8 Breathing waveform.

to be an oxygen-deficient situation. Exhaled CO_2 is produced as a by-product of metabolism, and excess CO_2 has a profound effect on the respiratory system, such that increases in the blood concentration of CO_2 are more powerful stimulators of respiration than changes in oxygen concentration. Maintaining appropriate concentrations of O_2 and CO_2 in the breathing air to assist in regulation of respiration is important to assure adequate energy production and for a person's health, comfort, and cognitive function. Maintaining appropriate oxygen levels is dependent on providing appropriate-quality breathing air; keeping CO_2 concentrations low (desirably below 0.5% for 8 hours of wear) is assured by preventing rebreathing of exhaled air. The potential effects and limitations on human tolerance imposed by exposure to increasing concentrations of carbon dioxide in the inspired air at rest and at a very, very high work rate are outlined in ISO standard 16976-1 [110]. ISO standard 16976-3 describes parameters associated with respiratory physiology [116].

4.3.4 Anthropometry and Range of Motion

One of the most difficult human factors to describe adequately for the purposes of PPE applications is the distribution of anthropometrics of the user population. It is one thing to attempt to describe the distribution of one or two anthropometric measurements in a select population to provide sizing guidance, but it is another to attempt to describe the entire head or body in any significant way. Hundreds of publications on population anthropometry of relevance exist; many focus on the head only or on the entire body and are often reported as 5th, 50th, and 95th percentile values. Various approaches to generating, compiling, and using anthropometric data have been followed. ISO 8559 [117] and NATO STANAG 2177 [118] suggest standardized body locations that should be measured or surveyed to obtain adequate anthropometric description for clothing design for a given population.

Numerous subpopulations exist based on age, gender, or ethnic origin, which need to be sampled representatively for the population to be fitted. For occupational use, this information tends to come from industrialized areas; published examples include Europe [119–121] (limited data for the world are also provided in some of these), the United States [122,123], China [124,125], and Korea [126]. Others are specific to the military user community (e.g., Canada [127] and the United States [128–130], various other military [131]) or to a first-responder community or other occupational subgroup [132,133]. Some studies specifically compare anthropometry among particular ethnic subpopulations [134].

Essentially, there are as many anthropometric combinations as there are people and hundreds of significant measurements that could be performed. Additionally, although in an individual user, many anthropometric measurements may be correlated with each other, this correlation information is lost once large databases of data are generated. Hence, it is impossible to appropriately assess the meaning of the distributions of individual measurements relative to one another; for example, the 50th percentile facial width measurement might occur only rarely in the same person as the 50th percentile neck circumference. Therefore, generating "average" anthropometry may be unrealistic without understanding whether any correlation between any particular set of measurements exists, and in what subpopulations. Software tools do exist that provide information on the statistical likelihood of occurrence of values for nearby measurements [135]. Three-dimensional scanning data are now available that retain overall correlated information, and in most cases the accuracy and precision of these data compare well with those obtained manually [136–138].

The most important issue that has had minimal systematic investigation relates to the significance of any given anthropometric measurement in PPE design. Any given measurement may have particular significance in designing for any or all of human performance, comfort, or protection, or (in a few cases) it may be of little significance. Further, the measurement could be significant for a given design of protective item but not for others; therefore, anthropometric data generated or mined specifically for designing one type of item might not contain the information necessary for another, even if the data generally cover the same body parts.

Most if not all of the work developing and assessing the validity of human test panels based on anthropometry in relation to protection has been performed for respiratory protection. In the late 1960s, respirator-fit test panels were developed by Los Alamos National Laboratory (LANL) [139,140] based on U.S. military anthropometric surveys (clearly representative of neither the civilian workforce nor of the current population diversity within the United States). The pass/fail criteria for the panel were intended to deliver the result that a single respirator model designed to fit 95% of the user population should have only a single failure for the 25 subjects tested. Over the course of subsequent work, this panel was evaluated for its ability to correctly identify important anthropometric features as predictors of fit, particularly for half-face respirators; its evaluation indicated that depending on the brand of half-face respirator and the gender of the wearer, particular facial measurements might be positively correlated or uncorrelated with the fit of the respirator (assessed by its protective performance); visualized leaks of half-face respirators were not correlated with facial dimensions used to generate the test panel; respirator leakage is strongly affected by nose and chin leaks; the gender of the wearer is a factor in how a respirator fits; and consideration should be given to many other facial dimensions when defining a respirator test panel and selecting a respirator for an individual wearer [141–145].

These results, in addition to others more recent [146] that assessed the current validity of the population distribution of the original data, indicated that a more representative panel needed to be developed for the workforce. One particular feature of these previous panels is that they have always been based on bivariate data sets: in other words, two (presumed) critical measurements were used and the panel of persons described by their values for those two measurements (e.g., facial length and width, or chest width and height). An illustration of a panel represented in this manner is given in Figure 4-9, derived from a more up-to-date analysis of the U.S. workforce [147] using a current anthropometric survey [122,148]. This bivariate approach to representing human faces is undoubtedly a bias of the limited ways we have of representing data on paper.

A new development proposed by the same authors [147] is the use of principal component analysis (PCA) to identify important anthropometric features as predictors of protection. Using principal component analysis, any number of head

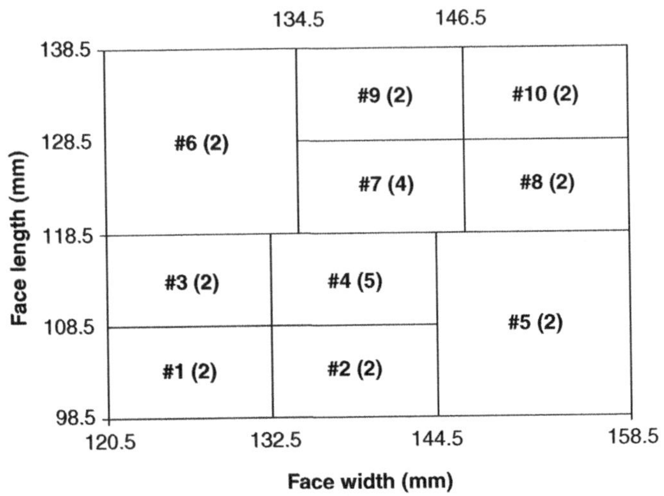

FIGURE 4-9 One example of a respirator panel based on facial length and width, as suggested by Zhuang et al. [147]. The values in parentheses indicate the number of persons from a 25-person panel that should lie within the dimensions specified.

anthropometric features, in a specified linear combination, can be used as predictors of performance.* Some of the dimensions found to be most important were the maximum frontal breadth, the bitragion chin arc, and the bitragion subnasale arc, but it is possible for the analysis to include many more. Based in part on this work, ISO has developed an anthropometric standard [149] for the development of respirators.

Many of these studies, including PCA analysis, focus on facial dimensions as being the most important presumed predictors of respirator fit. Nevertheless, it is important to remember that for both sizing and comfort of full-face respirators (as necessary for CBRN protection) and of helmet-based protective systems, the anthropometry of the entire head, and sometimes the neck, is just as important. There has been considerably less focus on this area in terms of providing useful standardized data; limited data that are available in this area address anthropometry for headforms for helmet fitting and testing [150–152]. However, there is a definite need for a proper correlation of head and facial sizing for the purposes of full-face and helmet-based respirator systems.

There has been little done in the area of systematically correlating sizing of protective clothing with protection. The general issues associated with designing PPE to fit the population properly is discussed further in Section 4.6.9. Correctly defining anthropometry is just as important for comfort as it is for protection, and further, the two are related. When people are uncomfortable, they are more likely

*One limitation of such an analysis is that there will be a single performance variable that is predicted (e.g., the protection provided by a respirator), and the outcomes will always be biased by the design of the respirator(s) chosen to perform the initial correlation as well as the performance parameter chosen.

to attempt to adjust or defeat closures, and excessive strain on the materials and closures if PPE does not allow sufficient freedom of movement may result in failure to maintain system integrity. ASTM F1154 [153] defines a series of exercises that are performed in a user trial to assess whether adequate freedom of motion is attained within PPE.

Hand Anthropometry. Hand anthropometry is critical to good glove design; an Australian user study ranked good fit higher than any other design factor for user satisfaction [154]. Various data on anthropometry and other relevant human factors exist. These include (1) conventional hand anthropometry of U.S. military personnel [155], (2) an anthropometry set developed for computer-aided glove design [156], and (3) comparison of certain engineering anthropometric and performance parameters between bare and pressure-gloved hands [157]. Key dimensions for sizing are generally felt to be hand width and breadth, which are easily measurable and correlate well with various other dimensions [156].

Foot Anthropometry. It has been demonstrated that female and male feet are significantly different [158]; female feet and legs differ in a number of shape characteristics, particularly at the arch, the lateral side of the foot, the first toe, and the ball of the foot. Daily-wear work boots for military or responder personnel may well be unisex. While protective overboots may be less sensitive to such differences, when protection is incorporated directly into the boot (e.g., combat or work boots), future footwear designers should take such differences into account.

4.3.5 Sensory Issues and Situational Awareness

The following sensory inputs and outputs are important to maintain: clarity of vision and field of view, quality of hearing, clarity of speech, and dexterity and tactility.

Clarity of Vision. Haze, distortion, and fogging, as well as absent or inappropriate vision correction, can all reduce the clarity of vision. Vision quality should be as close to 20/20 as possible. Distortion of vision is particularly disorienting when it is not stable, that is, if the feature that causes the distortion is moving relative to the eye (e.g., eyepieces that move in and out with breathing). Other undesirable characteristics that may not be captured by standard vision tests include halos or color distortion.

Field of View. The field of view (or field of vision) is defined as the area that is perceived in total by a fixating eye (or eyes). Figure 4-10 gives an example of a field-of-view diagram based on EN 136 [159]. The limits of the normal unrestricted field of view per eye are roughly (where 0° is the center of the field) 60° up (into the superior field), 75° down (into the inferior field), 110° to the outside (temporally, as located on left and right of the outside rings of the diagram), and 60° to the inside (nasally, lying in the dark gray area of the diagram). The field of view is also expressed separately for monocular and ambinocular (both eyes together) vision. Aside from the limitations to the peripheral view imposed by the eyepiece or visor, other components

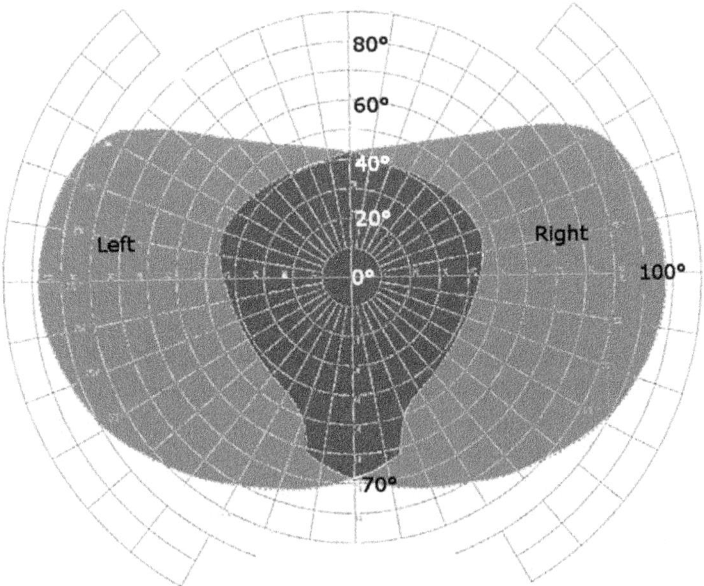

FIGURE 4-10 Field-of-view diagram showing unrestricted fields of monocular view for left and right eyes, with dark gray area the area of binocular visual overlap contained in both fields; ambinocular vision is the entire grayed area.

that can restrict the field include the respirator snout, canisters, hoses, helmets, and hoods.

Speech Intelligibility and Communication. Significant decrements in human communications capabilities (speech and hearing) can be observed when wearing PPE; results from military trials have shown performance losses of up to 60% in some measures [160]. Important parameters to consider aside from attenuation and overlaid noise are the frequency content and reverberation. The combination of local noise that obscures hearing from a person's own PPE (crisp and crinkly materials, supplied airflow/blowers, helmets, hoods), local background noise, and muffling and distortion of speech by the speaker's respirator can seriously degrade its quality by the time it reaches the listener. ISO is developing a standard for hearing and speech requirements [161].

Dexterity and Tactility. Wearing gloves can significantly increase the muscle activity, pinch strength, and discomfort in the hand and reduce the dexterity and touch sensitivity, and the duration of the activity required can result in different performance outcomes [162,163]. When investigating requirements for acquisition and development of the next-generation Australian NBC glove, the highest priority was proper fit, followed by dexterity, grip, chemical protection, and tactility [154].*

Skin Irritation. Components likely to come into contact with the skin will have texture and may have additives that can be leached out, and these must not have any short- or long-term irritating or toxic adverse effects [164] or give rise to dermatitic responses.

Odor. It is not unusual for materials to emit an odor, either when new or as a result of aging, that arises from various additives used in processing.

4.3.6 Integration, Compatibility, and Functionality

Personal functions that maintain health must be as unimpeded as possible. Normal bodily functions include drinking, eating, and voiding. The ability to clear nasal and visual lacrimation as well as vomitus can be crucial to maintaining a person's functionality. Additional important medical treatment functions include the capability to monitor breathing and heart rate, to administer drugs, and to perform cardiopulmonary resuscitation.

PPE must be compatible with all forms of ballistic and impact protection worn on the head and body; the functionality of both forms of protection must be maintained simultaneously. Compatibility with items that require the use of sights such as weaponry and night-vision goggles, as well as with heads-up displays is also crucial. Weapons firing can degrade protection, while wearing a respirator or hood can impede proper sighting of a weapon. PPE components such as air tanks, hoses, and blowers may be incompatible with seating arrangements in vehicles and aircraft. However, the availability of ancillary power and assisted air supplies when in these operating environments is advantageous. The Canadian first-responder standard [3], for example, requires that full systems, including ancillary items, be evaluated to assure integration in all aspects of performance. Overall, the impact of failure to maintain functionality and performance on operational outcomes must be assessed by user involvement on a case-by-case basis.

4.4 THE ENVIRONMENT

4.4.1 Climatic Conditions

The outside environment can have a significant impact on the comfort and capabilities of the wearer as well as on the survivability and functionality of the PPE. PPE may exacerbate physiological problems imposed by, for example, hot and humid environmental conditions. Alternatively, if properly designed and selected, it can assist in mitigating such problems: non-CBRN PPE is often designed to assist in

* Good fit, ranked first, would automatically address several comfort and dexterity and tactility aspects.

protection against rain or cold, for example. Achieving protection from CBRN agents in a full variety of outdoor environments is a considerable challenge; for example, some protective materials may perform poorly if wetted, and others may become brittle in the cold, affecting their durability as well as the freedom of motion of the wearer. It is important, therefore, to recognize the range of environmental conditions in which the PPE is to be used. Will adverse climatic conditions be likely, and what impact will they have on both the wearer and the PPE? Is it necessary to design separate PPE for different environments (e.g., hot weather, temperate, cold weather)?

NATO has described the various environments worldwide [76] for the purposes of test and design criteria, including annual and diurnal variations in temperature, relative humidity, and solar radiation around the globe, also considering the effects of altitude. The effect of storage under extreme temperature conditions is also considered. Atmospheric pressures and winds are mapped out, while levels and types of precipitation, ozone, sand, and dust are described in more general terms. Systems for use at altitude must protect against hypoxia (see, e.g., the NATO standard [165]).

4.4.2 Physical and Mechanical Stressors

The task that must be performed will often impose a number of physical and mechanical stresses on the PPE, and various external hazards can result in puncture, abrasion, tearing, fracture, melting, or other forms of failure of system integrity that will have an impact on protection and the likelihood of personal injury. Numerous standards (reviewed in Section 6.7 and Chapter 7) describe the levels of stress that must be survived by the PPE. For example, nuclear heat flash will cause thermal stress on the materials; NATO [4] has described the conditions that must be survived such that the body is not burned excessively.

It may also be true that equipment will suffer considerable mechanical stress during transport, perhaps more so than when worn. NATO [166,167] has described the types of mechanical stresses that can arise during transport by rail, air, sea, land, airlift, and transfer handling, as well as during storage on some form of conveyance while in an active warfare situation. NATO [168] has also standardized packaging requirements depending on the type of environment that is to be withstood in shipping and storage.

4.5 MATERIALS AND THEIR SELECTION

4.5.1 General Issues

A very useful reference is an article by Zhou and others [169]. The following issues are particularly important to consider when making material choices.

Function of Component or Subcomponent. Different functions of components within a protective item translate into different physical properties. A material that assists in forming a seal to skin may need to have particular elasticity characteristics; a reinforcing material must be durable; and a material may need to perform additional functions, such as blast protection. A material intended for air purification must be air permeable and porous.

Type and Physical Form of Agent. Different types or forms of agents require different material properties to protect against them.

- Airborne and liquid hazards of any type may penetrate porous materials, particularly at higher air velocities.
- Liquid chemicals can interact with materials in ways that are particularly challenging to material integrity by dissolving into or reacting with them; for more aggressive chemicals this also may happen for airborne forms.
- Chemicals are more likely to have immediate effects on skin if they penetrate and hence may have more stringent criteria for dermal protection than do biologicals and radiologicals, for which decontamination after exposure is an effective component of protection.
- Protection against biological and radiological hazards focuses on their aerosol state, although body fluids may also be considered.

Approaches to Cooling. Materials can assist or inhibit cooling:

- Air-permeable materials, if they can be made suitably protective, provide the simplest means of permitting convective or evaporative cooling for clothing.
- Moisture vapor–permeable materials permit some evaporative cooling.
- Reflective materials reflect heat from outside and may be used in some firefighting applications.
- Air-impermeable materials permit cooling only by convection through imperfectly sealed closures or by forced-air cooling; the use of impermeable materials generally raises a significant thermal management problem.
- Phase-change or thermally reactive materials have the potential to cool the wearer for brief periods by absorbing body heat.
- Layering of separate materials (rather than laminating) generally increases resistance to thermal transfer in both directions by introducing insulating air gaps.

Hand Properties, Conformability, and Comfort. Hand properties pertain to factors such as the texture (stiffness, smoothness, softness), elasticity, and drape of the material; protective clothing is most comfortable when it feels like daily-wear clothing. Also, soft and stretchy materials in general conform and seal better to other surfaces, forming more reproducible, comfortable, and reliable closures. Materials should not cause skin irritation, either trivially due to surface roughness or more seriously due to the presence of irritating or sensitizing chemicals that can leach out of the material and through the epidermis. Such chemicals might result from polymer processing or from nondurable surface treatments.

Survivability. To retain its integrity, any material chosen must be able to withstand the conditions of use and exposure to agent. Chemical agents tend to dissolve many polymeric (plastic) materials, resulting in blisters and soft spots in the material that are prone to failure. Also, chemicals may degrade optical surfaces such as lenses or faceshields, causing pitting or crazing. Thick, hard, heavy, stiff materials are most durable; this is clearly exactly the opposite of what one would wish to use in PPE from the point of view of usability and wearability except for protection against trauma.

Decontaminability. If insufficiently hardened materials are used, an agent cannot be removed on decontamination and a contact hazard will remain. Smooth air-impermeable nonporous materials are in general most easily decontaminated, whereas rough surfaces and certain types of surface charging can result in difficulty in removing agent. Decontamination in general is not effective against agent absorbed into porous materials (in fact, decontamination of liquid agent with solvent can result in enhanced penetration of the agent into the material). The result is that a significant residual vapor hazard may remain after decontamination of porous materials. Therefore, outgassing (desorption) characteristics are of as much importance in the choice of materials as is the potential for absorption.

4.5.2 Fibers and Textiles

Cloth materials are used in combination with other materials in protective clothing material systems because of their comfort as well as the obvious extensive familiarity of the industry with relevant manufacturing and construction techniques. They are distinguished here from the polymers and plastics discussed in the next section as being constructed from fibers, whether woven or nonwoven, that are manufactured in bulk sheets or rolls for subsequent construction into clothing items.

Filament fibers can be constructed from natural or synthetic materials; the latter may be textured to give physical properties similar to those of natural fibers. When different fibers are blended together, they may impart multifunctionality to a fabric—for example, by adding a conductive fiber to a yarn. To produce nonwoven fabrics, fibers and other structures, such as thin films, are joined using various mechanical or chemical methods. Traditional woven fabrics are produced by interweaving fibers at right angles, and a variety of properties can be introduced by varying the type and density of the weave, thickness, and type of yarn. Knitted fabrics have more porosity and stretch, meaning that they are less likely to be able consistently to prevent penetration by an agent, while often being more comfortable and insulating.

Textiles are not usually themselves highly protective but are often used in the construction as the outer and/or inner layers of material combinations or laminates, with an additional more protective material sandwiched within. Sometimes, the cloth fibers themselves may have some CBRN protective properties; examples include active carbon cloths or carbon-impregnated stretch fabrics, fabrics treated with a reactive layer, and filtering fabrics. To add properties that are not inherent to the weave or the yarn, textiles may be finished by either mechanical or chemical means. Mechanical processes can, for example, change the surface texture when calendaring, and the weave density can be changed by heat treatment. Chemical processes are used very commonly to impart multifunctionality to a textile by applying a chemical treatment to the fabric's surface.

Important properties when it comes to the selection of cloth materials for PPE are durability, resistance to shrinkage, thickness, air permeability, hand properties, and the ability to take treatments such as liquid repellents, reactive coatings, dyes, or printed patterns (such as camouflage). Comfort layers to be worn next to the skin must be designed to be nonirritating, and may enhance comfort by wicking moisture away from the skin (e.g., by using highly water absorbent knits). The ability of material to withstand agents is obviously also important. Thicker, more tightly woven, less air-permeable materials will resist penetration by any physical form of agent. As an outer layer, the cloth may be treated for liquid repellency, or alternatively may be designed to wick liquid drops to spread them out so as to present a reduced magnitude of hazard to any given location.

Cloths should be resistant to degradation by all of the chemical classes of interest in the application; for example, nylon may dissolve in organic solvents, making it unsuitable as a chemically resistant material. Other possible hazards must also be considered, for example cloths constructed from certain polymeric bases may be inclined to melt in heat, causing a significant skin contact hazard, making them potentially unsuitable for use in an environment where fire is a significant risk (e.g., firefighting and damage control parties, aircraft and combat vehicles).

Barrier materials (discussed in the next section) may be strengthened or made more fire and flame resistant by layering them within fabrics constructed from composite yarn with a fiberglass core [170] or a fire-resistant woven textile layer [171], respectively. Synthetic fibers with excellent mechanical, thermal/combustion [172], and/or impact performance may be added to impart specific properties [169,173]; examples of high-performance polymer fibers include polyamides such as Kevlar (Dupont) and Twaron, Nomex, PBI (polybenzimidazole), ultrahigh-performance polyethylene (spun from ultrahigh-molecular-mass polyethylene), M5 (polyhydroquinone–diimidazopyridine), PBO [poly(paraphenylene benzobisoxazole)], and Zylon. Inorganic fibers made from carbon (including carbon nanotubes), glass, and various oxides can impart similar properties [169,173].

4.5.3 Barrier Materials

For the purposes of this discussion, barrier materials are considered to be designed to be impermeable to everything; in other words, agents as well as all other gases, liquids, and solids are kept out, and they are wind- and water resistant. Materials in this category include conventional polymers such as thermoplastics or elastomers, singly and in combination, as well as advanced engineered materials such as nanocomposites.

Conventional Materials. There is a wide variety of such materials available for use, and they have an equally wide variety of physical properties, such as durability, elasticity, melting temperature, and resistance to permeation by chemicals (by their nature all are impermeable to aerosols). Weaknesses in properties for any given polymer can be addressed by combining them in various manners. Lamination together of polymers with different agent-resistant characteristics can provide broad-spectrum protection against permeation, while outside layers of a different polymer or cloth may be used for comfort or to provide enhanced strength and durability. In addition, polymers may be blended to achieve unique properties. Elastomers are used where stretch is appropriate to achieve more universal or comfortable fitting.

It is generally necessary to test the absorbency and chemical resistance of plastics being considered for equipment that requires chemical hardening. For example, perfluorinated materials are resistant to adsorption, and depending on microscopic structure, to permeation; polyalkenes (polypropylene and polyethylene) are quite resistant to agent degradation; and PVC [poly(vinyl chloride)] is very absorbent, although it may be used in blends with success.

Reference material that tabulates the resistance of various materials to chemicals in terms of both permeation and degradation is available from manufacturers of materials and PPE items, and in compendia such as the *Chemical Protective Clothing Performance Index* [174]. Although somewhat dated (last revised in 1999), the *Performance Index* lists all of the protective materials in commercial use at the time of issue and evaluates their performance in over 350 models of protective clothing against a large number of classes and mixtures of chemicals. In some applications, resistance to degradation by radiation may also be important, and information on this topic is also available [175]. NATO describes paints that can be used to make a surface resistant to contaminants and decontaminants [176], and sacrificial and strippable outer layers are possible.

It is important to note, however, that although this type of information can be used as a first step for material selection, all systems must still be tested, usually with live agent and actual decontamination procedures. It is also true that new materials are constantly being developed and marketed whose performances are not well known: for example, composites and nanomaterials.

Nanocomposites. Another approach to producing materials that can improve burden involves producing thinner or lighter-weight materials that can maintain high protection levels. Nanocomposite materials [177] can alter the structure of polymer membranes by the use of additives that reduce the opportunity for chemical permeation. Nanoclay materials [178] consist of minerals forming platelets of material that when dispersed into a polymer, form what is called a *sterically constrained mesoscopic structure*. Because the platelets themselves are impermeable, the permeation path length for agent through the polymer is greatly increased, as agent must work its way around the platelets (Figure 4-11). The resulting doped polymer retains many of

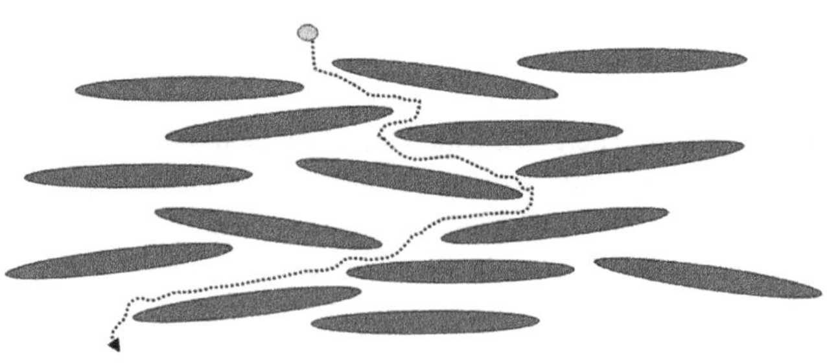

FIGURE 4-11 Structure of platelet-based nanocomposites, illustrating the tortuous permeation path.

the properties of the original polymer, including flexibility and optical clarity [179–183]. Gusev and Lusti [184] have discussed the rational design of such materials, and examples are given by Duncan and Pedersen [185] and Grove [186].

Conventional materials also may be hardened to permeation by **surface modification**. For example, ion-beam treatment of polymers profoundly reduces the diffusibility of solvents into the polymer by changing its surface structure [187].

Radiopaque Materials. These materials [188,189] are now marketed for radiation-protective suits; for example, Demron is a laminate that contains a center polymer film that incorporates organic and inorganic salt particles that contain large electron clouds that can block some lower-energy emissions, including x-rays and gamma rays [190], as well as low-energy particles. Future generations of such materials may be designed to enhance the amount and energies of radiation that can be blocked.

Applications. Some of the more important applications in PPE that polymeric barrier materials may have are the following:

- **Optical surfaces.** Polymers are used to provide optical surfaces in helmets, hoods, and respirators, as they are lightweight, clear, and formable. As such, the polymer chosen must be particularly resistant to physical damage of any sort and clouding from contact with agent. Polycarbonate is impact resistant but may need to be surface-coated for scratch resistance. Polyurethane is a possible choice where flexible materials are needed, but still has performance limitations in the application [191].
- **Gloves, boots, and facepieces.** These must be constructed from flexible, elastic materials, and at the same time must be durable. Single-use gloves are common for laboratory use, but when considerable resistance to chemical permeation and durability is required, as in CBRN protection applications, thicker rubber materials are usual. For use in facepiece materials, or where rubber seals to the skin are expected to be airtight, the ability of the material to conform to skin surface is important. Different rubbers vary widely in their permeation and absorption characteristics; fluorinated rubbers (Viton) and halogenated butyl rubbers are usually the most agent resistant, and silicon rubber is the most permeable.
- **Clothing materials.** The types of polymers used in hazmat-style garments typically consist of relatively thick multilayer laminates to impart all the necessary properties of chemical resistance and durability to abrasion and puncture simultaneously; as such, the hand properties of these materials are usually not the best for comfort.

4.5.4 Selectively Permeable Materials

Selectively permeable (SP) *materials* allow certain chemicals to permeate through them while holding back others, based on either size or solubility selectivity.

Moisture vapor–permeable (MVP) *materials* are a particularly class of SP materials whose selective permeability relates to allowing moisture vapor to pass through them. MVP materials are currently used in, for example, sportswear and firefighter turnout gear to permit enhanced evaporative cooling while maintaining good wind- and/or watertightness [192]. MVP materials have clear applications in the CBRN protection field as well, although there are some potential limitations. To be useful as materials that reduce the physiological burden while providing some CBRN protection, MVP materials should have the following characteristics:

- Their permeability to moisture vapor should be as high as possible.
- Penetration or permeation by the target agents should be below physiologically relevant levels. Many such materials can provide effective protection against penetration by aerosols; a more difficult challenge is providing protection against chemical permeation, or penetration by vapors while maintaining high levels of moisture vapor permeability; chemical resistance of the various available materials varies greatly [193].
- Moisture vapor permeability as well as the protective capabilities should be reasonably independent of the relative humidity and moisture content of the membrane and of the surroundings.
- Durability must be such that its integrity can be maintained under normal conditions of use. The thicker the material, the lower the MVP value, and therefore durability may be difficult to maintain at the same time as high moisture transport rates.
- The material should have appropriate thickness and hand properties.

This class of materials in general holds out radiological and biological aerosols well, as well as resisting penetration by liquids to some degree while permitting evaporative cooling; this makes these materials a good candidate for protection against biological and radiological agents. Such materials are not, however, necessarily appropriate for comprehensive CBRN protection, as they are not necessarily resistant to chemical permeation or vapor penetration. Hence, particular examples may have a particular suitability for some types of protection and not others.

There are two general types of MVP materials, microporous and monolithic. Figure 4-12 illustrates the conceptual differences between the two types. The microporous structure on the left permits diffusion of water vapor from humid air next to the skin (at the bottom) through the porous channels to the outside (at the top). On the other hand, the monolithic, nonporous structure on the right permits diffusional permeation of water molecules through its elastomeric structure, consisting of hydrated, intertwined polymeric chains. These two types are discussed in more detail below.

Microporous* membranes (Figure 4-13) have very small pores that inhibit the flow of liquid because of the pores' small size (on the order of micrometers,

* Note that the term *microporous* as commonly used for these materials does not conform with the definition used in Section 4.2.2; by the previous definition, the pores in the materials described in this section are macropores.

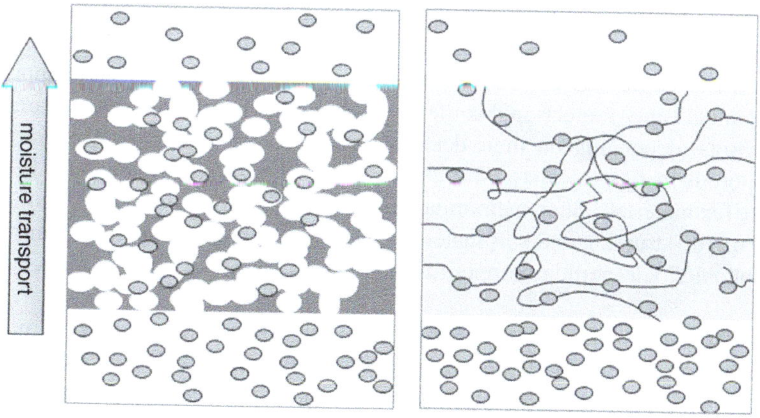

FIGURE 4-12 Comparison of microporous (left) and monolithic (right) moisture vapor–permeable materials.

less than that of a small water droplet at 0.1 mm) and the material's intrinsic *hydrophobicity*.

> **Hydrophobic materials** repel water, causing it to bead up, and attract oils and many solvents, causing them to spread. **Hydrophilic materials** do the reverse.

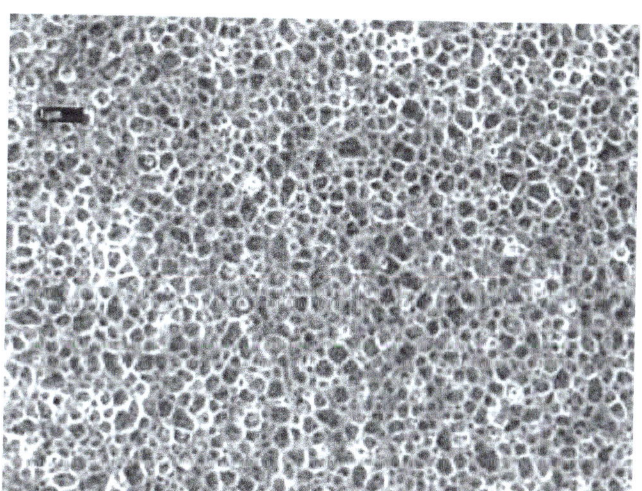

FIGURE 4-13 Electron micrograph of a microporous material. (Reproduced with permission of American Association of Textile Chemists and Colorists, from Obendorf, S.K., AATCC Review, July–August 2010, © 2010; permission conveyed through Copyright Clearance Center, Inc.)

While holding out polar liquids, microporous materials allow vapors (including moisture vapor) to pass easily through the porous network, whose channels are substantially larger than the size of a water molecule. This makes them inappropriate for chemical vapor protection, but they should be effective at holding out particulates or aerosols, depending on their detailed structure. The best known example of a microporous membrane material is expanded polytetrafluoroethylene (ePTFE), used in GoreTex materials. Such membranes are not themselves very durable, owing to the rather fragile structure and soft materials of which they are composed. Combination hydrophobic/hydrophilic microporous membranes have also been developed [194].

Monolithic Materials. Other, much less porous hydrophilic membrane materials may achieve selective moisture vapor permeability, albeit at a lower rate; examples include certain hydrogels and elastomeric films [195]. To achieve high moisture vapor permeability, the polymer should be composed of many intertwined chains containing polar groups, which may be hydrated by the addition of water. The water of hydration is present in reasonably high concentration, and individual molecules are somewhat free to diffuse between the chains forming the structure, giving rise to high moisture vapor permeability. Water permeates the material on the higher-humidity side, and as it diffuses through, the water of hydration on the lower-humidity side is pushed out. Examples that have been commercialized include polyether–polyurethane and polyether–polyester materials. Durability of microporous ePTFE membranes can be enhanced by laminating them between monolithic MVP layers [196] (resulting in somewhat lower moisture vapor transmission compared with that of ePTFE alone).

For microporous membranes, moisture vapor permeability is usually very high and almost independent of relative humidity, since the porous structure is not affected by the water concentration. In contrast, for polyether-based monolithic continuous membranes, high moisture vapor transmission rates occur only once the body is already sweating heavily so that the material itself is thoroughly hydrated [197].

New MVP materials include sulfonated aromatic polymers that have a high moisture vapor permeation rate at various humidity conditions [198] and blends of hydrophilic materials that can be injection molded [199]. SP materials based on interpolymer complexes [200] of poly(vinyl alcohol) and polyethyleneimine have been developed; the hydrogen-bonded network may provide a barrier that relatively hydrophobic molecules such as sulfur mustard and soman cannot easily penetrate. Additionally, the polyethyleneimine component may act as a chemically reactive barrier.

Two different hybrid SP/active carbon-adsorbent materials for CBRN protection (Chempak Sorptive fabric and Spiratec Hybrid) have been introduced; they combine two protective layers, an SP material layer over an active carbon layer. The advantage of this combination is that any vapor that does manage to enter the system either via permeation or through closures may be absorbed by the active carbon inner layer, and the material as a whole resists liquid penetration.

Other types of SP materials may eventually be developed that will provide additional capabilities to resist chemical permeation and penetration. For example, a material that allowed oxygen through selectively could be of benefit in breathing air purification, but existing materials cannot yet provide gas flows sufficient for the application. **Micro- and nanosieves** are perforated membranes formed from regular structures either in molding [201], by using lasers or ion-beam etching [202], or by chemical design using various phase separation processes [203–206] and could have future applicability. This has been discussed more completely by Van Rijn [207].

4.5.5 Filtration Media

Conventional filtration media are constructed from fibrous materials laid down in mats whose fibers are typically in the micrometer size range and higher. Nanofibrous materials are now under development to take advantage of different filtration behavior on this size scale. Other types of more structured media could have pores intrinsic to their chemical structure.

Conventional fibrous media are generally constructed from a nonwoven mat of fibrous material (see Figure 4-2). Filter fibers may be constructed from organic materials, such as cellulose from paper or cotton, or inorganic materials, such as glass. The important characteristics that determine such a filter's ability to effectively remove particulate in an airstream include fiber density, fiber diameter, and mat thickness, all of which determine filtration efficiency, pore size, and pressure drop and resistance to flow. Filtration occurs from front surface to back; in other words, the majority of particles are removed by the first fibers they encounter. This means that if larger particulates are encountered, a filter may become clogged before its full filtration capacity is used up. On the other hand, filters become more efficient as they clog, even as the resistance to airflow increases.

Electret filters enhance the filtration efficiency by taking advantage of electrostatic attraction. While the overall charge on the filter is close to neutral, individual fibers are either positively or negatively charged and therefore attract oppositely charged particles (as well as inducing charge in neutral particles). Corona or triboelectric charging mechanisms are used to charge the fibers. This additional capture mechanism can result in a higher filtration efficiency for a given pressure drop, particularly in the area of interest around the traditional MPPS range of 0.1 to 0.3 μm [208], and the MPPS is in fact shifted to a smaller size (e.g., 0.05 μm) [209]. The electrostatic mechanism tends to pull particles onto the fiber from all directions, resulting in a more uniform distribution of adhered particles and slower clogging. It also has a finite capacity; once the fibers are neutralized by adhering particles, the filtration enhancement is lost [210]. The charge on the fiber may be lost as a result of other environmental factors, such as impinging radiation, liquid solvent, and time [210]. Removal of electrostatic charges on electret respirator fiber media by isopropanol treatment is known to shift the most penetrating particle size from 40 to 60 nm to 250 to 300 nm (typical of fibrous filter media) [211]. However, it is not yet clear how severe the environmental exposure needs to be to degrade the electret properties [209] (Filtration by granular beds can also be enhanced by electrostatic charging [212].)

Nanofibrous layers (Figure 4-14) may have some filtration advantages, although they are fragile and difficult to fabricate. When fibers are in this size range, they have an enhanced surface-to-volume ratio, yielding more effective filtration per unit

FIGURE 4-14 Polyurethane nanofibers spun onto a conventional nonwoven material. (Reproduced with permission of American Association of Textile Chemists and Colorists, from Obendorf, S.K., AATCC Review, July–August 2010, © 2010; permission conveyed through Copyright Clearance Center, Inc.)

weight. In addition, the flow dynamics change around such small fibers, with less turbulent or "stick" flow, resulting in "slip" flow conditions [213]. This reduces the flow resistance and at the same time enhances the capability of diffusion to become a significant capture mechanism for smaller particles. Reports have indicated that relatively thin nanofibrous layers can afford complete filtration protection against particles with diameters within the range 1 to 5 μm [214] and provide significant improvements in the range 100 nm to 1 μm [215].

Theoretically, nanofibers can be constructed from any material that can be formed appropriately. To create nanofibers, a number of processing techniques exist, such as drawing, template synthesis, phase separation, self-assembly, and electrospinning [216,217]. After formation, such fibers must be well supported on other materials, due to their fragility. Electrospinning can be performed using polymeric fluids that must have adequate viscoelasticity (usually controlled by an appropriate combination of molecular mass and concentration of the polymer in solution) and conductivity in order to be electrospun (i.e., to form uniform fibers) [218]. Various materials are now commercially available [215,219]. Commercial nanofiber centrifuge technology is now available that increases the production rate by orders of magnitude [220]. Nanofibers also have the potential to have additional beneficial properties; for example, superhydrophobic nanofibers have been produced [221,222]. Novel nanofibers constructed from alumina have shown excellent filtration efficiency of small viral particles relative to their low pressure drop [223].

Filter Construction. To maintain their structural integrity, fibrous filtration media may need to be supported between other stronger layers that have a larger pore size. In

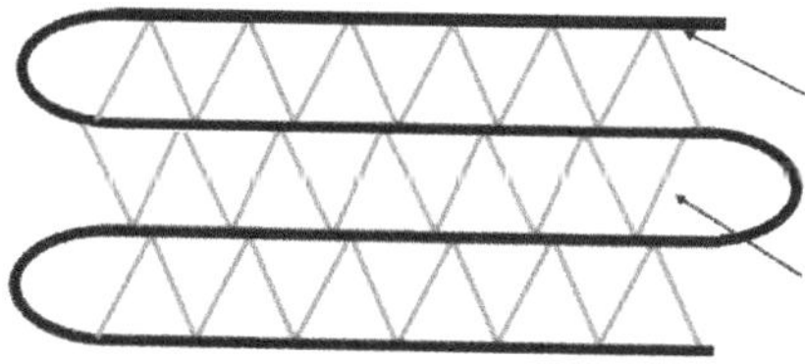

FIGURE 4-15 Construction of a pleated filter medium.

addition, such a construction may aid in preventing early clogging, as larger particles are filtered out in the outer supporting layer, acting as a prefilter, rather than in the inner micro- or nanofibrous layer. Airflow should be directed to ensure that the entire useful surface of the filter medium is utilized; if the air enters the filter housing through a small hole, it will tend to flow more through that portion of the medium directly opposite the hole. Therefore, designs that direct flow radially out before flowing through a flat medium, or like a cyclone over the interior of a cylindrical medium, have advantages, spreading the flow over a larger usable area of the filter material.

The depth of the filtration layer may be increased by layering together mats of material or by using batting. Although this increases filtration efficiency due to a higher probability of encounter, it also increases flow resistance. The increase of flow resistance sometimes has benefits in that some particles may be filtered more efficiently at lower flow velocities, when interception and diffusion are predominant capture mechanisms. However, where high airflow is important to aid in breathing or cooling, this design approach is less desirable.

Alternatively, filtration efficiency and flow can be improved simultaneously by offering the largest external surface area of the filtration medium possible—in other words, rather than increasing the depth of the material, increasing the amount of medium available while maintaining constant depth. Such an approach increases filtration efficiency while reducing flow resistance and has the advantage of adding more medium in such a manner as to reduce the likelihood of clogging. This type of construction can be performed by pleating thin foldable media (Figure 4-15). The surface of a filter medium may need to be treated with liquid repellent coatings to reduce the likelihood of filter wetting by bulk water or by condensed moisture or other vapors in the air; filter wetting results in undesirable effects such as alteration of porous structure, fiber diameter, particulate adhesion efficiency, and resistance to flow.

Efficiency of Filtration. Generally speaking, the term *filtration efficiency* is used to describe the performance capability of the types of filters used for CBRN protection. Filtration efficiency is simply the fraction or percentage of particles removed by a filter from the airstream, usually measured by counting those remaining downstream of the filter element. It must be defined under certain types of conditions and may vary as a function of flow rate, particle size, and particle type. Various standards organizations and national bodies have defined standard classes of materials used in filtration elements. For example, the National Institute for Occupational Safety and Health

(NIOSH) and the Committee for European Norms (CEN) employ different test protocols and have different performance requirements for filtering facepiece (FFP) respirators. For example, NIOSH requires a minimum of 95% and 99.97% filtration efficiencies for N95 and P100 filtering facepiece respirator materials, respectively, while CEN requires 94% and 99% efficiencies for FFP2 and FFP3 respirator materials, respectively. The classes and test methods are discussed in more detail in Section 6.8.3.

4.5.6 Adsorbents

Adsorbents useful for CBRN protection are capable of removing chemical vapor contaminants from an airstream by physisorption, sometimes followed by chemisorption. The speed with which air passes through the adsorbent affects the likelihood that adsorption or reaction will occur. In general, the less time the air takes to pass through, the less likely it is that the hazard agent will have encountered a suitable adsorption or reaction site as it passes, and the more likely it will be entrained with the air, passing through unaffected. Above a certain critical flow through such a medium, defined for a given bed depth, breakthrough can be almost instantaneous. For respiratory protection, adsorbents must be able to purify the air more efficiently and at higher flow rates than in clothing. This is because for dermal protection, higher concentrations are usually permissible due to dermal toxicity being less than inhalation toxicity, while flow rates are lower as they are caused by ambient wind and motion rather than breathing.

Air-purifying elements intended to remove chemical vapors for respiratory protection are often constructed from beds of granular meso- and microporous particulate adsorbents, with the air gaps between the grains providing the connective macroporous network. Although this is an inexpensive means of generating the adsorbent structure required, it is relatively irreproducible. In addition, the bed integrity is potentially disrupted by rough handling. Furthermore, full utilization of the carbon surface may be prevented by diffusional resistance through the mesopores before the useful surface is reached. For clothing, adsorbent bed depth is very short relative to that in a respirator air-purifying element (millimeters rather than centimeters). Typically in this case, the adsorbent is not present in granular form but rather in some other form that can be constructed more effectively into clothing materials by layering between other cloths, as it must be flexible and durable.

Active (or **activated**) **carbon** is by far the most commonly used adsorbent for CBRN protection. Although relatively inexpensive, it is an extremely effective adsorbent for vapor-phase chemicals because of two properties: its porous structure and its chemical composition. Active carbon is effectively capable of developing an infinite number of porous structures with different densities, pore size distributions, and networks, and its surface is modified relatively easily to add various chemical moieties that add specific reactivities. Active carbon is generally manufactured by *carbonizing* (burning) and *activating* an organic substance such as wood, coal, or coconut shells. Carbonizing forms the initial porous structure, by burning off volatile organic gases such as water, carbon dioxide, and low-molecular-mass organic molecules, leaving a primarily carbon-based framework. Activating by further heating in the presence of an "activating" agent such as a metal salt or steam converts the carbon to a microcrystalline graphitic structure and increases the surface area available for adsorption. The activation may also add chemical moieties such as –OH, which assist in attracting molecules through hydrogen bonding. The area of the active surface of the carbon type used in CBRN protective filters is on the order of 1000 m^2 (about the size of a football field) per gram of carbon [224].

Active carbon is a very versatile material, in that like carbon-based molecules in general, it can be constructed into many different forms that suit the application. Active carbon is available inexpensively in powdered and granular forms, often used in respirator canisters, but to impart particular desirable characteristics it can also be manufactured in other structures, such as fibers, beads, and monoliths. Various forms of active carbon can be sandwiched between cloth layers to form clothing material systems that protect against high-molecular-mass chemicals such as persistent chemical warfare agents. Many of the original materials used by the military decades ago included a layer of foam that contained loosely bound powdered active carbon. Current materials typically contain either active carbon fiber–containing cloth, which is formed by carbonizing a woven polymeric precursor such as polyacrylonitrile, or alternatively, beads of active carbon laminated in place between two cloth layers.

Active carbon can also be incorporated into monolithic structures where a fixed solid structure is relevant (such as a respirator canister). These structures could possess advantages over structures composed of fibers, beads, or granular carbon, with more controlled, stronger networks of porous structure. They can be constructed from various precursor structured templates: for example, by carbonizing a structured polymeric material [225,226], casting from an inorganic template [227,228], or binding together granular or powdered materials [225,229], including honeycombed composites, with clay [230].

Despite its remarkable ability to remove a large number of vapor-phase hazard agents from an airstream, active carbon has a number of limitations in performance:

1. **Low-boiling agents.** As described in Section 4.2.2, active carbon removes vapors very efficiently by a physisorption process even at short bed depths, but only if they are relatively high-boiling compounds. Impregnated active carbons can remove low-boiling vapors by chemisorption, but this is a much less efficient process with lower capacity, very limited compared to its overall surface area. In general, the most effective means of increasing capacity is simply to provide a proportionally larger amount of adsorbent, and to increase effectiveness a greater bed depth and/or lower flow rate is required. The active carbon in clothing materials is not impregnated, for two reasons:
   - The low-molecular-mass conventional CWAs are not particularly dermally toxic, meaning that impregnants are not necessary to remove them.
   - The very short transit time of the contaminated air through the material at high wind speeds means that there would be insufficient time for reaction with impregnants to occur given the very short bed depth of the carbon layer.

- This means, however, that low-molecular-mass dermally hazardous TICs are not well removed by active carbon clothing materials and hence will not be protected against effectively by this type of material system.

2. **Water and other contaminants.** Atmospheric water degrades the protection provided by active carbon in two ways: It adsorbs to the surface via physical forces, taking up surface that is then less available for chemical agents; and water reacts with some impregnants and hence removes their subsequent availability to react with chemical agents via chemisorption. Hence, all active-carbon-containing filtration materials potentially have a limited shelf life in air, particularly once their packaging has been breached, and once wetted with water they will be ineffective. Other common and not overly toxic airborne contaminants (e.g., gasoline) may have a similar effect, either adsorbing to the surface or reacting with the impregnants, making air purification less effective subsequently when hazard agents are encountered.
3. **Canister penetrants.** Penetrants such as perfluoroisobutylene and disulfurdecafluoride are compounds that do not stick to conventional activated carbons. Specially designed penetrant-protective carbons containing organic amine impregnants can protect against organofluorines [231].
4. **Liquid protection.** The capability of these clothing material systems to protect against chemical agents will depend primarily on the liquid repellency of the outer cloth layer, the air permeability of the system, and the amount (in mass per unit area) and type of activated carbon layer.
5. **Particulate protection.** Typically, active carbon materials themselves provide little protection against particulates; some other mechanism of protection must be used if such protection is needed.
   - However, active carbon fiber materials have some capability to remove particulates in addition to their adsorbent properties, and by combining these two functionalities they may have some advantages in certain applications.

Zeolites. Zeolite adsorbents have been used in adsorbent-based air purification systems to achieve broader removal capability. Cation-containing zeolites can adsorb molecules on the basis of their electrostatic interactions with the metal ions, while hydrophobic silica zeolites preferentially absorb organic solvents [106]. Zeolite adsorbents constructed from granular material have been commercialized for the removal of various types of organic compounds from airstreams. Specific size-based separation using zeolites incorporated into membranes has been achieved [232]. Monolithic zeolite structures are also possible.

4.5.7 Reactive Materials

Reactive materials can be incorporated into a protective device or material layer to more effectively remove the hazard posed by an agent once it has come into contact with the reactive surface. For reactive materials to be useful for protection, they must satisfy a number of requirements:

1. The reactive groups must react with the target hazard agent, and preferably, as many different classes of and specific target agents as possible:
   - It may be possible for the reactant to generically deactivate most biological agents, although there are some differences between the various groupings. Most are susceptible to oxidizers, for example, although spores are resistant to many chemicals and may require more specific targeting.
   - Reactions with chemicals tend to be more specific, and not all hazard agents can be targeted with the same reactant, although oxidizers are again very effective generically.
   - There are no chemical reactions that will remove the hazard posed by radiological agents, as their hazard is generated at the nuclear level.
2. The reaction must render the target hazard agent less hazardous:
   - Simply capturing the hazard agent on a surface may reduce the hazard.
   - More often, the reaction is targeted to breaking down the hazard substance; the by-products of this breakdown must be nonhazardous or considerably less hazardous than the target agent.
3. The reaction must be sufficiently fast to be effective compared with the length of time the hazard agent is in contact with the material:
   - If the contact time is potentially short—for example, as the hazard agent passes over or through the material—the reaction must be nearly instantaneous.
   - If the hazard agent is captured by the material, the time for reaction can be longer; in this case, it is not only the immediate hazard to the wearer that is being reduced, but also the subsequent hazard either during removal or prior to disposal of the PPE.
4. The reactive groups must be stable to environmental conditions of exposure and wear:
   - If the item is designed to be worn continuously for some period, and possibly laundered and reused, the reactive groups must withstand water, sunlight, detergents, and other possible environmental contaminants prior to the hazard encounter.
   - For a single-use item, the reactive groups may be preserved by appropriate packaging and storage.
5. The reactive material must itself be used in such a manner as not to be hazardous to the wearer:
   - The reactive groups may be relatively benign to the wearer because of their specific chemistry.
   - The reactive groups may be unable to affect the wearer because of the relevant route of entry; for example, skin is considerably less active a route of entry,

and it may therefore be possible to use certain materials in clothing that could not be used near the eyes or respiratory tract.

- The reactive groups may be immobilized and separated from contact with the wearer such that they do not come into contact with the wearer: for example, layered between clothing materials or held in an APE.

6. The reactive groups should have sufficient capacity to be useful against hazardous quantities of an agent:
   - Some reactive groups may be regenerated by treatments during laundering.
   - Catalytic groups that promote reactions without being consumed have a greater capacity than do reactive groups that react one to one with agents and are consumed in the process.
7. The reactive material must be accessible to the agent along with any coreacting species:
   - For catalytic chemical protective coatings the solubility and diffusion of reactants such as O_2 and H_2O in the matrix may be critical.
   - The surface containing the reactive species must match the hydrophobicity, hydrophilicity, and charge of the agent to be deactivated such that contact is likely to occur.
8. The reactive groups must be able to be attached to the surface of the PPE material or fibers:
   - Some reactive materials are added as surface coatings.
   - Other reactive materials are chemically attached (via a covalent bond) to the PPE material: for example, via reaction with polymers or incorporation into the polymer during the polymerization step.

The ideal reactant for the purposes of deactivation would react instantaneously with most possible types of hazard agents, preferably without being consumed in the process; however, because of this high reactivity, this type of chemistry is potentially very susceptible to degradation by environmental exposure and is likely to be hazardous to the wearer.

Reactive Chemistries. Various reactive chemistries can be used that generally fall into one of a few classes: These include oxidizers and reducers, and acids and alkalis. Oxidizers and reducers act by either taking or giving electrons to other species in the course of a reaction, respectively, while acids and alkalis either give or take protons, respectively. Within these classes, specific chemistries may be more effective at reacting with certain agents than with others. Novel fibers and incorporation of reactive chemistry have been reviewed by Edwards et al. [233]. The reactive chemistries noted below represent a general overview without their suitability for incorporation into PPE systems necessarily being implied.

As mentioned previously, impregnants are reactive atoms or groups: for example, metal species such as particular oxidation states of silver, manganese, molybdenum, or zinc, or groups such as chlorine or amines that are added to the surface of an adsorbent such as active carbon. A number of relatively gentle decontaminant solutions use oxidizers related to oxones [234] or oximes [235], which have been shown to decompose most chemical and biological warfare agents with varying speeds.

In addition to the types of reactants noted above, chemicals that catalyze reactions, such as certain enzymes or photoactive species, have also been investigated; in this case, the reactant might actually be the oxygen or moisture in air, while the catalyst causes the decomposition reaction to occur much more quickly than it normally would. As described later, there are numerous strategies for killing vegetative organisms by disruption of their life cycle which may involve very specific chemistry designed for the organism or family or organisms. Spores and toxic chemicals are a more significant deactivation challenge, and typically much longer contact times between the reactive material and the agent are required for effects to be seen. It is possible to conceive of other means to promote decomposition reactions (e.g., electrical discharges) that could be incorporated into protective equipment; however, to date no such concepts have been put into practice.

Chemical Detoxification by Catalytic Systems. Incorporation of enzymatic catalysts into various supporting materials such that they retain their activity is an ongoing exercise. Various methods for attachment of enzymes to carriers have been used, including adsorption, covalent attachment, entrapment, encapsulation, and coating [236]. Diisopropylfluorophosphatase deactivates nerve agents and has been immobilized into polyurethane coatings [237] and onto fabrics [238]. Polyoxometalate catalysts have been investigated extensively for their applicability to self-decontaminating coatings. Polyoxometalates are transition metal oxygen anion clusters that are stable to oxygen and can have many different elemental compositions that lead to a tailoring of their properties. A particular strength of one subset is their ability to catalytically decompose sulfur mustard [239]. For example, polyoxometalates supported on porous carbons have been reported to form selective and recoverable heterogeneous catalysts for the rapid room-temperature oxidation of thioether analogs of sulfur mustard [240].

Photoactive Species. Photocatalysis is the acceleration of a reaction through the use of a light-activated species; in this case, the desirable reaction causes the breakdown of a hazard agent. A commonly used photocatalyst is titanium dioxide; a thin homogeneous layer of titanate nanotubes impregnated with tungstate salt can use solar energy to catalyze breakdown reactions. Using this catalyst, liquid nerve agents and simulants, as well as sulfur mustard, were degraded in a few minutes in the presence of sunlight [241,242], and textiles have been functionalized [243]. Titanium dioxide can also be used in conjunction with nanofibers or nanoporous media [244], and various similar oxides can be particularly active as nanoparticles [245].

Antimicrobials. Various types of antimicrobial compounds are available:

- **Silver compounds.** Silver ions and silver oxide exhibit broad-spectrum biocidal activity toward several bacteria, fungi, and viruses. The silver cation deactivates cellular enzymes by coordinating to electron-donating groups, particularly sulfhydryls and thiols [246]. Silver cations also interact with cytoplasmic constituents, resulting in general coagulation, and may also interfere with cell replication. SilverClear is a commercial treatment using silver crystals that has demonstrated broad antibacterial activity [247]. A silver bromide nanoparticle–polymer composite [248] shows many of the same antimicrobial properties as other silver supporting materials, yet is said to be easier to produce and less toxic to humans; the system employs a silver salt that can easily be embedded and is easily dissociated. The synthesis of the material is accomplished by the precipitation of silver bromide, which forms nanoparticles within the polymer. It has been tested against surface, airborne, and waterborne bacteria and hence shows promise as a bactericidal coating. Silver nanoparticles have been incorporated in chitosan-modified cotton with good antimicrobial activity [249].

- **Copper compounds.** Copper oxide fabric treatments have been commercialized [250–253] and shown to have potent antimicrobial activity against fungi, mites, bacteria, and viruses. Copper's broad biocidal activity is similar to that of silver [246].
- **Other oxides.** Nanocrystalline oxides such as MgO, CaO, and Al_2O_3 have demonstrated reactivity toward several functional chemical groups of molecules adsorbed onto their surfaces. Pesticides and nerve agent simulants have been shown to be degraded [254] when these oxides were incorporated into fabric treatments. Multifunctional textiles have been produced using electrospun polyurethane/zinc oxide nanocomposite fiber webs layered on cotton substrates, yielding ultraviolet protection and antibacterial functions [255].
- **Quaternary salts.** Long-chain quaternary ammonium and phosphonium salts exhibit broad-spectrum biocidal action by dissolving into and disrupting cell membranes. As examples, they have been incorporated into cross-linked copolymers [256], normal [257] and hyperbranched electrospun polymers [258], and dendrimeric polymers [259], and have been designed to self-concentrate to the surface as a hydrophobic additive in hydrophilic polyurethane polymers [260].
- **Halogens.** Triosyn resin is a polymer of a quaternary ammonium salt whose counterion is a triiodide group; this group can release iodine, which then attacks the structural proteins and the membrane lipids of microbes [261]. The efficacy of iodine in respirators has been demonstrated against various bacterial species, but results on viral efficacy are contradictory [209,262–264] perhaps dependent in part on temperature and humidity conditions. Iodine has been found to be more effective than silver in wound dressings, is faster acting, and because it has significant broad-spectrum biocidal activity [265] is less likely to result in resistance. This would imply that it might have a broader application in biocidal materials for biological protection. Haloshield is a chlorine-based *N*-halamine textile coating that has also been commercialized [266].
- **Combination and other.** Combined *N*-halamine/quaternary salt polymers also exist, with each of the functionalities having different specificities in terms of biocidal action [267]. Antimicrobials have been spun into nanofibers [268] and

incorporated into nanoscale inorganic materials [269] such as amorphous silica, titania, and colloidal silver. Cliniweave [270] is an antibacterial fabric treatment that acts by disrupting membrane transport function (metal chelation), inhibiting ATP (adenosine triphosphate) synthesis, and disrupting the cell wall coupling enzyme; the presence of three different mechanisms is claimed to reduce the likelihood of developing resistance.

4.5.8 Smart Materials

Smart materials often demonstrate relatively sophisticated molecular design, such as shape memory or controlled chemical release, or might simply have some form of miniaturized electronic sensing built in. In general, they fall into three categories [271]:

- Passive smart textiles sense the environment.
- Active smart textiles sense and react.
- Adaptive smart textiles (the most "intelligent") sense and react in more than one way, or reversibly, changing behavior depending on circumstances.

Examples of triggers that could actuate a response in a smart material include:

- pH
- Oxidation and reduction
- Solvent exchange
- Ionic strength
- Any form of electromagnetic irradiation (e.g., light, high energy, microwave)
- Temperature
- Physical deformation or pressure
- Magnetic field
- Electric field

Passive sensing textiles are useful only when followed by an action on the part of the wearer; for example, to prompt the wearer to perform decontamination, PPE could be composed of a sensing textile which signals that an encounter with a hazard substance has occurred via a color change. **Active** smart textiles might direct a heating or cooling system to engage when the temperature beneath the clothing is outside the comfort zone, or they might trigger the release of a neutralizing chemical in the presence of a hazard. **Adaptive textiles**, which are just a special case of active textiles, have the most potential, since they themselves provide the change between normal and enhanced states when needed, potentially yielding optimization of factors such as physiological burden, energy consumption, and protection. Criteria such as flexibility, water resistance, launderability, and durability must be maintained, meaning that the simpler the material system, the more acceptable it is likely to be.

Some particular examples of developmental and currently marketed approaches to smart materials include:

- **Adaptive ventilation materials** open when high humidity resulting from sweating is encountered and close when sweating stops, due to U-shaped holes that are pushed open by a hygroscopically expanding polymer film [271]
- **Adaptive cooling materials** [272] change phase in a given temperature range and absorb heat during the transition, resulting in cooling [273]
- **Color-changing materials** include active camouflage [274]

4.5.9 Nanomaterials and Other Protective or Next-Generation Materials

The most revolutionary work to improve materials for protective applications is occurring in the area of nanotechnology. By engineering materials to provide particular properties that may not be easily available in current-generation materials, some of the next advances in protective equipment design may be made possible. The entire field of nanoengineered materials, examples of which have been described in previous sections, shows great promise for the future. In this section we also discuss various specialty materials that provide useful functionalities, other than CBRN protection, to be incorporated into PPE.

Nanomaterials are really numerous different types of materials that all have features present on a nanoscale. Of those that are pertinent to PPE, examples include fibers made from nanocomposites or carbon nanotubes, or with nanoparticles adhered to the surface, and nanoscale coatings on other materials. Adaptive materials are likely to be nanomaterials, as their behavior is likely to be engineered at the molecular level. Some of the properties not already discussed that can be altered effectively include repellency, wicking, durability, electrical and thermal conductivity and insulation, adhesion, fire and flame resistance [275], and transparency.

Flame and Fire Resistance. The addition of carbon nanotubes has been demonstrated to yield increased durability, fire and thermal resistance, and conductivity to various materials [276], and select hardened coatings can increase fire retardancy [277]. Flame-resistant materials such as Nomex resist the spread of flames and degradation of the material (i.e., ideally are self-extinguishing) and do not char or melt [275,278,279]. Heat-shielding materials generally reflect heat away to reduce the likelihood of melting or ignition. These approaches are usually achieved through special coatings but can include metallic layers as part of a composite material.

Ballistic Protection. Materials for ballistic protection, including various aromatic polyamide fibers, such as Kevlar, Twaron (Enka), and ultrahigh-modulus polyethylene, are used for ballistic protection [280].

Sealing and Fitting. **Superadhesive material** concepts are being investigated that may have applicability to closures. Synthetic modified surfaces shaped like the adhesive structures in gecko feet [281] are under development; such materials could be used as effective reusable adhesives for seals and closures. Transparent acrylic thermoplastic elastomers are now described as having adhesive properties [282]. Thermally activated **shape memory polymers** [283] may also be useful for sealing and forming custom closures.

Superelasticity. This type of elasticity (pseudoelasticity) is a shape memory property not requiring temperature to reverse changes in shape brought about by stress, which may have benefit in assisting in the formation of seals or three-dimensional structures. For example, superelastic eyeglass frames press against the head with constant stress, making fit easier to obtain [284].

Repellency and Self-Cleaning. **Superhydrophobic materials** are under development in which the surfaces are altered to mimic natural superrepellent surfaces that are found in the plant world (the "lotus-leaf effect") [285,286]. These surfaces have particularly "hairy" morphology, and the ultimate goal is to provide repellency toward both polar and nonpolar low-surface-tension fluids. Fabrication methods include particle deposition, sol–gel techniques, plasma treatments, chemical vapor deposition, in situ polymerization, and casting techniques [287]. The rough surfaces are formed in the process by phase separation, growth, rough coating, or an intrinsic nanoparticle-based structure. The surface to be treated can be rigid, such as a hard plastic, or flexible, such as a polymeric or fibrous material, or may be superhydrophobic itself, such as electrospun fibers [288]. These surfaces can be further enhanced by combining with other coatings, and advances are being made in this area [289]. Fluoropolymer coatings may add extra repellency to these already repellent surfaces [290]. Their high surface area can also be used to effectively carry reactive coatings, such as photoactivated singlet-oxygen generating agents, and up to a 1000-fold increase in availability of the photoactivated agent has been demonstrated by this means [289]. Superhydrophobicity has been combined with low reflectivity [291]. It is also noteworthy that the ability of superhydrophobic surfaces to repel water also gives them a passive antimicrobial effect (as contaminated water will not adhere to the surface).

For the purposes of **antifogging** and **self-cleaning**, water repellency or hydrophilicity can be appropriate, along with reactivity. **Highly repellent coatings**, which are organic compounds containing silicon atoms, have been produced by chemical vapor deposition to add hard antifog coatings to complex surfaces [292]; self-cleaning materials with both hydrophilic and oleophobic properties have also been proposed [293]. Various sol–gel types of materials that take advantage of nanotechnology have promise for coatings of this nature [277]; they may be hydrophobic, hydrophilic, or photocatalytic (usually containing TiO_2). Hard coatings that are employed for other applications may also be modified to add hydrophobicity; sol–gel technology can generate scratch-resistant hard coatings, and modifying such coatings by adding silica particulate filler can result in a lotus-leaf effect in such a hard coating [294].

Transparency. Novel transparent materials may also have a significant impact on the design of lower-burden PPE, particularly respirators where new concepts for eyepieces, visors, and even the entire respirator could benefit from the use of transparent materials. The area of transparent nanocomposites that may be used as protective layers in armors may have applicability as well to PPE, providing hard, transparent components [295,296]. Prototype transparent soft barrier materials have been developed that show solvent diffusion resistance comparable to that exhibited by flexible high-barrier materials [297].

Conductivity and Energy Management. Condutive materials have a wide variety of potential uses: Antielectrostatic buildup, sensing, and transmission of information and energy are the most promising for CBRN PPE. For reduction of **electrostatic effects** to prevent discharges that have the potential to cause ignition and effects on sensitive equipment, resistivities should be reduced as far as reasonable. Conductivity can be added, for example, by laminating conductive layers, spraying, ion plating, or sputtering a conductive material onto another material layer, incorporating conductive fillers or fibers into another material, or adding conductive fibers or yarns into a fabric [169]. **Energy-harvesting materials** may react to light or movement [298] to generate energy, which can then be passed to another part of the system to power devices; they would typically be a film on another material layer. **Conductive heating materials** containing carbon nanotubes can be incorporated as films or impregnated into textiles [299].

4.6 SYSTEM DESIGN

4.6.1 Introduction and Background

This section is directed specifically at the PPE system designer or integrator and begins by reiterating the importance of the overall principles of solid requirements development and life-cycle management processes to assure appropriate and robust outcomes of the design process that can be put into long-term practice. Many of the concepts that have been described in earlier sections of this chapter will then be brought together and related directly to specific design issues at the system level, suggesting which materials and design elements may be appropriate in achieving specific PPE capabilities as well as describing their limitations.

One of the challenges in the field of personal protection is the fact that for years, the protective ensemble has been managed as a collection of items that have largely been specified, designed, and built independently rather than as a protective **system** designed from scratch with a unified approach. This approach has begun to change but remains a challenging issue because of the requirement for coordination between many different players during the entire life cycle of the system, and the concomitant time and expense. This coordination can become problematic at many stages of the process, even after the system has been put into service. If, for example, the shelf life of one item is shorter than that of the others so that the item must be replaced sooner, many organizations would insist on tendering the item's replacement. Requalification of an entire system based on a change in a single item or subcomponent is an onerous task and one that organizations may be unwilling to undertake. Hence, the replacement subcomponent might be selected without a proper system-level evaluation of its impact. It is also rare for a system to be constructed and put into service in concert with a well-defined concept of operations and use. In practice, even if these concepts have been defined in advance, constant rethinking may be required as the limitations of the available technologies are imposed onto the user's "wish list" of performance capabilities and modes of operation.

Given the expense and time involved in appropriate development of CBRN protective systems, it is counterproductive to attempt to shorten the design cycle by bypassing steps in the process, as PPE may be produced that does not in fact have the needed characteristics for overall mission success. Therefore, as a general rule, the approach taken should include a long-term plan for life-cycle management that includes the predesign and procurement phase, which can generally be performed well in advance of an actual procurement program. The most important component of this phase is to ensure that all user organizations have a well-developed and frequently exercised concept of CBRN operations so that when the next requirement for new PPE comes along, the organization is already prepared and knowledgeable. This includes documentation of all the activities and ancillary equipment with which the PPE must work in concert.

Second, it makes more sense for PPE design and development to be performed in stages, with a large component of modeling incorporated into the process. Modeling not only includes computer-based simulation, but also physical modeling of the PPE components wherever possible or practical. One excellent example of the benefits of the physical modeling approach has recently been reported in the development of a single-use pocket-size rapidly donned escape respirator [300]. The many restrictions that were placed on the design based on this concept of use were accommodated by the use of handmade models in wearer trials prior to computer modeling and design and tooling. Where this type of approach is impractical, rapid prototyping can be of value. As computer simulation tools become more powerful, it becomes possible to predict parameters such as thermal strain and protective performance for entire systems, although this field is still in many respects in its infancy.

Finally, it has been tempting for designers to identify promising new technologies and to redesign protective items or systems to suit the technology first and the needs of the user second. A certain amount of this is both inevitable and desirable in the early stages, since every design is a series of compromises and it is difficult to predict in advance what these will be for a given technology without actually implementing it. Nevertheless, the technology can end up driving the process without a thorough assessment of the gains introduced and the inevitable trade-offs that will result.

4.6.2 Early Design Considerations

Particular design options may eliminate certain possible concepts of operations, or may result directly from important user requirements or limitations. Thus, both user requirements and the implications of design choices need to be well understood at the outset. A few examples of the interrelation of design concepts and user requirements are given below.

Each PPE item or system may be single- or multiple-use. If multiple-use, the PPE might be disposed of after contamination or may need to be reused after decontamination. Single-use equipment has the advantage of fewer constraints imposed by the requirement that it continue to protect after a variety of exposure conditions, but it is practical only when the supply chain for the equipment is reliable and rapid. For example, on deployed operations the military prefers reusable equipment and generally requires decontaminability since the number of readily available spares is limited by a number of logistical considerations and the required duration of use may be lengthy; by comparison, the training and cost required to provide a robust decontamination capability is viewed as more easily supportable. On the other hand, first responders may prefer disposable equipment, as storage of spares is more feasible and the likelihood is lower that the equipment will need to be reused before resupply could be completed.

The required duration of stay in the hazard zone may also affect whether PPE needs to be switched "hot." For example, the air-purifying element or air supply of a respirator can be depleted in a relatively short period of time (sometimes less than 30 minutes), making exit and reentry impractical. In this case the system should be designed to permit safe switching in a contaminated area. Some specialist wearers may be so overburdened by multiple PPE components layered over one another that designers should start by looking for opportunities to combine particular functionalities in a single design component or PPE item; for example, aircrew systems are under development that combine whole-body immersion [301], cooling [302], and CBRN protection [303]. Despite the obvious advantages, all of the difficulties of trying to combine what are often very disparate requirements into a single material or design remain, and this type of approach must be carefully considered for feasibility.

The Soldier Integrated Headwear Systems technology demonstration project developed concepts of an infantry headwear system including CBRN protection and considered all the integration, technology, and trade-off issues simultaneously during the early design process. A combination of approaches, including modeling, mock-ups, and component design were used to begin to evaluate these concepts [304]. Capability requirements were developed and associated priorities assigned very early in the process [305], followed by evaluation of particular technologies to deliver capability; the combined capability/technology pair was ranked during the design process. In this case, general priorities were established that pertained to capabilities for the following operational roles: warfighting (most important), CBRN operations, and peacetime operations (least important), with the more detailed capabilities within each role then prioritized accordingly.

Requirements groupings were split into:

- Whole head: overall protection, thermal comfort, range of motion, load, physical comfort, fit and adjustability, power and data, and detectors
- Vision: vision protection, local awareness, visual displays, and enhanced vision
- Respiratory: CBR protection and breathing
- Speech: voice communications and computer speech input
- Hearing: auditory protection, local awareness, auditory displays, and enhanced hearing

The project used an analytical hierarchy process to prioritize requirements and the likelihood of available technology to meet them, thus ensuring that the product would meet the most important functionalities of the wearer and that trade-off decisions could best be made [306].

It is crucial early in the process to consider the full range of human characteristics that may affect performance in the intended user group. As one example, accommodating the full range of anthropometry of the user group should be considered up front in the design, as it has a huge impact on comfort, functionality, and protective performance, with the design options usually imposing particular trade-offs that may not be acceptable to all user groups. Designs that rely on a good fit to the wearer for comfort and protective performance usually must be provided in many sizes, but ensuring proper sizing and availability of the correct size to each user then becomes a significant logistical burden, and individual issue of equipment is usually required. On the other hand, one-size-fits-most concepts are logistically very attractive, since PPE can be shared much more effectively and issued without special sizing requirements, but this approach has other drawbacks. For example, totally encapsulating suits that are made oversized to have minimal fit issues are burdensome as a result and have to rely on overpressure air to maintain protection, while hood-style respirators that rely on a single size of stretchy neck seal are usually very uncomfortable to the wearer.

In the sections that follow we amplify on particular design approaches and considerations in relation to delivering specific protective and human performance objectives.

4.6.3 Maintaining System Integrity to Prevent Penetration and Leakage

Any PPE can be penetrated at a location where its integrity has been breached: due, for example, to puncture of materials or poor performance of seals or closures. A system must have integrity when it is new, and it must be able to maintain this integrity in use and throughout its life cycle. Various issues that may affect integrity are addressed here.

Materials must be sufficiently durable to withstand puncture, tearing, and abrasion under the conditions of use. For clothing, reinforcement materials can be placed where necessary on wear areas such as knees and elbows, and for other items, thicker or more resistant materials may used: on the soles of boots, for example. Next-generation concepts toward hardening of materials may be of importance here, permitting reductions in thickness or less need for reinforcing materials.

Good fit of the overall ensemble or protective item is important for optimum protective performance. When an ensemble fits and forms well to the individual, there will be little leakage where the closures seal to the body, particularly as a function of movement. If garments are loose-fitting and closures not tight, motion will cause a *bellowing effect* in clothing, with air being pumped in and out wherever possible. Bellows action has been observed experimentally by Schlieren photography [307].

The **bellows effect** can be understood by analogy with a fireplace bellows. As the bellows handles are pulled apart, air is drawn in, and as the handles are closed, air is expelled. When PPE items (such as suits) are pumped in the same manner, air moves in and out of the closures as a result of the motion of the wearer. The larger the reservoir of air, the more impermeable the materials from which the clothing is constructed, and the leakier the closures, the larger this effect will be.

Bellowing can be reduced by:

- Designing tightly sealed systems
- Using air-permeable materials
- Using sufficient overpressure
- Designing equipment to fit close to the body

Operationally, bellowing can be reduced by reducing the amount of movement.

Fitting well to the individual can be very difficult to achieve because of the highly variable human anthropometry involved: Logistics dictates fewer sizes, while good equipment performance dictates more sizes. Adjustable size concepts may reduce the number of sizes required or improve the quality of fit, although such systems should remain as simple as possible. Another approach is to use seal materials that are highly formable to the individual, such as encapsulated gel technology, which is being investigated for use in respirator face seals [308].

Some important features of **closure design** are as follows:

- Closures must be designed to minimize penetration of hazards to the extent required to achieve adequate protection; this does not necessarily mean that the tightest possible closures are required, depending on equipment design and protection requirements.
- Extra resistance to penetration may be provided by seam sealing and use of liquid-tight zippers, particularly relevant for highly protective systems designed using air-impermeable materials.
- Closures that are overly complicated are often ineffective; if users require too much time, training, or assistance to seal a system properly, putting PPE on incorrectly is much more likely in use; for example, a respirator harness adjustable at six points is more likely to be adjusted incorrectly than one that adjusts at only four or two points.
- Poorer protection may result from tightly sealing or butting one item of equipment against another, as it may result in dislodging of both when either one is dislodged. For example, tight sealing of a hood against a respirator using a rubber seal may cause the respirator to pull off the face as a result of head or neck movement if the hood is too tight on the body.
- Other equipment with which the item is designed to integrate must be considered throughout the design process to ensure that closures remain effective throughout use. If highly effective closure performance is desired, it is likely

that only select items that have been designed and validated to work together can be used together in the system.

Overpressure and directional airflow can be used to minimize the effect of airborne penetration, since flow through leakage points will tend to be out rather than in.

- Blowers that provide the overpressure also give the benefit of cooling in many circumstances.
- However, blowers can be quite bulky, fragile, and demanding of power, as well as requiring significant logistics and maintenance support.
- Much smaller fans and blowers are now becoming available that may be used to provide supplemental air flow or overpressure when combined with good design.

4.6.4 Preventing Penetration and Permeation of Liquids Through Materials

Passage of liquids through materials and closures is enhanced by prolonged contact or pressure, as well as being more likely the larger the contact area. The likelihood of **penetration** and **permeation** of liquids through materials is generally reduced through the use of:

- Appropriate sealing means around seams and material junctions
- Liquid repellent coatings
- Impermeable and hardened barrier materials
- Reinforcing layers
- Reduction of air permeability

From the point of view of design, different requirements for liquid resistance can have a very significant impact on design. The requirement for highly resistant materials may be reduced for particular user groups or in particular parts of a PPE system if they are unlikely to come into contact with gross liquid contamination. The nature of the liquids under consideration is important; repellent coatings usually work differently against hydrophilic and hydrophobic liquids, for example.

The significance of rapid and effective decontamination should not be underestimated; removal of the hazard significantly reduces the likelihood of significant effects on the wearer by reducing the amount of permeation and the likelihood of penetration. Hence, designing for liquid protection must consider ease, speed, and efficacy of decontamination procedures when used on a PPE system, and realize that operational procedures and design can both be altered to achieve optimum protective performance.

4.6.5 Preventing Breakthrough of Chemicals by Using Adsorbing or Reactive Elements

It is apparent that air-purifying elements to remove chemical vapors must be simultaneously highly air permeable (offer low resistance to airflow) and highly effective at purification. Any air-purifying element can be made more effective by increasing its effective capacity for a given total flow of air impinging on the element. This can be accomplished by a variety of means. In the case of vapor adsorption, efficiency can be increased by increasing the effective surface area:

- Increasing the total amount of adsorbent is effective but a last resort.
- Providing a longer bed depth increases the likelihood of capture of the vapor as the flow of air increases.
- Providing a large total exposed cross-sectional area of adsorbent bed to the airstream reduces the velocity of the air through the bed and hence increases the likelihood of adsorption.
- Engineering the adsorbent at the micro- and nanoscales can provide maximal exposed surface area.
- Designing the element so that all parts of the bed see equal flow results in the most effective use of the adsorbent bed.

The latter two concepts also potentially increase the rate of, not just the capacity for, adsorption, which is often the limiting effect at high incident airstream velocity.

When used in clothing, adsorbent materials afford an additional benefit in that chemicals that have penetrated a system, whether through breaches in system integrity or by breaking through the materials, may be removed from the inside out; that is, the adsorbent may scavenge the vapors from inside the PPE. Because dermal permeation of chemicals is relatively slow and doses that cause effects are relatively high, this mechanism can be quite effective in maintaining high-boiling vapor concentrations below harmful levels. Keeping the adsorbent layer close to the skin increases its efficiency in this regard [309,310].

Effect of Airflow. The use of realistic maximum airflow rates and patterns to set the design requirement is important. Overestimating the potential flow will make the design that much more difficult, while underestimating the potential flow can result in a design that will fail catastrophically, since the amount of breakthrough of vapors through adsorbents is not linear as a function of flow. At a high enough flow rate, almost instantaneous breakthrough of the entire vapor challenge will occur through an adsorbent bed.

The airflows through a protective item may be a result of breathing (for a respirator), or for clothing, incident wind caused either by normal air movement and movement of the person, or additionally, various types of vehicular motion, such as traveling in an open aircraft or transport vehicle, or being in the vicinity of helicopter rotor wash. For air-permeable adsorbent-containing material systems, high incident wind can have a dramatic effect on protective performance [311], reducing dermal protection by orders of magnitude. In this case it is particularly important to understand how the challenge concentration and duration will also be affected by these conditions, as both are likely to be reduced at the same time as protection is being reduced. It is also important to understand how the design of the clothing will affect the face velocity (speed of the air at the surface of or through the material). Increasing the resistance to flow by reducing air permeability, or reducing the air gap by moving the material closer to the skin, will significantly reduce face velocity and probably increase removal efficiency at the same incident wind condition.

Nature of the Chemical. The relationship between the chemical structure of the vapor to be removed, its toxicity by various routes of entry, and the capabilities of various adsorbents and reactants to remove it from an airstream must be well understood. For example, low-molecular-mass low-boiling vapors are poorly removed by many adsorbents; this is a fairly fundamental property because it is related to their tendency to form condensed phases, the same property that affects the likelihood of adsorption onto a surface. Because of their generally small size, size-based separation possibilities are limited; air must be able to pass through the purifying element relatively freely. It is therefore important to understand whether removal of such low-boiling materials is important; for example, many have limited dermal toxicity (in part because of their tendency to evaporate rapidly before they can dissolve into skin to any significant extent). However, most chemicals cause some respiratory or eye effects, because of the low barrier to gas dissolution and diffusion of both these routes of entry; further, the low-boiling chemicals may be present in higher challenge concentration because of their higher volatility. Therefore, to improve their capabilities with respect to these low-boiling chemicals, reactive species are often introduced into respirator canisters. Here again, all of the considerations listed above for adsorbents are important, and it is noteworthy that reaction must generally be preceded by adsorption even if for a shorter time, so that the reaction has time to proceed. Chemical reactions are rarely as rapid as needed, and therefore the agent must be at least briefly immobilized near the reactive groups.

The fact that adsorption can occur, albeit somewhat ineffectively, for low-boiling compounds should not be neglected for its possible negative consequences as well. As discussed previously, such ineffective adsorption may cause a chemical to appear to be removed from the airstream over the duration of a test designed to assess performance; however, subsequent migration through the bed may occur if reaction does not occur, resulting in breakthrough at a later time after the challenge vapor has been removed, particularly if clean airflow continues (e.g., continuing to breathe through a canister after leaving a contaminated area). Therefore, adsorbent selection, reactant selection, and bed design must all take this possibility into account.

Reactive species that are added to PPE elements may require coreactants to be effective, or alternatively, their reactive efficacy may be inhibited by other species present in air. For example, water is a ubiquitous airborne species that can enhance or inhibit such reactions, particularly when it is coadsorbed onto adsorbent surfaces [312]. Since humidity levels can vary hugely, any system must be capable of functioning regardless of water concentration in the surrounding air. Since for CBRN applications, removal of many different chemicals is required, adsorbent beds may contain multiple adsorbents and reactive species. Assessing how these multiple components interact with each other is an important part of design; in canisters, beds from different adsorbents or containing different impregnants may be layered or mixed, depending on the stability of each in the presence of the other.

4.6.6 Preventing Penetration and Reaerosolization of Aerosols

Particulates may be prevented from penetrating systems by the use of barriers (i.e., maintaining system integrity, as described in Section 4.6.3) or can be removed from airstreams by filtration.

Particulate Blocking. Any air-impermeable barrier material will stop particulates. Selectively permeable materials such as those described in Section 4.5.4 can be an improvement in terms of thermal burden, allowing air or moisture vapor to pass through the material, permitting some degree of convective or evaporative cooling.

- Blocking materials must be combined with good closures to prevent bellows effects.

Filtration of Agent. No filter will have 100% filtration effectiveness, and therefore, to properly understand when its effectiveness drops, it is important to understand its performance characteristics. For example, in general, filters exhibit a minimum in filtration efficiency at some particle size in the submicrometer range (see Section 4.2.2), and filters may be more effective at different flow rates, an effect that may vary as a function of particle size. The construction and materials from which the filter is composed will also determine filtration effectiveness, which may vary as a function of additional factors such as charge on either the filtration medium or the particulate hazard [313]. Finally, even after capture, aerosols can reaerosolize after filtration when the particulate is not thoroughly adhered onto or embedded in the medium.

- Glass, cellulose, and electret fibrous materials have all been used in particulate filters; in CBRN respiratory protection in general, the choice of material has been driven by the need to achieve high degrees of protection with the lowest possible breathing resistance. Commercial high-efficiency filters are generally fiberglass or electret in nature, with the electret properties providing extra filtration capability at the MPPS without an increase in breathing resistance. Electret properties are susceptible to decay over time in storage.
- Fabric that does not have extra particulate filtration capability can be penetrated by aerosol particles which are less than about 5 μm in diameter; therefore, impermeable materials or filtering fabrics may be employed for those aerosols that are considered to be a significant dermal hazard.
- Reaerosolization of hazardous materials may be reduced by including a capture material. Reaerosolization is more likely from surfaces to which the aerosol is not tightly adhered, or from air-permeable materials due to the backflow of air. On the other hand, loosely adhered material is more easily removed by washing during decontamination.
- There is some evidence that nanofiber filters have considerably reduced pressure drop across the filter for a given filtration efficiency for certain particle sizes [215].

Removal or Deactivation of Agent. As for vapor protection, certain types of materials may scavenge an agent during and after penetration so that its impact is reduced. In all cases, sufficient contact time between the agent and the scavenging material is required.

- A reactive component can be added aimed at deactivation of C or B agents (self-detoxification) after their capture.
- Adhesive materials on the interior of the PPE could potentially also be designed to scavenge particulates in order to capture them after they have entered.

4.6.7 Protecting from Non-CBRN Hazards

It is the norm that any CBRN protective concept must also protect the wearer from other common hazards, and in some cases, from less common ones.

Physical Degradation and Durability. Many different types of projectile, sharp, or abrasive objects can result in degradation of the materials of which the system is composed, resulting in subsequent loss of protective performance, or even more immediately, can damage wearers themselves. Protection against piercing or abrasion can be provided by reinforcing or thickening the various material layers, or by choosing intrinsically strong materials such as polyaramids (e.g., Kevlar). Materials should resist fatigue resulting from bending or abrasion. Fabrics can be more durable when the fibers are larger and are free to move when stressed [314], and flexible polymers may resist puncture and flex fatigue better than stiff polymers do. Fluoropolymer coatings can also enhance the durability of outer layers.

Ballistic Protection. Ballistic protection should be layered over CBRN protection, as CBRN layers generally fail to protect once they have been breached. Protective undergarments can be a good choice to combine with ballistic protection as the ballistic layers provide substantial protection against air and liquid penetration, and the undergarments integrate better in combination with the overlayer. Materials are described in Section 4.5.9.

Impact. Impact protection is typically incorporated into headgear, which may be integral with the respiratory protection (e.g., in aircrew helmets). A low-density material such as polystyrene foam may be provided as an impact-absorbing layer.

Heat and Flame. Fire at a CBRN event would not be an unexpected consequence of explosive dissemination, and many user groups (e.g., naval damage control parties and firefighters) have a primary role in fire suppression. Additionally, the extreme heat of a nuclear explosion is often considered as a nuclear hazard that must be taken into account for the military. When heat and flame are significant potential hazards, materials that can melt or ignite easily are inappropriate choices. The fire-resistant polyaramide Nomex may be chosen to be included as a woven component in the outer layers. Traditional thermal insulation materials can help protect the skin from heat as well as providing an additional barrier to liquid penetration; protective air pockets are created through constructions such as double-woven or knitted structures, or batts of loose fiber wadding quilted together between two layers of fabric. Cotton-containing layers are a good choice to be worn next to skin, as they are melt-resistant.

Signature Reduction. In some cases, the person wearing PPE could be facing an adversary that is the actual or potential perpetrator of the CBRN event. In this case, there are a few important considerations to assure that the PPE does not enhance the wearer's detectability. To reduce the visual or infrared signature, the outer layer of the system should be able to take an appropriate camouflage or color. Materials should not increase the auditory signature; crispy or crinkly materials can give away a person's position. Similarly, the respirator should add as little noise as possible; overpressure systems yield the sound of blowers and/or moving air through valves, and audible alarms are inadvisable.

Signature Increase. Although in the large majority of cases, signature increase would be undesirable, in the case of rescue personnel it may be desirable to make a person more detectable—improving recognition when visibility is obscured, or in the dark. This can be achieved through the use of bright colors, reflective, fluorescent, or photoluminescent materials, tracking devices, and audible alarms that sound when a person is down or low on air.

4.6.8 Optimizing Human Performance

Thermal Stress and Cooling. Various currently available approaches that are being pursued to reduce thermal stress among PPE users are outlined here. Some of the design factors that can help in maintaining acceptable thermophysiology of the wearer include:

- Thermal insulation (high to reduce impact of extreme heat or cold, low to enhance air permeability; see below)
- Air permeability (high to enhance sweat evaporation and convective cooling)
- Water absorption (high to minimize discomfort from sweat buildup)
- Ventilatory airflow (high to enhance sweat evaporation and convective cooling)
- Moisture vapor permeability (high to enhance sweat evaporation)
- Heat-activated cooling reactions (high to enhance cooling)

Many of these concepts require energy (via a power source or stored chemical energy), and therefore their implementation is currently limited.

Air- and moisture-permeable clothing fabrics can be utilized to take advantage of convective and evaporative cooling that will result from passive exchange of air and/or moisture through the materials. This type of approach may be particularly valuable when designing daily-wear and uniform concepts [315]. Reducing encapsulation where possible increases normal ventilation processes.

Microclimate cooling provides cooling directly to the wearer rather than cooling their environment and is accomplished through some cooling device, often powered [111]. This type of approach may have value for specialized users who have access to power (e.g., within a vehicle) or who need to operate in bulky PPE for relatively short periods of time (e.g., explosive ordnance disposal technicians). Cooling may be offered to the whole body or can be targeted to a large surface area, such as the torso, or to an area where the circulatory system is close to the surface, such as the wrists. It has been observed that intermittent microclimate cooling (e.g., 2 min on, 2 min off) is as effective as continuous cooling, resulting in the possibility of extended cooling duration for the same power requirements, or reduced power requirements for a given duration [316].

There are a number of approaches to providing a cooling medium that have been patented [317–319] or are sold commercially, including blowing ambient air, effective to remove moisture and sometimes heat depending on ambient conditions and work rate, or chilled or ice water, effective until the circulating water equilibrates with the environment, often about 15 to 30 minutes.

Present-day liquid-cooled systems generally consist of an undergarment-style tube suit constructed from hood, jersey, and pants, with coolant flow approximately uniform over the body; their heat removal capacity approximates 500 to 600 W when wearing protective clothing, but the absolute limit is predicted to be above 1250 W, greater than the required amount of over 1000 W produced during maximal exertion [111]. Automatic or guided control of flow over different body regions is possible to improve effectiveness.

Vests are manufactured that contain phase-change materials that effectively absorb heat and in so doing, change their structure; this concept, like that of cooling water, operates for a relatively short duration until the phase change is complete. In principle, this same concept could also be provided to other body locations.

Systems that use supplied or powered air can provide cooling by redirecting the exhaust air over the body. One study has been performed which indicated that the effect of the respirator tested [a hooded powered air-purifying respirator (PAPR)] on discomfort was related primarily to slightly increased work of breathing and, at least when worn without additional heavy PPE and at the work rates tested, it did not contribute directly to increased sweating or heat stress [320]. This is perhaps not surprising, as the powered air system would contribute to cooling by convection and evaporative heat loss. It is more difficult to ascertain how much the respirator contributes to thermal burden once the test subject has reached a state of uncompensable heat stress; it is apparent, however, that it may be possible to compensate for some of the stress imposed by remaining PPE items by using cool supplied air.

To reduce moisture within PPE and increase comfort and effectiveness of cooling, possible approaches include the use of desiccants as well as wicking materials that pull moisture away from skin (e.g., cotton glove liners), and increasing the surface area for evaporation. Operational measures that can also be used include the development of heat indices; these are means of monitoring and/or providing guidance to the wearer on their thermal status and how work rate should be altered to maintain health.

Real-time monitoring of important physiological status indicators such as heart rate, core temperature, and breathing rate is now becoming possible through advances in miniaturization, monitoring, and telemetry. Important requirements for routine implementation would be that they be easy to use (noncontact, with no preparation or requirement for medical knowledge), rugged and reliable under conditions of use, of minimal size and weight, capable of being powered, and available at low cost.

Thermal stress can also be minimized by rehydration; people wearing protective clothing can lose 1 to 2 $L \cdot h^{-1}$ to sweating when exercising [321]. Any hydration system needs to be able to allow people either to drink freely while working, at lower rates, or provide fluid during breaks at a relatively high flow rate to replenish lost water. People will also rehydrate themselves more reliably when provided with cool water than with warm water for drinking [322]. Finally, concepts that provide just-in-time protection, minimizing the length of time before the PPE is put on or encapsulated provide the wearer with a start state as close as possible to physiological norms [323,324].

Oxygenation. *Dead space* within a respirator (on the inside of the exhalation valve) is the space that contains exhaled air that is capable of being rebreathed. This space needs to be kept as small as possible to ensure that the oxygenation levels of the breathed air are sufficient and that expired CO_2 is not rebreathed, which is why essentially all respirator styles have a nose cup to contain the exhalation air. Some resistance to exhaling the air into the remainder of the facepiece (for full-facepiece designs) needs to be introduced, typically through unidirectional inhalation valves; another option is a nose-cup design that pulls away from the face on inhalation and seals to the face on exhalation.

Filtering facepiece respirators (described in Section 5.3.5) that have an excellent fit to the face can yield relatively low oxygen concentrations within the respirator (around 15 to 16% [325,326]) because of their large internal dead space. There is insufficient information overall on the effect of wearing CBRN and filtering facepiece respirators on blood oxygenation and CO_2 buildup over long periods of activity and whether there may be significant gains in human performance to be achieved through redesign.

Comfort and Physical Burden. Many aspects of comfort involve reduction in the amount of energy required to perform a task, which will minimize thermal stress, as described above, as well as energy use. Several factors cause the wearer of PPE to exert more energy than when in the unprotected state; these include increased work to move, to breathe, and to maintain balance.

The main contributor to work of breathing is resistance to inhalation and exhalation, which are a function of:

- The properties of any air purification media used, such as packing density and porosity of particles or fibers, filter thickness, and bed depth surface area on the face of the medium.

- The history of use: clogging of filters is possible.
- Valves that force unidirectional flow: inhalation valves prevent backflow of air out of the facepiece, which would degrade the quality of the air purification medium, or out of the nose cup, which would result in rebreathing of air, while the exhalation valve prevents inhalation of contaminated air from the outside.
- The flow rate: higher flow rates (caused by high breathing rates) increase resistance in general by increasing turbulence.
- Leakage into the respirator: higher face-seal leakage means lower breathing resistance by (undesirably) bypassing the air purification medium.
- The design of the respirator facepiece: design assists in directing flow and minimizing turbulence.

Therefore, the design of the medium and the respirator should be considered in order to optimize flow, decrease resistance, and decrease likelihood of clogging. Breathing flow rate cannot be controlled while leakage into the respirator must be consistent with the value required for the protective performance desired. The center of gravity should be maintained as close as possible to normal to minimize muscular fatigue and strain. Additional load carriage elements, such as backpacks, can compound problems that result from carrying other ancillary equipment associated with PPE. Overall system design to include PPE and all ancillary items taking all of these factors into account should be carried out. Weight must be balanced around the body, keeping it low to prevent overbalancing. Transferring weight onto muscle groups best able to support it through the use of frames or exoskeletons should be considered.

Reduction of weight and resistance to movement are accomplished using thinner, flexible materials; stiff, heavy materials are also undesirable as they are uncomfortable in contact with skin. Specific features can be introduced to improve ease of movement, such as in the Canadian NBC glove, where pleating at the knuckles improved its comfort and dexterity performance relative to an unpleated counterpart [154]. Comfortable fit will similarly be achieved with either highly formable materials or a large selection of sizes designed to the anthropometry of the user population (see Sections 4.3.4 and 4.6.9). Materials in contact with skin in particular should have good hand properties in general, some component of stretch often being desirable, and, if possible, contain a wicking layer to remove sweat from the skin (e.g., glove liners).

General Functionality and Ergonomics. PPE is often to be worn for extended durations, and therefore clothing components must often satisfy some aspects of normal use for work clothing. Issues include the ability to be appropriately colored or treated, color-fastness and resistance to fading, resistance to shrinkage on laundering or differential shrinkage between joined materials, and the capability to be sewn or seam-sealed; these are all important in regard to the materials and manufacturing methods to be chosen.

ISO has standards published and in development in the area of ergonomics that can be used to improve design [327–329]. General ergonomic issues that may not already have been addressed under human factors and performance above include how quickly, easily, and reliably items and systems can be put on and removed, including, for NATO, time to transition between various dress states (open to closed) [24]; this includes forming closures, integrating items with each other, and so on. Some sizing features may be adjusted in advance (e.g., mask harness settings, to reduce donning time). Systematic studies on how to design functionality into protective clothing and peripherals have rarely been undertaken. One such study [330] looked at the addition of wearable input (control) devices and the relevant criteria that would improve functionality.

Sensory Impact. Quality of vision is probably the most important issue with PPE. The field of view must be as large as possible, in particular allowing good binocular vision directly ahead and good downward vision to allow for good footing. These should be achieved simultaneously while the person has his or her head aimed in an upright and forward direction. If possible, good peripheral vision is desired at the same time, although this is ranked slightly lower than the other two. Insofar as the respirator is concerned, this configuration is most likely to be achieved with a visor rather than individual eyepieces, and with a small snout to the respirator, with the canister or hoses low, off to the side, or behind. Typically, the requirement to be able to bend the respirator facepiece for compact stowage in a carrying case has made a single visor design problematic, but newer materials have made this design more achievable (using a flexible polyurethane visor [331]).

The visor or eyepieces must be made from an impact- and scratch-resistant material with low haze and distortion, antifog and glare, while being as lightweight as possible. The respirator or headwear system may also be a convenient means to protect or improve vision quality by allowing the mounting of, or integrating, sun, nuclear flash [332], and laser protection, or infrared goggles, via outserts or inserts.

Fogging results from the condensation of minute water droplets onto the inside surface of an eyepiece; the water source in a respirator is either the moist exhaled air or sweat evaporated from the face, which then condenses on the eyepiece due to a lower ambient temperature outside the PPE than that within the eye space (at around 32 to 37°C). Antifog coatings may be hydrophilic, absorbing moisture to prevent the formation of droplets, or act as wetting agents, in which case the excess water forms a thin layer that runs off the lens due to gravity. Hydrophilic coatings have a limited capacity before they will be saturated, but are regenerable (i.e., once they dry, they are again effective). In either case, the coatings must be insoluble in water so that they do not leach off the surface, and hard such that they do not scratch; they are generally polymeric in nature.

Aside from the use of antifog coatings, fogging is also minimized by designing the facepiece such that fresh breathing air sweeps over the eye space. This can be achieved either by directing all the incoming air over the eye space before it enters the nose cup, or by directing a fraction of supplied overpressure air into the eye space, as well as by ensuring that exhaled air cannot enter the eye space to any significant degree.

Glare results from reflection off surfaces and is managed by antireflective coatings. These coatings may need to be antireflective not just in the visible spectrum but also to minimize the signature in other areas. **Haze** tends to occur as the surface quality degrades over time, or if a surface is not perfectly smooth. Hence, hard, smooth, durable coatings are important.

A significant portion of the user community requires **vision correction**, and for many of those users, eyeglasses are the correction means of choice. Respirators should be designed from the outset to be compatible with vision correction. The most common means of vision correction is using a glasses insert that mounts onto the eyepieces or nose cup. Achieving the correct focal distance, and maintaining the stability of the lenses such that they do not move when a person breathes or is active (which, aside from degrading the quality of vision, can cause motion sickness) are important. The corrected field of view should be as large as possible and the total field maintained to the extent possible.

The **auditory** impact of a system in terms of signature has been discussed in Section 4.6.7, and many of the same factors can affect a person's auditory capabilities. Materials or design features that are noisy or muffle the ears are usually undesirable except where hearing protection devices are necessary for noisy environments, in which case alternative appropriately filtered electronic auditory inputs are usually desirable. **Speech transmission** is enhanced through the use of mechanical or electronic amplification: for example, by a passive speech diaphragm located near the wearer's mouth or by a microphone. The nose cup shape, material, size, and mounting are all important in maintaining quality of speech to minimize reverberation and frequency distortion. Mechanical amplifiers and microphone pickups may be directional, meaning that they must be carefully located for effectiveness. A linear decrement in speech intelligibility as a function of the background white noise level reaching the listener has been observed [333] for a diaphragm-containing respirator. In the same study, varying speech diaphragm size yielded no difference in performance, meaning that smaller diaphragms can be contemplated (down to 30% of the standard area evaluated of about 13 cm^2) [333]. Other studies showed no advantage of using a microphone over a diaphragm for telephone communications [334] and with moderate background noise (60 dBA) [335].

Visual and auditory inputs can be enhanced through a variety of means: for example, by a specially designed headwear item that uses sensing, signal amplification, and potentially artificial intelligence to present information to the wearer through means such as heads-up displays and audio feeds. The materials chosen for PPE construction should be free of odors that might be significantly unpleasant for the wearer, particularly those materials used for the facepiece and for air-handling or purification systems. Treatments such as amine impregnants and plasticizers are common culprits.

Psychological Well-being. Aside from all of the previous issues that can affect well-being, there are a few additional issues that can affect human performance, such as:

- Claustrophobia, which will be alleviated with good quality of vision and field of view, and good air quality
- Inability to identify others, which will be alleviated with large transparent visors, good quality means of communication, or special identifiers built into the PPE [336]

4.6.9 Optimizing Sizing

The first issue in optimizing sizing is to ensure that the appropriate metrics for the user population that are relevant to the PPE being developed or selected are available; these include knowing the user population's makeup (age, gender, level of fitness, overall anthropometry) and the activities that will be performed when wearing the PPE. Then the optimum number of sizes will need to be established such that comfort and functionality are maintained. At some point, an appropriate representative panel of users will need to be used to evaluate the success of the sizing system. As discussed previously, although it may seem that the best system would have many possible sizes, in fact too many sizes can be as much of a problem as too few. Too few sizes, of course, results in poor performance of the PPE with respect to functionality and comfort. Too many yields the likelihood that the system for issuing the correct size to each person will break down as a result of difficulty with maintaining inventory, having sizes available where needed at the scene, or correctly assessing the proper size in the first place.

Respirator Sizing. A proposal has been made to develop a respirator sizing system based on principal components analysis in which different facial aspect ratios are targeted rather than simply "size": for example, five sizes, based on average, tall thin, short wide, short thin, and tall wide faces. ISO standard TS 16976-2 follows this approach, including international anthropometric data; the standard also includes standard headform facial dimensions for each size proposed [149]. Other issues that may need to be addressed include various difficult-to-fit features, such as sunken temples and narrow or small chins. Separate men's and women's sizes could be appropriate.

One-size-fits-all adjustable sealing options would be the least logistically burdensome, but this requires new design approaches not currently available or practical; sealing at the neck might be a more flexible option if issues with comfort and restriction of circulation could be overcome. Individually customized seals through the use of facial scanning would provide the most robust sizing capability. Overpressure and directional flow reduces the reliance on proper sizing to some degree, although at high work rates, protection may fail.

Clothing Sizing. Aside from comfort issues, sizing clothing too small can result in closure failure; items that are designed to overlap may gap instead. Motion will exacerbate interface problems, causing jackets and ankles to ride up, for example. It is thus to be expected that sizing clothing too small is more likely than sizing too large to lead to significant problems in both comfort and protection. Nevertheless, sizing too large will lead to its own problems: more bellowing of outside air through the material and the closures, more difficulty in forming snug closures due to excess of materials, and a large air gap between clothing and skin that in some designs causes poorer protection due to an inability of reservoirs within the clothing to deplete the hazards [337]. Automated body measurement systems can make selecting the correct size of clothing more reliable; for example, the Canadian Forces use a system that scans a person in the standing and sitting positions and uses the images to select the correct size [338]. The information is stored in a database to ensure that the correct size will be issued in the future.

Sizing of Other Equipment. Other equipment already issued personally may not fit well when CBRN protective PPE is donned beneath, and issuing more than one size to each person is usually impractical. To simplify such issues, PPE should be designed as thin as possible, with few pressure points, and where possible, other equipment should have some sizing slack to accommodate PPE beneath when necessary. Particular issues involve helmets and shoes.

4.6.10 Other Design Issues

Decontaminability. The ability of equipment to be decontaminated for the purposes of reuse depends on its "hardness," as described in Section 4.2.1; in this case, its ability to withstand decontamination solutions (often, chemically harsh) and procedures (often, involving high temperature) is the important requirement affecting selection of materials and design. Complex mechanical or electronic systems are unlikely to be decontaminable and survive, unless well protected, and decontamination of expensive, yet sensitive equipment is a thorny issue. The best approach to designing a system for safe reuse is to prevent its contamination in the first place, using some sort of sacrificial disposable layer over top. NATO has discussed many of these issues [339].

Many organizations feel that for PPE, the only decontamination requirement is the ability to perform expedient immediate decontamination sufficient to permit safe removal, since it is extremely difficult to assure sufficient cleanliness for safe reuse. For immediate decontamination, the important criterion is that the equipment must remain intact during the decontamination process, protecting the wearer from both the hazard and the decontamination procedures until removed. The decontamination process must work with the material and design choice.

- For example, clothing constructed from air-permeable materials is poorly resistant to liquid-based decontaminants, and although it may be possible to use such materials in a careful application method, a better approach might be suction-based decontamination to pull the hazard away from the clothing and the person.
- PPE systems to be decontaminated using pressurized spray-based systems must be constructed using liquid-impenetrable materials and closures, and any human-mounted electrical systems must be liquid-tight, powered off, or removable during liquid-based decontamination.

Equipment Survivability, Integration, and Maintainability. Many aspects of equipment survivability are encompassed by those design features already described that prevent penetration or breakthrough of materials. However, durability of the entire system is also important and is best evaluated in item- or system-level evaluations.

Integration with other routine equipment and functions is also a significant consideration for user-level trials. In general, the overall PPE silhouette should be kept as close as possible to the body. PPE should not interfere with important entry and escape functions (e.g., from vehicles, downed aircraft, or confined spaces). Hoses, hose connections, blowers, and so on, can be a significant problem for air supplied or powered air systems, as they are prone to damage from snagging, bending, impact, or trapping beneath a person, and therefore proper mounting and layout will assist in reducing the likelihood of damage. Weapons sights and goggles often require that respirator eyepieces be kept close to eyes. Air-purifying elements such as canisters may interfere with normal weapon-sighting positions. Keeping canisters as small as possible and away from the eyes and shoulders will assist in minimizing this issue. Regardless of the best design, extra training may be required to maintain firing accuracy.

Minimizing the burden on a person who is carrying the equipment before it is opened is as important as weight and silhouette when it is worn. Minimizing packaging has a benefit, although it may also make protection of the item within the packaging more difficult. PPE should not create extra pressure points when worn with load-bearing equipment or helmets, and therefore locations such as the scalp, shoulders, back, and waist areas should be examined to avoid seams, seals, tubing and hoses, and connectors where they could cause problems. Keeping a slim silhouette to gloves and boots not only improves desterity and mobility, but ensures that other equipment that requires interfacing with the hands and feet can be used properly, including weapon triggers and various types of hand- or foot-actuated controls.

Multiple Hazards. It should also be recognized that trying to protect against multiple hazards simultaneously may be beneficial or detrimental to the net performance of a system. Table 4-2 includes some examples that compare the potential effects of including or neglecting certain types of protection against physical hazards on CBRN protection and/or human performance.

Fashion. Black et al. [340] have identified a number of "fashion factors" that if properly taken into account, are more likely to result in user acceptance:

- Self-perception and identity
- Cultural identification with recognizable social groups
- Fashion currency; awareness of relevant fashion and lifestyle trends
- Feelings and emotions, including comfort and well-being
- Tradition and innovation: impact of emerging technological and fabric trends
- Appropriate form, style, materials, and color (overall concept and silhouette)
- Cut, style, and proportion (i.e., not "old-fashioned")
- Manufacturing processes and detailing
- Functionality and fitness for purpose

Based on personal experience, the importance of human nature in relation to the foregoing issues should not be neglected; for example, Canadian soldiers invariably

TABLE 4-2 Result of Including or Neglecting Protection Against Other Hazards on CBRN Protection

Physical Hazard Encountered	Flame/Fire	Static Charge	Ballistic/Blast
If hazard is encountered with no protection	May degrade subsequent CBRN protection due to material damage and loss of system integrity	No likely impact on subsequent CBRN protection	Will degrade subsequent CBRN protection when loss of system integrity results
If protection against the hazard is provided	Antiflame/fire treatment of material may be incompatible with other treatments and degrade comfort; can be contained in separate overgarments	No impact on CBRN protection; however, many flame- and fire-protective solutions themselves require the inclusion of electrostatic discharge mitigation	Normally contained in additional equipment; may improve CBRN protection by addition of layers, may degrade CBRN protection due to system integration issues, decrease comfort, and cause decontamination problems

Source: [4].

"lost" their carefully designed and geeky-looking combat glasses that integrated with their respirator, preferring to go without vision correction, while newly designed protective combat uniforms when presented to the user were judged first based on the appropriate placement of pockets.

Cost. Finally, although in life-support equipment cost is often considered secondary to performance, it should not be neglected. Minimizing the number of parts and sizes and improving serviceability will lower costs throughout the life cycle of the equipment. On the other hand, because CBRN PPE is generally expensive, maintenance rather than disposal of the entire item is to be preferred. Examples of ways to improve maintainability include design of a respirator with replaceable visors, valves, and so on, or renewal of consumable protective functions such as reactivity in clothing materials by recharging with solutions during laundering.

4.7 MODELING PERFORMANCE AND HUMAN PHYSIOLOGY

4.7.1 Introduction

Historically, there has been relatively little effort in the area of modeling as it relates to protection provided by this type of PPE, particularly at the system design level.

Nevertheless, the value of such modeling has become more apparent in recent years, with increased activity in this area. The particular value of modeling is threefold:

- When successful, it can provide considerable benefit in understanding and improving fundamental design requirements issues as well as detailed design concepts.
- Once a model has been developed and validated, it can permit exploration of considerably more design space far more quickly and effectively.
- The mere attempt to model provides valuable insight into what aspects of the physics, chemistry, and biology of protection and human physiology we do not yet understand sufficiently well to truly assess the success of a design once a PPE system is assembled. This helps to drive further research and data acquisition in the field.

Modeling can be both mathematical and physical, and the most successful approaches combine the two; in other words, a protective system (including the person wearing it) is conceptualized and reduced to simpler components in order to understand their importance. The important interrelationships that may be modeled are indicated in Figure 4-16. All of these interrelationships are important; nevertheless, it should be apparent that one interaction that is of primary interest is how the hazard from CBRN agents can be reduced by design of PPE. Remarkably, it is rare that this particular facet has been modeled in any detail. It is conventional to use the interaction between the hazards and the PPE material as a much simpler surrogate—a model, in fact, for the performance of the PPE itself—but in doing so, many important performance aspects are neglected. In fact, further interaction between the materials and the design is of equal importance in determining the effectiveness of hazard reduction. Environmental factors such as wind, temperature, and humidity are well known to

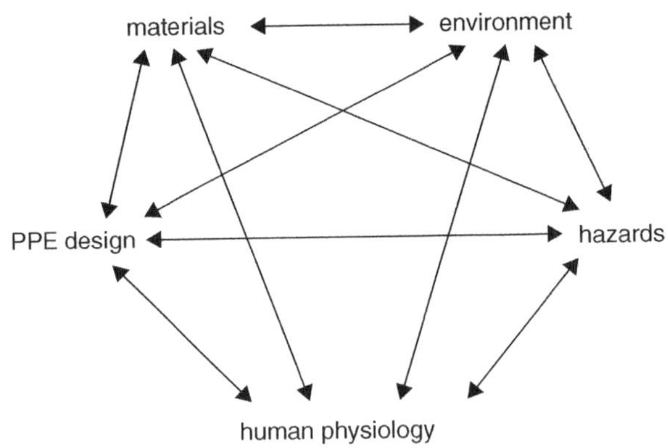

FIGURE 4-16 Interrelationships that may be modeled in order to understand PPE performance and relevant human physiology.

have considerable effect on both PPE performance and on the person wearing the PPE in terms of thermophysiological strain.

4.7.2 Protection Performance Modeling

The effectiveness of protection depends on the following parameters:

- The hazard: its physical form, chemical nature, and magnitude
- The environment: temperature, wind speed, humidity, and so on
- The person: his or her susceptibility to an agent due to physiological differences, and their work rate
- PPE materials: their ability to hold out an agent by penetration and permeation, their capability to be sealed to other materials, their effect on ventilation rate, and their ability to trap, destroy, or retain an agent
- PPE design: the ability to keep interior concentration of agent minimal—for example, by minimizing ventilation by contaminated air and/or maximizing flow of uncontaminated air

Each of these parameters interacts with the others to determine the total effectiveness of protection, and therefore any realistic model must take into account all of them explicitly by inclusion or by justifying their omission as a factor when they are not believed to be a major contributor. Some examples of existing pertinent models follow.

Airflow and Ventilation. One option is to conceive of people or their limbs as simple objects such as cylinders, and of protective clothing as sleeves or containers of material that surround them. This simplified approach can be modeled more easily both physically and mathematically than can a person wearing full PPE. This permits computational fluid dynamic modeling of airflow [341–343] and simple predictions of protection performance based on deposition of aerosols or adsorption of vapors [337,344]. For example, in the case of a cylinder, the influence of air permeability, the wind speed, the diameter of the cylinder, and the distance between clothing and cylinder surface can be investigated [344]. More extensive modeling has been performed using different models for different scales [345], including direct numerical simulation for flow underneath the clothing, microscale direct numerical simulation for determining the relation between textile porosity and permeability, and flow and heat transfer predictions by direct numerical simulation and Reynolds averaged simulation.

Dynamic Adsorption on Active Carbon Beds. Numerous models have looked at the prediction of breakthrough of vapors through activated carbon beds. Among these, the Wheeler–Jonas model is the most widely used [346].

Aerosol Protection and Filtration. Models of filtration at the fiber or material level are numerous, and a review of these could fill an entire book [347–355]. Despite the plethora of filtration models, their incorporation into more complex systems has been sparse. Modeling of aerosol protective performance by taking into account the physics of aerosol filtration and deposition for a cylinder [356] as well as its contribution to protection in a full system constructed from filtering materials [356] has been carried out.

Liquid Protection. Models of liquid repellency have been proposed and validated experimentally [357,358].

Full System Protection. Modeling of dermal exposure to aerosolized chemicals has estimated the skin uptake of solvents [359], taking into account the presence of clothing and respiratory protection. Various moving mannequin test platforms have been developed as physical models [360,361].

4.7.3 Human Performance and Physiology Modeling

The additional interactions between PPE and human physiology and performance have begun to be considered. Simulation methods for human physiology have been reviewed by Fiala et al. [362].

Human Performance. A variety of simulation methods have been used to describe human performance for the purposes of military outcomes [363], and the effect of PPE can be incorporated as a form of advantage (favorable survival if CBRN agents are deployed) or disadvantage (due to increased burden or reduced operational effectiveness), depending on the scenario.

Flow Modeling. As described in Section 4.2, flow modeling is also particularly useful for looking at thermal transport and cooling [345]. A CFD model of respiration in a half-height mannequin, including aerosol aspiration, has been developed [364].

Thermophysiology. The Fiala human physiology and thermal comfort (FPC) model is a framework of linked models that predict human thermophysiological and thermal sensation responses to various environmental conditions [362]. A combination of CFD and FPC models can be used to predict thermoregulation [365]. A combination of simulation models, computational schemes, software architecture, and computer-aided design (CAD) software systems has been used to quantify the thermal performance of clothing and its impact on the thermal biology of the human body in various wearing situations (Figure 4-17) [366]. The model has been validated using human trials and is intended to permit dynamic iterative design and engineering of clothing. Another model that also takes into account the balance of such factors as pressure, geometry, heat, and movement with the outside environment has been developed [367].

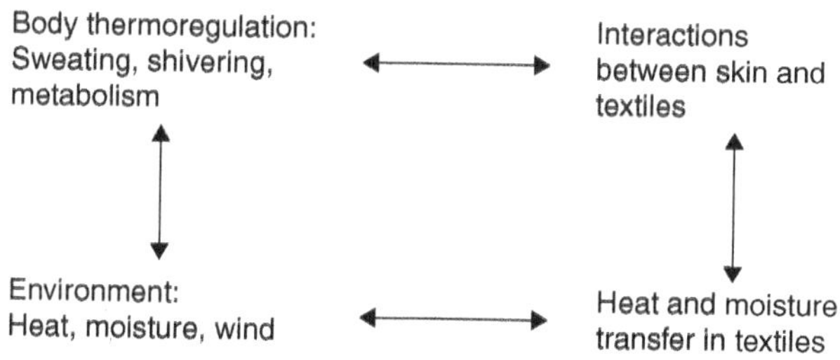

FIGURE 4-17 Components of thermal modeling of clothing. (Based on [366].)

The use of mannequins for thermal physiology modeling can be advantageous. The Hohenstein Institute in Germany has a thermal mannequin for evaluating clothing insulation as well as a skin model for evaluating the effect on skin temperature of fabrics and sweating [368]. The U.S. Army Research Institute for Environmental Medicine has developed the Heat Strain Decision Aid tool based on their biophysical models [369] and has a sophisticated thermal mannequin [370]. A sweating thermal mannequin that can be exposed to chemical challenges is under development [371]. Other mannequins that can be used to apply the FPC model are being developed [372,373].

Comfort. Manufacturability has also been modeled in a desire to streamline the construction prototyping process [374], and this model may also assist in comfort design. Models of fit are also being developed; powerful computational tools combined with pressure measurements have been used to best fit the complex geometries of the human face with a respirator seal to achieve comfort as well as protection [375].

4.7.4 Toxicity Modeling

The ultimate determinant of protection is how much agent enters the body in such a manner as to cause effects (i.e., reaching a target organ). Although this area has received considerable attention due to its relevance to an unprotected person, there is scope to develop better laboratory and theoretical models of human physiology that could help to increase our understanding of how hazards interact with and are taken up by the body. As discussed in Chapter 2, the effect of these extremely hazardous substances on human physiology is often poorly understood. Animal models have been used extensively as human surrogates (recent relevant examples being inhalation of spores for a realistic indoor release using a swine model [376] and ricin toxin aerosol using a mouse model [72]), or artificial membranes as skin surrogates (e.g., chemical permeation [377]), but there has been less work, either laboratory or theoretical, taking the models closer to the human.

Respiratory Toxicity. The greatest hazard arises from airborne agents gaining entry via the lungs. It is understood that because vapors have uniform physical characteristics, their behavior is also quite uniform under most conditions. The behavior of aerosol agents is much more complex. However, because of its relevance to the nuclear industry, radiological particulate deposition in the lungs, and concomitant dose delivery to the body, have been modeled extensively and generally incorporated into the ICRP's models in publications 30 and 66 [20,378,379]. The inhalation process and distribution of various-sized particles entering the nose and mouth have been modeled [380].

Dermal Toxicity. A good model to understand human *skin permeation* is the live skin model, in which skin removed during surgery is used in permeation cells [381]; the location of the hazard material within the skin layers can be isolated, the rate of absorption through the skin monitored, and recovery and removal by various washing or decontamination means investigated. The significant disadvantage of this method is the difficulty in obtaining suitable skin samples, which are also likely to be biased in type or origin (often removed in cosmetic surgery). Once realistic artificial or cultured skin models are available [382,383], studies in this area will be advanced considerably.

Alternative approaches to assessing skin permeation exposes humans to relatively nontoxic dermally absorbed materials [384], and skin barrier properties can, in part, be inferred from the behavior of these other chemicals. Local absorption (where the material is retained in the skin and does not enter the bloodstream) is much more difficult to monitor than systemic absorption by these means, and animal models are used more commonly [385]. Overall dosimetric models of dermal uptake of vapor-phase chemical warfare agent [82,83] have also been developed. Deposition of aerosols onto the body has been measured and modeled fairly extensively [386,387]. Dosimetric estimation models have been developed from these models [388–390]. Dosimetry of aerosolized solvents for spray painting has been modeled [359,386].

Ocular Toxicity. There is little information on the vapor toxicity of agents via the eyes, select chemical warfare agents being the exception based on both animal and human exposure; eyes are a highly complex physiological system that may be the target organ or a route of entry into the body, and are extremely challenging to model. Semiempirical modeling of deposition of particulate onto the eye has been described [391].

5 Protective Equipment: Concepts, Components, and Systems

In this chapter we describe the various concepts and styles of equipment that are currently available to protect the routes of entry against CBRN agents, as well as the fundamentals of how they protect.

5.1 TERMINOLOGY

At the outset it should be mentioned that standardization of terminology is not easily achieved, and generally, terminology varies depending on the jurisdiction, and even within a jurisdiction. Standards devoted exclusively to terminology exist, on protective clothing as published by the American Society of Testing and Materials (ASTM) [392] and ISO [393], and for respirators by the Japanese Institute of Standards (JIS) [394] and (under development) by ISO [395]. Europe has a vocabulary standard for personal eye protection [396]. NATO has a general vocabulary standard for CBRN [397]. Title 29, U.S. Code of Federal Regulations, Part 1910.134, also describes much vocabulary related to RPDs.

General vocabulary on textiles and their properties can be found in:

- ISO 3572:1976: Textiles – Weaves – Definitions of General Terms and Basic Weaves
- ISO 8159:1987: Textiles – Morphology of Fibers and Yarns – Vocabulary
- ISO 8160:1987: Textiles – Textured Filament Yarns – Vocabulary
- ISO 9092:1988: Textiles – Nonwovens – Definition
- ASTM D4920-08: Standard Terminology Relating to Conditioning, Chemical, and Thermal Properties

5.2 CONCEPTS OF USE

There are a number of aspects of the concept of use of PPE that result in various different subcategories within each type of item. In particular, these relate to the

Personal Protective Equipment for Chemical, Biological, and Radiological Hazards: Design, Evaluation, and Selection, First Edition. Eva F. Gudgin Dickson.

length of time and the number of times that the equipment will be worn. Equipment may be intended as:

- Single-use (disposable)
- Multiple-use; dispose after contamination
- Reusable after contamination

These differences have a significant impact on the design, capabilities, and subsequent cost of the item:

- Single-use equipment will often be less durable than multiple-use equipment, and if maintaining the integrity of equipment is critical for user safety, the relevance of single-use equipment in this application must be considered carefully.
- Designing for decontaminability is a particularly difficult challenge, and thus the user should be clear on this particular component of their concept of operations when designing or selecting PPE.

Hence, much equipment is intended to be multiple-use but not necessarily decontaminable for the purposes of reuse.

A related consideration is the level of training the user requires:

- Escape or other forms of single-use equipment may be intended for relatively untrained users, having a short intended use time, and offering the user limited operational capability. Such equipment is often not sized but rather "one size fits all," with perhaps some fitting features, and may be single- or multiple-use.
- Users who expect to perform more operationally complex functions of longer duration require training, and their PPE usually has more sophisticated features. Such PPE may be single- or multiple-use and will almost always need fitting and sizing.
- The areas that tend to be traded off when designing within these concepts of use are the following:
  - Efficacy of protection vs. physiological burden
  - Durability and decontaminability vs. cost

In the sections that follow we describe many types of PPE components and systems. Particular design issues are addressed. In some cases, equipment has been designed around particular standards, some of which are discussed in general in this chapter. The various relevant standards for testing and performance are discussed in more detail in later chapters.

5.3 RESPIRATORY PROTECTIVE DEVICES

5.3.1 Protective Concepts

To protect the respiratory tract, as well as associated routes of entry such as mouth and eyes if within the respirator, there are two important requirements: First, the breathing air must be free of hazards; and second, no hazards must be able to penetrate or permeate around or through the protective item.

The breathing air can be rendered free of hazards either by the supplied air or by air purification. **Supplied air systems** provide breathable gas from a supply that is:

- Worn by the wearer;
- Provided through a tether line from a tank or remote location with clean air; or
- Synthesized in situ by chemical reaction.

Air-purifying systems purify ambient air to remove and deactivate toxic materials by the following means:

- Particulates and aerosols are removed by adhesion to the filtration layer.
- Vapors are removed by adsorption and chemisorption onto the adsorbent layer.

For either type of system, toxic materials can be prevented from entering the protective item by providing a leak-free system; for some portions of the item, this may be relatively simple, whereas for others it can be a significant design issue: for example, sealing to the face or preventing in flow through the exhalation valve. Alternatively or additionally, leakage may be prevented by using a positive-pressure concept, in which air flows out through all potential leak paths at a sufficient pressure to prevent backflow. Materials must also be chosen to prevent permeation of chemicals over the intended period of use.

Some particular challenges of applying traditional respiratory concepts for use in CBRN protection lie in the following areas:

- The limited number of existing materials available that can satisfy the requirements for soft components such as facepieces, valves, and hoses, and transparent components such as eyepieces and visors
- The difficulty in maintaining protection and adequate clean air supply despite the high airflows demanded when breathing at exhaustive work rates
- Fitting the population to minimize leakages while not restricting vision or causing discomfort

It should be recognized here, as in all PPE use, that there are significant trade offs involved in selecting among the various concepts and that each has significant limitations that must be managed.

5.3.2 Components of a Respirator

Regardless of which style of respiratory protective device is used, all consist of at least a *respiratory interface*, which is the part of the respirator such as a facepiece or hood that connects to the wearer, and either an *air-purifying element* or an *air/oxygen source* and associated delivery components.

Air-Purifying Respirators (APRs). APRs contain, at a minimum, a respiratory interface (such as a facepiece, hood, or helmet) and an air-purifying element (APE) that may be a removable canister or filter element, or may be part of the facepiece, as in a filtering facepiece respirator. The air is drawn through the APE either by the wearer's inhalation action or with the assistance of a blower unit. The APE works to clean the outside air by various means described in Section 4.2.2. Powered air-purifying respirators (PAPRs) use a blower to provide increased flow and, potentially, overpressure.

Supplied Breathable Gas Devices. These contain, at a minimum, a respiratory interface and a means for supplying uncontaminated breathable gas to the wearer. *Self-contained devices* use pressurized breathable gas in valved cylinders, or generated by a chemical reaction during use, as an integral part of the RPD worn by the user. Although oxygen-generating devices are commonly used in escape applications for certain adverse environments (mines, submarines), they are not commonly used for this application, weight of and heat generated by the oxygen-generating system being concerns. Other supplied breathable gas devices use a gas source remote from the wearer derived from pressurized cylinders or drawn by the breathing action of the wearer from an uncontaminated area.

Typical Components of the Respiratory Interface. All respirators that cover at a minimum the entire face (most relevant to CBRN protection) contain certain types of universal components that assure functionality of the device. Various attachments and design features assure that the respiratory interface fits to the wearer and integrates appropriately with other equipment. The traditional full-face face-sealing respirator has some form of adjustable head harness. Alternatively, a seal to the neck or components to facilitate attachment to a helmet or hood system may be present.

Normal functioning of the wearer is assisted by other components. Eyepieces or visors that allow clear vision and integration with other equipment that may need to be brought up to the eyes are essential, and they may have other functions, such as ballistic protection. Some form of communication interface is necessary to permit clear speech, and in some cases to assist hearing. In most cases, a drinking facility is provided (with an accompanying CBRN hardened reservoir for drinking fluid).

Proper air management is critical to preserve the quality of the air within the respiratory interface, maintaining a directional path for incoming and outgoing air. One-way valves open only during either the inhalation portion of the respiratory cycle (inlet valves), allowing clean air to flow into the lungs, or during the exhalation portion (outlet valves), exhausting the exhaled air out of the RPD. The nose cup is present to minimize the "dead space" of exhaled air that is not exchanged completely on each breath, and to keep moist exhaled air from fogging up eyepieces, which have a clean air sweep across them as well. Inhalation and exhalation valves have some cracking pressure (i.e., they will not open if there is insufficient positive pressure against them). This cracking pressure will potentially be different in overpressure devices compared with negative-pressure devices. Finally, a leak-free means of attaching to the air supply or APE (canister) is required.

FIGURE 5-1 Tight-fitting military-style APR.

Styles of Respiratory Interface. There are two styles of respiratory interface to the wearer's head: tight fitting and loose fitting. **Tight-fitting respiratory interfaces** rely on providing an intact physical barrier between the device and the wearer. Nose cups, facepieces, and hoods that seal to the neck are examples of this style of respiratory interface (see Figure 5-1 for an example of a facepiece style of respirator). Such systems need to form a tight seal to the wearer's skin, usually on the face or occasionally on the neck. The seal itself protects against ingress of contaminated air. Typical APRs fall into this category; they are referred to as negative-pressure devices, as the air pressure within the device will become negative during the inhalation cycle. However, motion, speaking, or dislodging by other equipment can break the seal, in which case leakage may occur. This potential leakage can be reduced by using overpressure concepts such as self-contained breathing apparatus (SCBA) or tight-fitting powered air-purifying respirators (PAPRs).

Loose-fitting respiratory interfaces have no or a minimal seal anywhere to the skin and therefore rely on adequate breathable gas being provided at all times to accomplish an overpressure. This **overpressure** is intended to prevent hazardous substances from leaking into the area covering the important routes of entry: eyes, face, and/or mouth. Certain styles of hoods and helmets fall within this category (see Figure 5-2).

Nonencapsulating loose-fitting respiratory interfaces such as that illustrated in Figure 5-2 do not seal to the wearer's skin, and therefore they can only be used with devices that actively supply breathable gas to the respiratory interface; otherwise, there is nothing to prevent ingress of contaminated air and protection exists only in the presence of substantial overpressure. The amount of air supply required to maintain this overpressure depends on two factors: the breathing rate of the wearer and the resistance to flow out of the interface. As the breathing rate increases, an

FIGURE 5-2 Loose-fitting powered air-purifying respirator: hood style with powered air supply worn on the waist.

adequate air supply must be present on the inhalation portion of the breathing cycle so that there is still more air supplied than is being inhaled, maintaining positive pressure. With regard to design of the respiratory interface, the size of the "gap" between the interface, such as a hood, and the wearer's skin—the air exits through this gap—determines whether any overpressure can be maintained. If movement causes this gap to increase or causes a bellows effect with the interface material, exterior air may be drawn into the breathing region. In either case, for protection to be effective, the system must include a pressure demand valve that increases the flow to maintain the overpressure. For the relatively high protection levels required for CBRN protection, extremely high flow rates may be necessary as a result, and these may be difficult or impossible to provide.

Additionally, **encapsulating** (or **nearly encapsulating**) **systems** exist that may have no specific seal to the wearer but have few leakage routes either [i.e., the wearer is inside a giant plastic bag with a route for breathing air to enter (if necessary) and exit]. When worn with self-contained breathing apparatus (SCBA) inside the system, only an exhalation valve is needed, and overpressure is maintained within the entire system. An example of the general design of a loose-fitting nearly encapsulating

FIGURE 5-3 Nearly encapsulating suit worn with a PAPR: front view, and back view showing the APEs outside the suit. (Reproduced by permission of, and © Respirex International, date unknown.)

system is shown in Figure 5-3. A blower is worn within the suit, with the canisters outside; air is supplied from the blower to the head area. The breathing area may or may not be enclosed in a a nose cup; alternatively a neck dam may separate the breathing area from the body of the suit. Typically, the exhaled air exhausts down into the body of the suit and then through a valve into the atmosphere. Encapsulating concepts are not particularly affected by the wearer's anthropometry, and therefore in principle are to be preferred when trying to fit to the entire population to a very high degree of protection.

Air-Purifying Element Components. The air-purifying element or elements may be part of the facepiece, or a canister attached to the facepiece or a blower unit in the case of a powered air system. Canisters will have a housing consisting of an appropriately impermeable material, an inlet (designed to protect the contents from splash contamination or wetting by water during use, and decontamination), and an outlet that contains the attachment to the respiratory interface.

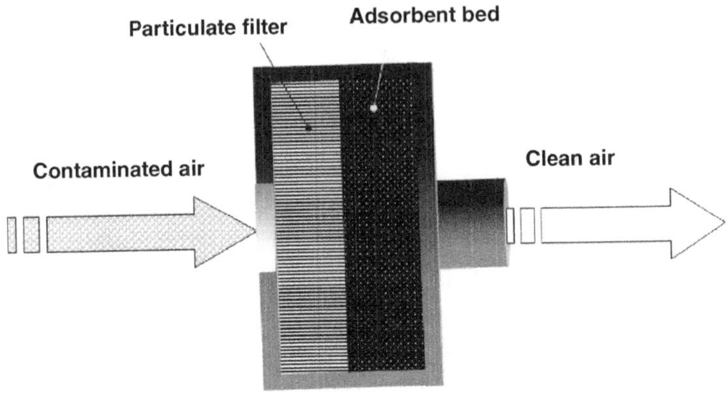

FIGURE 5-4 Schematic of the layout of a canister.

For general-purpose CBRN applications, the APE (Figure 5-4) will first contain a particulate filtration component to remove aerosols, followed by an adsorptive or reactive component to remove vapors. The aerosol-removal component will be in front of the vapor-removal component to minimize degradation of the adsorptive or reactive bed by encounter with aerosols and dusts. An APE containing only the particulate-removal element has potential application for B and R events, where no hazardous vapor is known to be present. The use of APE-containing devices is appropriate only when it is known that the APE has the capacity to remove all the hazards from the air and that the remaining air is suitable to breathe (i.e., has sufficient oxygen content). The hazard-removal capability of an APE is difficult to ascertain thoroughly, particularly for vapors. Its dynamic capacity (the speed with which it can remove the hazard) and theoretically, total capacity must be determined for each vapor hazard of interest, under a variety of realistic conditions of use (airflow rates, temperatures, and relative humidities in particular). This means that vapor-removing APEs have limited duration of use, which will differ for each hazard and condition of use.

Particulate removal components have fewer limitations; their capabilities vary most importantly as a function of the challenge particle size, surface tension (wet/dry), and charge, and not as a function of the specific agent. Thus, the performance can be more easily characterized generically for worst-case conditions. In most cases, capacity is limited by clogging of the filter, which is detected by the user as an increased breathing resistance, meaning that filter change is required. Attachment of the filter to the headpiece or hose is usually via a standardized connector [398,399].

Breathable Gas Supply Components. A breathable gas is supplied from a pressurized source. Possible gas sources include:

- **Gas cylinders:** designed to contain pressurized breathing gases at the highest practical pressures for efficient storage. The breathing gas may be purified air or various oxgen–gas mixtures.

- **Chemical oxygen:** provides breathable gas to the wearer by mean of a chemical reaction and includes a scrubber to remove undesired exhaust gases.
- **Compressor:** provides a continuous source of purified breathable air to the supply chain.

The gas source may be remote from the wearer and connected via an air-line or mounted on the body in the case of gas cylinders or chemical oxygen supply. The supply chain for the device may contain the following components:

- **Shutoff valve:** to allow or prevent the gas from entering the supply chain.
- **Pressure reducer:** reduces the high pressure to lower pressure levels that the wearer can breathe directly or that are used by a demand valve.
- **Demand valve:** there are two types, negative and positive pressure. In a negative-pressure demand valve, inhalation triggers the valve to open and exhalation stops the flow; during inhalation, the pressure may go negative. Positive-pressure demand valves are designed to maintain a pressure slightly above ambient inside the respiratory interface even during inhalation.
- **Continuous flow valve:** provides flow at a comfortable rate and pressure for the user in a continuous flow (rather than demand) mode.
- **Relief valves:** used to prevent overpressurization within the supply chain.
- **Transportation elements** (e.g., hoses and breathing bags): direct the gas along the supply chain.

Table 5-1 summarizes most of the relevant possible styles of respiratory protection for a CBRN application. More detailed discussion of the more common styles of RPD follows.

5.3.3 Self-Contained Breathable Gas Styles

Existing supplied breathable gas systems intended for CBRN use are essentially all self-contained breathing apparatus (SCBA) systems (see Figure 5-5) with a compressed air tank worn by the user using a harness, and connected to pressure regulators and supply lines, which are connected to a facepiece. This is an open-circuit apparatus, as the exhaled air is not recirculated.

The significant advantages of this concept are that:

- Breathing air is guaranteed in environments containing any concentration or type of hazard agent, or low oxygen.
- The supplied air generates an overpressure (at a minimum pressure guaranteed by the pressure-demand valve) that assists in preventing inward leakage.
- Breathing is assisted on inhalation.
- Defogging is facilitated if flow is directed properly within the facepiece.
- The person is not tethered to an air-line and therefore is free to move.

TABLE 5-1 Some Styles of RPD That Are Relevant for CBRN Protection

Design Feature or Capability	Type	Variants	Examples and Comments
Seal style to face or head	Half face	Tight-fitting single-use APR with built-in APE or replaceable canister	Particulate only could have relevance when worn with eye protection (e.g., filtering facepiece); particularly for particulate-only events (B, R)
	Full face	Tight-fitting reusable APR with replaceable canister	Particulate only, vapor only, or both
		Tight-fitting with supplied air	SCBA
	Hood	Tight-fitting single-use APR, with mouthpiece	Often used for escape concepts
		Tight-fitting APR with replaceable canister	PAPR likely
		Loose-fitting APR with replaceable canister	PAPR only
	Nearly totally encapsulating suit	Loose- or tight-fitting with replaceable canister	PAPR only
	Totally encapsulating suit	Loose- or tight-fitting	SCBA only
Manner in which air is supplied	Self-contained air	Demand	Uses less air and balances demand, therefore extending the air supply and optimizing protection
		Constant flow	Maintains overpressure, theoretically improving protection
		Open circuit	Air is exhausted to the outside
		Closed circuit	Air is recirculated by scrubbing undesirable exhaled gases, cooling, and supplementing oxygen

TABLE 5-1 ***(Continued)***

Design Feature or Capability	Type	Variants	Examples and Comments
	Air-line	Open circuit	Air is supplied under pressure from a remote source (e.g., tank, filtered compressor)
	Powered air purifying	Demand	Conserves life of canisters
		Constant flow	More common; maintains overpressure, theoretically improving protection
Particulate removal capability of APE	Nonoily particulate (e.g., biological aerosol)	99.97%	Called N100 per NIOSH
		99%	Called N99 per NIOSH
		95%	Called N95 per NIOSH; also R95 (somewhat resistant to oil)
	Any particulate	>99.995%	Called ULPA in Europe per EN 1882 (generally for collective protection/clean rooms)
		80–99.995%	Called HEPA in Europe per EN 1882
		99.97%	Called P100 or HEPA in the United States
		99.95%	Called P3 in Europe
		99%	Called P99 in United States
		95%	Called P95 in United States
		94%	Called P2 in Europe

FIGURE 5-5 SCBA worn over a protective coverall.

- Cooling will occur due to evaporation, combined with convection from a cool air supply (since air cools on expansion from the high-pressure tank).

An appropriately qualified SCBA system should be suitable for use in almost all CBR hazard environments insofar as protection is concerned, although there are other practical limitations, including the following:

- The air supply is very limited; even the highest-pressure tanks will yield only around 30 minutes of use when the wearer is working heavily.
  - Exiting a hazard area to resupply air can be lengthy and impractical.
  - A tank must be refilled by a licensed operator using clean air, and in a large event clean air may be at a significant distance.
  - Extra prefilled tanks are a significant logistical burden.
  - Enhancing air supply by carrying a spare bottle is an additional burden.
- Air tanks are bulky and weighty, even with the new generation of carbon fiber–reinforced tanks.

- Air supplies require warning systems to alert the wearer if air is low, and such systems are an additional expense and, when audible, restrict stealth operations.
- Requirements for air resupply, maintenance, and training are significant compared with simple air-purifying respirators.

5.3.4 Tethered Supplied Breathable Gas Systems

Breathable gas may be supplied via a tethered line from a plumbed-in air supply (within a building, for example), from a larger purified supply not worn by the user, or from a clean ambient air source remote from the source of contamination. The latter strategy would often be a poor one in a CBRN incident, as no such source may exist within a reasonable distance of the user, and some form of air purification would still be required to ensure that clean air is used. The use of a tethered line in general has limited applicability for roles where freedom of motion is required, although certain user groups might find applicability; for example, as described in Section 3.7.1, medical personnel, coroners, or pathologists who work in a fixed area might find this concept of use appropriate. An SCBA system may also be supplemented by a portable tank system with an air-line to extend the air supply.

5.3.5 Air-Purifying (Negative-Pressure) Respirators

The simplest forms of air-purifying respirators are those that have no form of assisted air supply; in other words, the force to drive the air through the purifying element is provided by the wearer's lungs. They are referred to as negative-pressure respirators because inhalation causes the interior of the respirator to be at lower pressure than the surrounding air, which will tend to force leakage in through any place where seals are not tight. Despite this disadvantage compared with overpressure respirators, they are simpler, with less logistical burden, and therefore remain a good choice where their use is permitted.

Commercially, a variety of styles of such respirators are available. They may be half-face, covering the mouth and nose only (Figure 5-6), or full face, covering the entire face, including the eyes. In terms of purification styles, they may be designed to remove only a particular subset of hazards. Particulate respirators may be rated to remove oily aerosols and/or dry aerosols, and are often constructed as filtering facepiece respirators (Figure 5-7). Gas and vapor respirators may remove only particular chemicals or classes of chemicals, such as organic vapor or acid vapor.

From the point of view of CBRN protection, as noted previously, most appropriate styles will cover the entire face at a minimum, although it is noteworthy that the use of particulate filtering facepiece devices for protection against contagious outbreak events is considered to be the norm [400,401], despite their demonstrably low protection capabilities against submicrometer particles [402,403] and generally poor fitting characteristics. It is noteworthy, of course, that surgical masks are also used routinely despite not being respiratory protective devices at all [400,404].

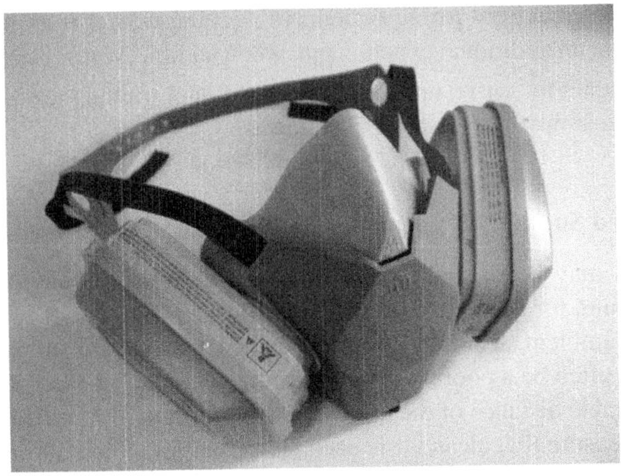

FIGURE 5-6 Half-face respirator.

There are various types of air purification capability in routine use:

- For particulate filtering only
- For vapor air purifying only
- A combination of air purifying and filtering

For CBRN purposes, vapor air purifying could only have applicability far from a chemical release event. Particulate filtering alone would have applicability to all biological and almost all radiological agents, while a combination of air purifying and

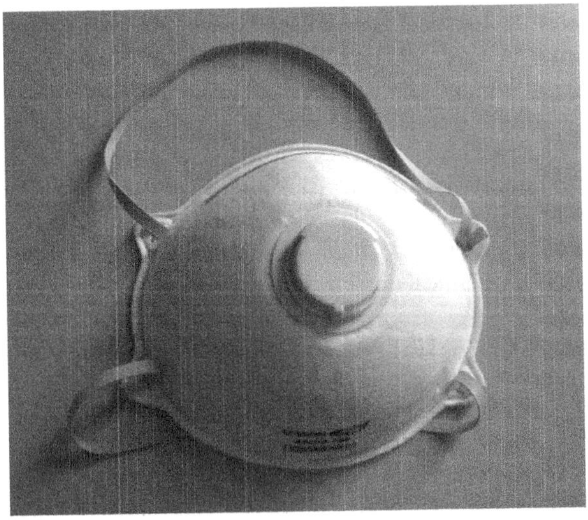

FIGURE 5-7 Filtering facepiece respirator with an exhalation valve.

filtering is appropriate for most chemical or unknown events where some combination of aerosol and vapor might be present. It should be reiterated here that there are CBRN environments in which *no* APR concept is suitable for use, and only an air or breathable gas supply is appropriate. These environments where APR use is inappropriate include:

- Oxygen-deficient environments
- Extremely high concentration environments (e.g., a release within a confined space or a large release outdoors)
- Vapor environments where the canister does not, or is not known to, remove the chemical
- Unknown environments (where the agent released is sufficiently uncharacterized in either nature or concentration that the capability of the system to protect against it cannot be assessed)

5.3.6 Powered Air-Purifying Respirators

Positive-pressure **powered air-purifying respirators** are generally supposed to be a significant improvement over negative-pressure APRs because of the following features:

- A blower, combined with an appropriate exhalation/exhaust valve combination in the facepiece, ensures that positive pressure is supplied to the facepiece, nominally forcing face seal leakage in an outward direction.
  - The exhalation/exhaust valve must have enough resistance to flow that it maintains overpressure within the facepiece.
  - The exhalation resistance can be reduced through the use of a pressure-compensating valve.
- Continual airflow provides cooling (via convection if there is a temperature gradient and via evaporation if there is a water concentration gradient) and also provides defogging.
- Air is not limited as it is in an SCBA tank.
- The positive pressure reduces the work of breathing by assisting in inhalation.

An example of a PAPR with a tight-fitting interface is shown in Figure 5-8. PAPRs that do not seal to the face (including loose-fitting PAPRs) also confer the advantage that:

- Facial hair, eyeglasses, and unusual facial features that might compromise a seal can, in theory, be accommodated.
- Fit testing and sizing are usually not required.

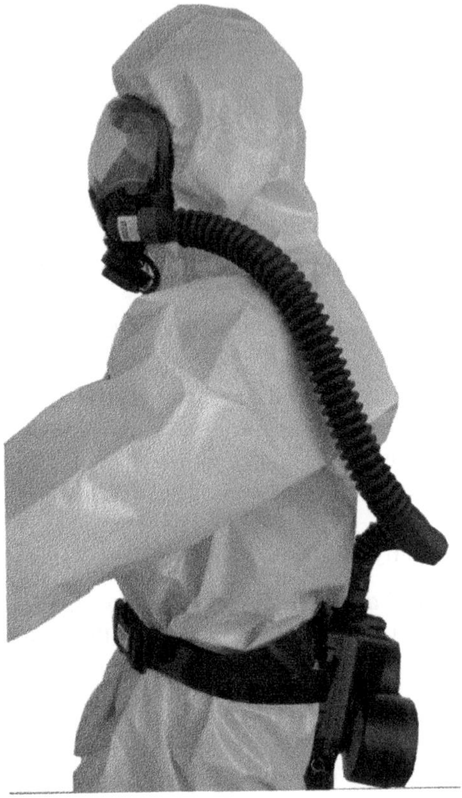

FIGURE 5-8 Powered air-purifying respirator with tight-fitting facepiece worn with a disposable suit; a blower with canisters is worn on the waist, with the air supply hose leading to the facepiece.

It must be stated that, in general, not all of these features necessarily result in overall improvements for the user. Some of the potential disadvantages or limitations of the PAPR system include:

- Continuous high flow through the canisters means that their lifetime is necessarily limited; the desired longer potential duration of use of the RPD due to physiological benefits thus requires a stock of canisters at the point of use.
- Canisters must be larger, more efficient, or doubled up to maintain required protection levels and durations, which leads to greater weight and bulk, particularly when including the blower.
- Battery life is often a limiting feature, and few systems have any end-of-service-life indicator; when batteries begin to die, blower power and protection fall off dramatically.

- Rechargeable batteries cannot guarantee a sufficient output after several rechargings; therefore, nonrechargeable batteries must be kept stocked and fresh at the point of use.
- A requirement may exist for resistance to flow on exhalation that may be higher than that in an APR, to maintain overpressure within the facepiece.
- Blower and hose construction are often less durable because these parts are not required to withstand high-pressure use, unlike SCBAs.
- The fact that PAPR blowers deliver high volumes at low pressures means that their capability to maintain overpressure within the facepiece is potentially not as robust as for an SCBA, which can provide substantial flow at a higher minimum overpressure.
- Loose-fitting PAPRs are particularly prone to overbreathing.

The latter two issues may be mitigated by wearing the PAPR within a nearly encapsulating suit, as described previously.

5.3.7 Emerging Concepts and Issues

Combined SCBA/PAPR respirators are now available (which generally can also be worn as an APR). The main advantage of this style is the extended wear times of the PAPR combined with an available clean air supply if agent concentrations become high, the canister cannot remove the hazard, or oxygen is depleted. Although this approach is attractive for organizations that have all three types of respirators on hand and would like to minimize the associated cost and logistics, the possible disadvantages of this concept should not be overlooked, as there is always a price to be paid:

- Increased weight and bulk will result from carrying both a blower and an air tank.
- Design compromises may result from combining both functionalities; for example:
  - If both hoses are attached separately to either side of the facepiece, the field of view will be affected and snagging or interference with the hoses is more likely, the facepiece is less likely to fit well beneath a visor, and the weight of the two sets of hoses may be more likely to cause dislodging of the facepiece.
  - If the two devices are placed in series (blower attached to the facepiece and compressed air tank feeding through the blower), the demand valve configuration may not function reliably.
  - It may not be possible for the wearer to switch back and forth freely between modes, for example, turning on and off an air tank worn on the back may be difficult for wearers by themselves, and for safety reasons the self-contained air is designed to override the PAPR so that in this configuration the PAPR would be usable only after the air tank was exhausted.

- It may not be possible to know when to switch safely between modes, so the PAPR functionality has limited applicability. Note that self-switching devices based on the detection of contaminants or oxygen depletion have been proposed [405]; it is, however, as difficult to imagine such a device assessing all possible situations successfully as it is to imagine that the user will be able to assess them successfully using other available devices.

It is possible for respirators to be **closed-circuit** (i.e., the supplied air is re-breathed) by using a chemically reactive system that resupplies oxygen and scrubs out excess carbon dioxide. Their particular advantage is that they can provide breathable air for a significantly longer duration (several hours). Although such devices have merit and are used for mine rescue, for example, no such devices yet exist that have been demonstrated as appropriate for CBRN use.* Additional disadvantages include associated weight and bulk as well as heat production from the chemical reactions.

Combined soldier headwear systems such as concepts proposed in the Soldier Integrated Headwear System technology demonstration project [304] may better integrate the many additional functionalities required in the head region. The various NATO target capabilities that were combined as foci in the concepts included lethality; mobility; survivability (CBRN and ballistic); sustainability; and command, control, communications, and intelligence. Two next-generation general design types that integrated functionalities differently were considered: a modular system that included an integrated removable respirator and a permanently encapsulated helmet with respiratory protection. An add-on system that used a conventional helmet and respirator was also considered.

Bioreactive particulate filtering devices that include a biocidal layer releasing iodine (described in Section 4.5.7) have been marketed.

Face-forming or adhesive respirators are under consideration that will remove much of the requirement for individual fitting and sizing [406]; superadhesive and shape memory materials may have applications here.

5.4 DERMAL PROTECTIVE EQUIPMENT (CLOTHING)

ISO [407] and European [408] standards describe general requirements for protective clothing. Many of the components described below have been developed to some ensemble standard, whether the component provides dermal protection only or is part of a full system including a respirator, and the applicability of these systems and their selection are discussed further in Chapter 7.

* NIOSH has asserted that positive-pressure closed-circuit self-contained breathing apparatus that use pure oxygen breathing gas should be limited to mines and mining atmospheres that do not involve exposure to open flames or high radiant heat, until they are proven safe otherwise.

5.4.1 Components

The types of items that can comprise dermal protective equipment (DPE) include jackets, pants, hoods, socks, undergarments, overgarments and coveralls, encapsulating suits, gloves, glove liners, and boots. How these components fit and work together with each other and with a respirator is as important in determining the effectiveness of the system as are their individual properties. Material choices range from inexpensive disposable options to highly chemical impermeable laminates or polymers to air-permeable active carbon material systems. Various peripheral items are discussed in more detail here, with the remaining clothing items included in Section 5.4.2.

Gloves are worn to protect the hands and wrists and, in general, can be expected to be exposed to higher levels of surface contamination than much of the remainder of the body, as well as requiring higher durability characteristics. Resistance to chemical permeation should be particularly high on the fingertips and palms for those persons who could come into contact with liquid chemicals. Increased chemical resistance is often provided by polymeric laminates and is counter to the requirements for tactility. Grip can be improved by adding texture; this is particularly relevant when using decontaminating solutions, which are often quite slippery. Allowing for good finger motion may require building in some extra features that ease motions; it is possible that the backs of the gloves could be constructed of different lighter or more permeable materials. An example of a military CBRN polymeric glove that has finger grips, accordion-like folds at the joints, and a cotton liner to absorb sweat is illustrated in Figure 5-9. The glove is also ambidextrous, which has a number of advantages in

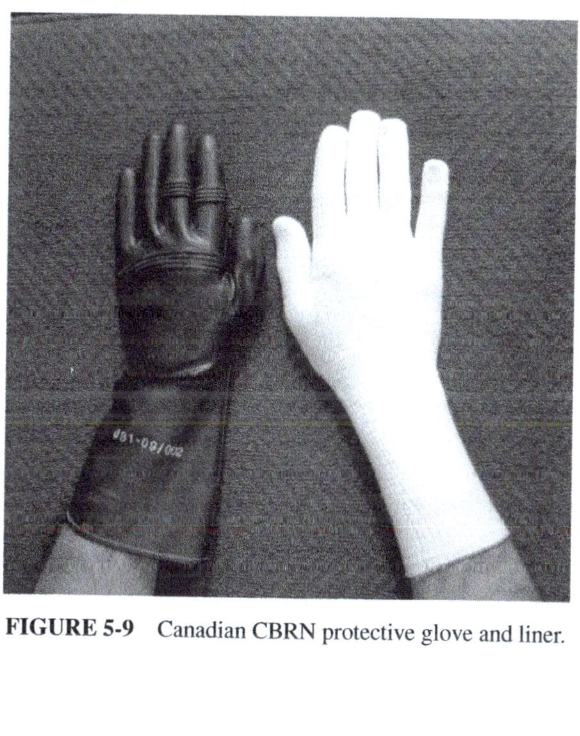

FIGURE 5-9 Canadian CBRN protective glove and liner.

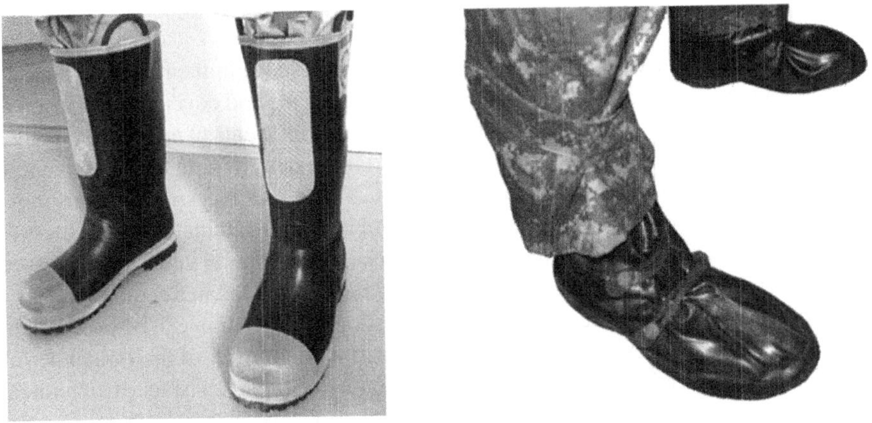

FIGURE 5-10 Hazmat-style overboots (left) and military overboots (right).

the areas of cost, logistics, and ease of use for the wearer. In a different approach, W. L. Gore has designed the Chempak Ultra Barrier Glove System, which includes a protective, permeation-resistant fluoropolymer liner worn inside a more fabric-like functional outer glove.

Similarly, **boots** worn alone or as or overboots protect the feet; the soles should be particularly resistant to permeation, and durability is an equally important issue. The traditional polymer boot or overboot is the most common approach (Figure 5-10); however, the standard-wear boot (e.g., work boot or combat boot) can also be designed with a protective membrane barrier included, or the protection provided in, or enhanced by, an interior layer such as a separate sock or integrated bootie.

Items of equipment not sold as CBRN DPE may also have a protective role to play; for example, ballistic vests, helmets, and visors can provide substantial CBRN protection by acting as an additional barrier layer.

5.4.2 Ensembles

For many years, the most prevalent protective ensembles used for CBR(N) protection fell into three general categories that are still available:

- U.S. Environmental Protection Agency (EPA) categories
- Single-use coveralls
- Military-style active carbon systems

These are described first, with more modern configurations following.

EPA Categories. The first category of system is the industrial spill response type, typically described using U.S. EPA designations of level A, B, or C; see Table 5-2. These reusable and decontaminable systems are intended to protect response crews

TABLE 5-2 EPA Ensemble-Level Descriptions

EPA Level	RPD	DPE	Intent of Use
A	SCBA	Fully encapsulating	Most hazardous environments
B (see Figure 5-11)	SCBA	Full-body splash protective	Significant respiratory hazard; dermal hazard only from liquid contact
C	APR	Full-body splash protective, or head, hands, and feet only, depending on hazard analysis	Respiratory hazard that can be removed by APR (e.g., below IDLH); dermal hazard only from liquid contact
D	None	Appropriate for general work environment	Minimal hazard

from contact with chemicals such as toxic and corrosive liquids (in the case of levels B and C) or to protect from all airborne substances when concentrations reach very high levels (level A). They have no associated performance standards.

The level A style of totally encapsulating protection, which can be used at even the high concentrations that can be obtained in confined spaces, is still the gold standard for CBR protection, provided that it is well designed from highly liquid-impermeable materials (as specified in NFPA 1991 [409]). Levels B and C have the respirator worn at least partly outside the DPE (Figure 5-11). There are no particular descriptors that ensure performance against airborne substances; the materials must be impermeable to liquids, and some degree of splash protection must be provided. Levels B and C suits are not necessarily well integrated with the peripheral items, as they are not evaluated as a system. In summary, these systems were developed primarily for open-air response to an accidental release of large volumes of liquid or gaseous chemicals, and because they have no performance specifications and need not have particularly effective closures, they are not necessarily appropriate for CBR protection unless they have demonstrated their performance as a system.

Any design that depends on an impermeable material combined with good closures to provide vapor and aerosol protection is completely dependent on the total integrity of the system. A single material or closure failure results in drastically reduced protection. Helping to compensate for this issue, extra dermal protection is provided in all level A systems, due to the fact that the clean exhaust air from the SCBA (contained entirely within the suit) fills the suit, and an overpressure valve keeps the suit under positive pressure. This same type of protection can be designed into level B and C systems by using overpressure and/or airflow through the suit to protect against ingress or flush it of agents; this approach may also assist in evaporative cooling. A separate positive-pressure air supply can also be used for the suit only, although this adds bulk and complexity.

Single-use (disposable) **protective coveralls** are used with whatever peripheral items and respirator are deemed expedient. This style is relevant when the dermal hazards are minimal (as in the case of biological materials) or skin decontamination

FIGURE 5-11 EPA level B ensemble.

is relatively efficacious even after hours of exposure (as in the case of radiological materials). Hence, a disposable coverall worn over normal clothing is often felt to provide sufficient protection and is there primarily to reduce the decontamination burden on the clothing as well as the skin. Durability is quite limited, with the majority of the protection provided by the peripheral items.

Both of the categories of equipment above imply a very high thermal burden because of the largely air-impermeable polymeric materials used. Working time within these systems is comfortably perhaps an hour at moderate activity levels, although it can be extended when necessary or in cooler environments. Duration of use may be limited by durability and/or working time.

Cold-War Military Systems. Many very early military systems were similar to the EPA-style level C systems. However, during the cold war, the military felt that they would operate around the clock in a contaminated environment. Therefore, the level of physiological burden that these systems provided was unacceptable, and an alternative approach was developed. The active-carbon-containing overgarments of the cold-war era were air permeable and worn with an air-purifying respirator, polymeric gloves, and overboots [324]. They were meant to be worn over combat clothing in cold climates, and more recently have often been worn as stand-alone systems (replacing the uniform) in warm climates. These ensembles were designed first and foremost for protection against militarized CW agents. This means that dermal protection was focused primarily on keeping highly dermally toxic liquid and vapor chemicals of moderate to low volatility away from the skin, which could be achieved through the use of adsorbent active carbon combined with a liquid repellent outer layer. The DPE provides some protection against aerosolized materials in general because all cloths are reasonably good filters of particles 3 μm and above. Active carbon adsorbents provide much less protection against high-volatility chemicals (many TICs), although these are rarely dermally active, and there are a few such compounds that have been identified as being of significant concern.

The air-permeable nature of an active carbon material, which yields considerable thermal burden advantage, means that certain types of vapors and aerosols can penetrate freely at sufficiently high wind speeds. Reducing the air permeability of the material system, or the size of the air gap between the layers, reduces penetration under these conditions. Materials that filter aerosols can provide extra protection if incorporated into the material system; active carbon fibers afford some increased aerosol protection relative to other forms of carbon, and extra fibrous filtration layers are also possible. Electrospun fibrous layers may offer a particular advantage here, as they are claimed to provide better filtration capability at higher air permeability.

Protection against nuclear events (N) in such systems was focused on ensuring that the materials were resistant to nuclear flash.

In some cases, these systems were adapted for first-responder use by making the materials more air permeable and less liquid protective. This was not based on any effective analysis of requirements but, rather, was primarily a drive toward comfort. Some of the more recently designed systems that can provide improved protection or functionality over those used historically are described below.

Protective Undergarments. As discussed previously, moving the barrier or adsorbent layer closer to the skin affords improved protection, keeping penetration low, with the small air gap effectively scavenged by the adsorbent. Protective undergarments or next-to-skin designs (Figure 5-12) are meant to be worn under other equipment. Outer clothing could consist of turnout gear, street clothing, or uniforms; undergarments are also a good choice beneath specialized equipment such as bomb disposal gear. Undergarment components can include hoods, jerseys and pants, socks, and glove liners. Typically, such garments contain active carbon and are stretchy for comfort and snug fit. The active carbon provides protection against the most dermally active chemical agents, while the outer layer can provide additional protection by acting as a barrier, reducing air permeability, or having a liquid-repellent treatment [309,310]. Overlap between components (e.g., ankle to sock) can be substantial without needing special closures; there is little movement and no significant gaps that would cause

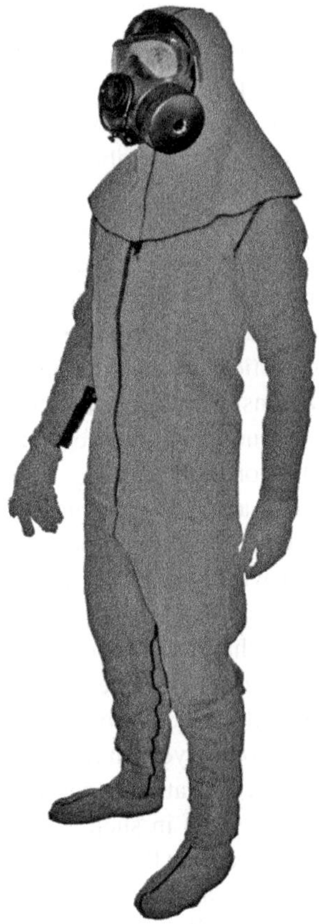

FIGURE 5-12 Protective undergarment system, including gloves, socks, and hood.

leakage; thus, protection can be excellent. However, thermal burden can be significant because of the multiple layers when the overgarments are included.

Recently, the first standards specifically designed for domestic response to terrorism involving CBR agents were published in the United States. **NFPA 1994 class 2 and class 3 systems** [410] have been designed and approved for use in CBR response. It is intended that class 2 ensembles are for incidents involving vapor or liquid hazards where the concentrations require the use of SCBA (above IDLH), while Class 3 ensembles are for equivalent but below-IDLH conditions. NFPA class 2 and 3 ensembles are illustrated in Figure 5-13.

Construction of a class 2 suit requires reasonably vapor-tight closures and impermeable materials to pass the requirements; this implies a certain degree of aerosol protection as well. A class 3 system has somewhat lower protection requirements, so the closures can be less tight. The materials must display a certain amount of evaporative heat loss through using moisture vapor–permeable materials, so the systems should yield a lower physiological burden. Each system is approved in its entirety

FIGURE 5-13 NFPA 1994 class 2 ensemble (left) and class 3 ensemble (right). (Photographed by Rick Bloomingdale, reproduced by permission of, and © Starfield-Lion, 2011.)

for dermal protection, including the respirator. Numerous approved class 2 and 3 systems are now available.

Other types of DPE have been investigated for their use as expedient protection for emergency escape. For example, **firefighter turnout gear** worn alone can act as a barrier to an agent; the materials are water repellent, bulky, and contain selectively moisture vapor–permeable membrane barrier layers. Gloves are usually durable with a synthetic or leather outer, and boots are chemically resistant. The systems are designed to be reasonably splash-proof overall, with closures to prevent water ingress. Therefore, there is protection intrinsic to such a system when worn with self-contained breathing apparatus in the standard response configuration. That level of protection appropriate for escape and rescue in a CWA release was demonstrated for these systems by the U.S. Domestic Preparedness Program in the late 1990s [411]. It was deemed that incorporation of protection directly into turnout gear could lead to improved response capability in the case of a known terrorism event, leading to the **NFPA 1971 turnout gear (CBRN option)** style of equipment [412]. The equipment is truly dual-functional; it is meant to be worn as turnout gear for structural response, but with an additional demonstrated capability to keep out CBR agents through a combination of an appropriately protective barrier layer in the material system and protective features designed to be deployed to keep out airborne substances. Commercial systems are available.

5.4.3 Emerging Concepts

Several areas are of particular interest in improving the performance of dermal protective systems.

Military Systems. Improving human performance by reducing the physiological burden is a major goal. The notion of lighter-weight protective uniforms [315,413] for the military, designed to be worn in place of more burdensome overgarments or stand-alone DPE when the perceived threat or hazard is lower, is an attractive one. A protective uniform concept would provide protection somewhere between that of a nonprotective uniform and that of a full-up ensemble that could provide protection for days against a wide variety of hazards. It has the properties of a uniform when worn in the open state, but can be closed up for more complete protection. A lower burden can be achieved, for example, by reducing the weight of the materials, improving their hand and comfort properties, or increasing the air permeability; it is important that this lower burden be achieved in both the open (daily wear) and closed (protective) states. The concept of use for such a uniform would be more limited, meaning that some hazards might need to be avoided, or protection could be of shorter duration, with the intent to use the uniform as an escape option. Nevertheless, the built-in "just in time" protection concept means that decisions about when to don the more burdensome overgarment or stand-alone concepts do not need to be made too early—a person has some protection all the time and remains in a relatively positive physiological state throughout (both prior to, and after closing up for protection). Some of the challenges with this type of design include the durability required for it to be worn as a uniform while maintaining protective capabilities, and retaining the correct hand properties of the materials, to keep them as uniform-like as possible. A number of militaries are updating their protective clothing systems to modernize and achieve a lower burden, including those of Canada, the UK, the United States, and Germany (Figure 5-14).

It was noted previously that class 3 NFPA 1994 systems use moisture vapor–permeable materials, and some military concepts also include them. If total heat loss values can be made sufficiently high, these concepts have the potential to be competitive for comfort with more highly protective air-permeable systems while providing broader-spectrum protection. An example of a military-style protective system incorporating MVP materials is shown in Figure 5-15.

NFPA 1951 includes the **CBRN technical rescue protective ensemble** concept for use by emergency services personnel assigned to or involved in search, rescue, treatment, recovery, decontamination, site stabilization, extrication, and similar operations at CBRN incidents; at the time of publication, no certified ensembles yet

FIGURE 5-14 Two examples of military systems designed to be worn as stand-alone uniforms. Left: The IdZ system, under development for the German Federal Armed Forces. (Reproduced by permission of, and © Blücher Gmbh, 2011). Right: The CB^{plus} uniform concept (worn in the closed state), developed by Defence Research and Development Canada.

exist to meet this standard. Similarly, ensembles specifically certified for use in particulate CBRN hazard incidents do not yet exist. Both NFPA ([410], **1994 class 4 ensembles**) and Canadian first responder PPE standards ([3], **Z1610 C4 ensembles**) suggest the appropriateness of such an ensemble for response to a biological or radiological event. ISO also specifies requirements for particulate protective clothing (**ISO type 5** [414]); however, the testing is not based on a fully integrated system (i.e., respirator, gloves, and boots can be substituted) and is also based on inward leakage and not deposition, and therefore this set of requirements provides undefined particulate protection levels.

The Canadian Z1610 standard [3] additionally includes a wide variety of possible ensemble configurations designed for specific capabilities within a CBRN response incident, described in more detail in Section 7.3.1. Various potential respirator styles are matched with suitable dermal protection capabilities as entire systems, suitable

FIGURE 5-15 Protective system using an MVP barrier. (Photographed by Rick Bloomingdale, reproduced by permission of, and © Starfield-Lion, 2011.)

for the possible roles and locations within an incident. Approval of systems against the standard is under way.

Similarly, the U.S. National Institute of Justice has published a standard for **law enforcement ensembles** [415], described in more detail in Section 7.3.1. Respiratory protection is not covered, nor is protection from ballistic threats, explosives, or ionizing radiation. The combination of respiratory protection style and clothing yields various configurations (LERL 1 through 4). As of this date, no systems are approved under this standard.

PPE for **radiation protection** useful for generic CBRN applications is still in its infancy. One system approved under the NFPA 1994 class 2 category (Figure 5-16) includes the partially radiopaque material Demron, described in Section 4.5.3. Demron is the only impermeable CBRN fabric that permits heat exchange, enabling the wearer to be cooled externally without having to penetrate the suit. The same materials have been incorporated into bomb suits and personal protective armour [416].

FIGURE 5-16 Radiation shielding suit and shield. (Reproduced by permission of, and © Radiation Shield Technologies, date unknown.)

By combining different functionalities, the burden can potentially be lowered, particularly for those who must have multiple types of protection. The **advanced NBC protection system for aircrew** combines protection against whole-body immersion and CBRN agents, as well as active cooling and CBRN protection [301–303].

5.5 FUTURE CONCEPTS TO IMPROVE PERFORMANCE IN USE

In this section we suggest some generic ways in which performance in use can be improved in the future, both through design of the PPE itself and by developing better concepts of use or indicators. Some of these have already been mentioned in earlier emerging concepts sections.

Fitting. Equipment that does not fit properly will neither protect properly nor will it be suitably functional. Therefore, designing equipment that is one size fits all with some adjustability, that is easy to fit, or that can be preadjusted and then fixed will reduce the logistics associated with this problem. Custom-fitted and custom-produced

PPE is also a possibility with current scanning and rapid three-dimensional printing capabilities. Real-time sensors that detect failures in fit are a possible solution, for example, integral pressure sensors in closures that detect when a seal pressure falls below a certain value. Approaches to rapid determination of air-purifying respirator fit in the field are under development [417,418].

Monitoring Protective Capacity. When equipment is worn or stored too long, it may fail to protect, due to either exceeding its normal capacity or to reduced capacity due to environmental exposures. Hence end-of-service-life indicators [419] or residual-life indicators [420,421] for protective performance would be of significant benefit. Since most protective materials have different protective capacities for different agents, a one-size-fits-all approach is conceptually and technically quite challenging.

Donning and doffing. Equipment may protect well when it is worn, but take too long to put on properly, be put on too late, removed before completely decontaminated, or difficult to remove without contamination transfer. Smart protective equipment that is worn all the time and closed up automatically based on a remote sensor or environmental trigger would have benefits in dealing with the donning issues. Ease of removal through the use of quick releases and reduction of the likelihood of contact with contamination during removal through peelable outer layers should be considered. Localized exposure/contamination indicators on the exterior of the PPE would make the decontamination and removal procedure more foolproof: for example, a color-change material that indicated chemical agent contamination of surfaces. Self-decontaminating materials would also reduce the likelihood of contamination during and after removal.

Multiple Hazards and Functions and System Integration. CBRN equipment is expensive and annoying to implement and wear. It is also highly dependent on having system integrity in order to perform adequately. Combining protection against other hazards with CBRN protection has the advantage that the user is more likely to wear the equipment if its primary function is something other than CBRN protection; in addition, when protection against other hazards, such as ballistics, moisture, or fire lies outside the CBRN layer, its integrity is more likely to be preserved in use. Integrating CBRN protection into the next-generation soldier system by integrating with the helmet and other protective layers as well as planning for reduced burden on the wearer by dual-use items such as CBRN protective footwear, or clothing requiring no extra outer layer, will increase the likelihood that the PPE will be used when it is needed and that it will function as desired. Increasing the breadth of the CBR hazards protected against by the PPE—for example by using an "all-hazards" air-purifying system that allowes only selected gases such as oxygen and nitrogen through or by using long-life lightweight closed-circuit breathing apparatus—would be of huge benefit.

6 Performance Evaluation and Standard Test Methods

In this chapter we outline how to develop performance evaluation methods, including how and when PPE should be evaluated. Relevant standard test methods are reviewed.

Evaluation of protective systems has evolved rapidly over the past 10 to 15 years. Previously, many of the properties of protective systems were evaluated largely at the material level under laboratory conditions. A much better understanding now exists of how to measure the performance of protective equipment in a realistic manner. It is becoming routine to include evaluations that characterize the performance of an entire protective ensemble, under conditions resembling field use; material-level tests are also evolving to permit an understanding of how they relate to behavior in systems, with modeling helping to bridge the gap.

With an understanding of the requirements as discussed in Chapter 3, and of how materials, designs, and people perform under different conditions, it is possible to develop appropriate conditions for evaluation that represent performance under typical as well as extreme conditions. The military has had standards for the protective performance of CBRN respirators and clothing for decades, focused primarily on general-purpose use by infantry and aircrew in cold-war scenarios of use; performance standards for specialist users or more contemporary operating environments are now under discussion. Standards for civilian use are relatively recent and are still being developed. At this stage in our understanding, it is not wise to assume that any protective system or method is sufficiently well understood that it would not benefit from constant scrutiny and potential reevaluation using new methods and the approaches for defining requirements outlined in Chapter 3.

6.1 TEST SELECTION AS DETERMINED BY LIFE-CYCLE PHASE

Phases in the Life Cycle. Testing that is performed on a particular PPE item or system evolves and changes over the life cycle of a material, as different sets of performance issues are identified. Five major phases in the life cycle of the design of an item of PPE are relevant: research and development, qualification, preproduction, production, and storage and deployment.

Personal Protective Equipment for Chemical, Biological, and Radiological Hazards: Design, Evaluation, and Selection, First Edition. Eva F. Gudgin Dickson.

Research and Development (R&D). This phase, which evaluates the suitability of particular protective concepts for consideration as systems to be qualified, is performed and led by the group developing the equipment. Materials and systems are developed, evaluated, and compared with existing systems. Often, a particular novel concept has been introduced, and test methods may need to be modified or developed to accommodate new requirements. A subset of critical performance tests will usually be identified that the new concept must pass in order to be screened through the R&D phase. The tests at this stage need not necessarily be completely realistic but, rather, should generate interpretable results that permit comparison among concepts. Certain types of performance parameters are generally important enough that they will be investigated here: examples might be protection against one or two selected airborne, liquid, and/or aerosol challenges, durability, or other properties critical to the particular performance challenge that is being addressed. Whole-system testing against at least one type of airborne challenge should be performed to evaluate prototype designs. To perform such testing, both the ability of a material to be incorporated or manufactured effectively into a protective system and the implementation of any appropriate system design elements (particularly closures) will be evaluated.

Selection and Qualification. This stage, whose responsibility lies with the purchaser, requires the most comprehensive set of evaluations possible. Every aspect of the performance of the system must be considered, including all worst-case but realistic conditions of use that could limit or degrade performance, including numerous possible issues related to degradation resulting from normal wear and storage life. If an item of equipment meets a given set of standards, and a certification organization for the standard(s) is available, purchasers need only confirm that the standard requirements are appropriate for their application. If standards are unavailable or met voluntarily, the organization requiring the equipment bears the responsibility for ensuring the thorough execution of the entire phase. The cooperation of the manufacturer may be involved, or third-party testing may be performed (usually considered to be the gold standard).

Although the test program may only involve a single protective concept that has been selected from several alternatives and is ready for qualification, it is good practice that multiple-concept items or systems be evaluated at this stage in order to permit selection among them. The materials, styling, sizing, and overall design of an item or an entire system must be evaluated in a multitude of ways, using lab-scale and human testing, some of these evaluations being carried out in the field or under simulated field conditions. If an individual item is being evaluated, it must be considered in combination with every other item that might be used in combination with it, both protective equipment and ancillary equipment that will be used at the same time. If this phase is being performed at the end of a development activity, selection and qualification should be performed with prototype systems of multiple sizes, based on initial stated requirements; alternatively, if commercial systems are being considered, production items may be used.

Hundreds or even thousands of tests may be needed, each with multiple replicates. Such a qualification program must be well designed and managed effectively to yield the best result, with clearly planned decision points regarding how to address potential test failures. Procurement specifications can be prepared and finalized at the end of the activity based on the performance targets achieved, recognizing that it is common that no system will actually meet every performance target specified by the requirements. Decisions regarding whether a particular requirement is merely desirable, as opposed to required, may need to be made at every point in the process, and clear and constant communication among all the interest groups involved in the qualification activity is essential.

Preproduction. This phase, which is the transition to mass production, is executed by the organization(s) developing and/or manufacturing the equipment. A significant selection of critical evaluations, along with examination of manufacturing integrity, must be performed to confirm the manufacturing capability and consistency, since translation to mass production may lead to a significant number of quality control issues not encountered previously. Production of materials may have been scaled up, manufacturing moved from manual to automated processes, and the number of sizes being manufactured increased, all requiring reexamination of critical performance parameters. Concept packaging and storage conditions must be established at this phase, often using accelerated aging studies in which extreme environmental conditions are applied.

Production Quality Control. Many of the critical performance parameters examined in the qualification and preproduction phase are of relevance here, and a small subset of these will probably continue to be evaluated throughout the production cycle. Requirements for this phase may be set by the purchaser, particularly where a long-term procurement contract is in place. An overall quality check must be performed regularly, some of which will be accomplished by visual inspection. Issues may be identified in various manufacturing steps, such as sewing and seam sealing, packaging, or proper sizing. Performance tests usually focus on protection and system integrity and might consist of, for example, one or more challenging but reproducible material (and sometimes seam) tests and at least one type of system-level integrity test.

Storage and Deployment. Responsibility for execution of this phase may fall on the purchaser, the manufacturer or seller, or a combination of both, depending on the terms of the purchase and local regulations. After manufacture, items must be stored in accordance with the guidance developed in the preproduction phase. Although a shelf life may have been estimated based on accelerated aging studies, performance will still need to be validated by sampling of stored items, preferably from several different storage locations, at designated intervals. Similarly, items intended for multiple wearings should be sampled at intervals to determine whether the estimates of service life developed during qualification and preproduction were correct after actual storage and use.

The tests that must be performed will be selected based on the types of materials and items in question. Material tests may be sufficient to identify issues such as loss of structural integrity, reactivity, or permeation resistance resulting from degradation of materials. Inspection followed by item- or system-level testing might be required to identify or quantify certain issues resulting from stiffening, polymer or material degradation, wear, or shrinkage, particularly where fit issues may contribute to the result.

6.2 ISSUES THAT MAY PREVENT EFFECTIVE EVALUATIONS

PPE for CBRN incidents must be designed to protect against any possible form of hazard agent that might be encountered under many different adverse conditions of use. This is a daunting task insofar as design is concerned, which becomes even more challenging when comprehensive evaluation methods must be established. Before approaching the problem of designing or selecting test methods for PPE systems, it must be understood that although it may seem to be simple enough to tell "whether something protects," this is, in fact, a very complex statement. All test methods are an approximation of reality: The only real test is to put a person in the equipment, expose him or her to the hazard agent under realistic conditions of use, and wait to see how the equipment and the person perform. Given that this is not a likely approach, every test method is designed to try to simulate this performance in some way, and each test result requires some assumption or interpretation to relate it to performance in use. It may take several layers of test methods and/or assumptions to predict one simple performance parameter, with hundreds of such parameters being of interest.

In the next few sections we outline some of the difficulties and pitfalls that may arise during the adoption and use of test methods.

6.2.1 Translating Requirements to Effective Test Methods

The first step in the approach to selecting or designing evaluation methods is to acquire a thorough understanding of all the conditions of use. This should be done following the processes outlined in Chapter 3. For CBRN use, because of the potential extreme nature of the hazards encountered, it is desirable to have a considerable level of certainty about the performance characteristics of the ensemble; although it is not necessarily the case that the ensemble will protect under all conditions, it must be known and planned for if it does not.

One difficulty that may arise in this process is lack of continuity or communication among the various groups involved in the process. For example, the evaluations may be performed by independent groups or agencies that have specialized capabilities for testing but that are not involved in any of the initial design and test selection process. In particular, the details of the requirements development may be confidential for particular types of users because of the sensitive nature of the threat and operational information, or the materials and design details could be proprietary. Under such circumstances, it is difficult for testers to perform an educated analysis of the results obtained, and they are not able to advise designers and users when the methods are inappropriate or the results do not make sense. Therefore, it is important for everyone involved in the evaluation process to understand as much as possible about protective performance fundamentals so that the best tests are designed and selected in consultation.

6.2.2 Scope of Evaluations

It must be recognized that different protective system concepts and materials protect differently against different types of agents; this means that enough agents and test conditions must be selected to be truly representative. Further, there is no single worst-case set of agents and test conditions that will evaluate equally all examples of PPE; some agents may be easily protected against by one type of PPE system but a difficult challenge to another. Hence, the test designer must have some a priori knowledge of the types of PPE and conditions of use that the tests are designed to evaluate, and this may mean that for a new technology, a series of exploratory tests must be performed in order to understand its strengths and weaknesses. No truly representative set of tests can be developed that will always evaluate all PPE systems (including those not yet developed) correctly under all conditions, since the next new concept in protection will be developed *precisely because* it protects differently from any that have been designed previously. This can occasionally prove an obstacle to adoption of a new protective technology, as the existing, accepted test methods and protective standards may not evaluate it appropriately.

The design process may begin with new materials intended to solve particular design deficiencies. These materials are likely to be manufactured initially in small quantities, with the result that only small-scale or bench testing can be performed to attempt to screen candidate materials. However, lack of correlation between bench tests and system tests is not uncommon and can make the screening process unreliable. As a result, it is important to assess whether bench tests correlate with system-level results as often as possible.

6.2.3 Standard Test Methods

It is apparent that standard test methods are a necessity in being able to correlate and compare data obtained at different times or by different organizations. The requirements development process within a given organization or user group can lead to standards for test methods that may remain in place for years, and these methods may migrate from one group to another (e.g., military test methods have often been adopted by civilian standards organizations). The difficulty that can arise is that the process of reasoning that led to selection of a particular standard test format or set of conditions, and in particular the assumptions involved, are often not documented; as a result, these methods may sometimes be used inappropriately after a time until new knowledge and thorough reevaluations reveal the problems and new methods are devised.

6.2.4 Time and Cost

When devising a test method or test program, the issues of time and cost cannot be ignored. Evaluation methods that involve toxic materials are challenging and costly. Evaluations that require the use of full protective systems may take months to years to complete, particularly in PPE development programs, as large amounts of prototype materials are manufactured, designs are translated into systems generally of multiple sizes, and numerous elaborate evaluations are performed. Hence, there will always be a programmatic drive to reduce the cost and complexity of the evaluation program and the test methods used. The most important efficiencies that can be achieved in this area relate back to the necessity to understand to the best extent possible the actual requirements for the system and conditions of use. Using methods that best relate back to these factors and that deliver results as early as possible in the program so that problems are caught quickly obviously has definite benefits in preventing unsuccessful concepts from proceeding too far through the program.

Appropriate screening approaches that can be employed in stages will reduce the specific cost of executing the tests but may significantly increase the overall time needed to complete a project; in addition, screening approaches that are based on less complex or simulated analyses may yield results that are not good predictors of system-level performance. Sometimes a single test or combination of tests can be found to be a relatively good predictor of performance in other tests, which then need only be applied to a smaller subset of concepts to confirm results at the qualification stage. Finally, using a system-level approach to procurement can result in significant cost savings, because there is no risk that items will need to be reselected when they do not work together; generally, support and maintenance costs are reduced by having a single support team.

6.3 SELECTION OF TEST CONDITIONS

6.3.1 General Limits

There are literally tens of thousands of possible CBRN agents (some of them hazardous only in large volumes), each of which in theory could be encountered in various physical forms and in a wide variety of conditions of use. A particular difficulty is to condense all these possible conditions of use into a manageable set of tests and test conditions that reflect all of the worst-case performance characteristics of the PPE ensemble as well as encompassing typical performance conditions. It is apparent that a relatively tiny set of possible combinations of challenge agent, concentration, physical form, and environmental conditions must be selected for evaluation purposes, and the rationale for the process of elimination should be documented carefully for future reference.

Limits on the range of possible variables can first be defined based on specific user requirements, resulting in a potentially different subset of relevant test conditions being chosen for each user group. Threat data may also be used to shorten the list of possible hazards. When selecting test formats and test conditions, it is usual to consider worst-case conditions of use derived from the high-level requirements. As already noted, it may be difficult to predict in advance with new materials and systems what the actual set of worst-case conditions is, particularly when it comes to protective performance. Which particular combination of agent, temperature, relative humidity, wind speed, and so on, is the most difficult to protect against? In addition, it can be difficult to know where to stop when combining possible hazards. Some combinations

TABLE 6-1 Examples of Routine Environmental Exposures That Could Degrade Protection

	Environmental Exposure		
Exposure Sequence	Very High Wind	Water	Environmental Contaminant
Simultaneous with exposure to CBRN hazard	Open, moving vehicles or helicopter rotor wash	Seawater, sweat	Organic compounds such as gasoline, oil, alcohol, cleaners; present as liquids or vapor; smoke; dirt; body fluids
Prior to exposure to CBRN hazard	Not relevant	Seawater, sweat	Organic compounds such as gasoline, oil, alcohol, cleaners; present as liquids or vapor; smoke; dirt; body fluids
Subsequent to exposure to CBRN hazard	Open, moving vehicles or helicopter rotor wash	Decontamination solutions particularly applied with high pressure; heavy rain/high seas	Decontamination solutions, particularly those applied with high pressure

may be possible but sufficiently unlikely, or difficult for a protective system to survive, so they may be excluded from consideration. For example, protective clothing may realistically be able to protect a person against thermal injury from fire or chemical injury, but not both presented at the same time or in rapid sequence. Such issues that fall outside the range of the capabilities of the equipment must be documented instead as possible limitations on use to be recognized and, if necessary, dealt with procedurally.

The likelihood of exposure to multiple conditions either sequentially or simultaneously that could affect protective performance needs to be considered. In Table 6-1 we list some of the factors that can degrade protection provided by many materials when encountered before, during, or after exposure to the CBRN hazard, and these considerations may require certain choices of preconditioning or test conditions. Limits will often need to be placed on the scale and number of possible performance evaluations, based on the availability of the equipment to be evaluated and other practical considerations. Therefore, tests must be prioritized and the number of replicates and control samples required must be considered carefully.

6.3.2 Test Scale and Test Platform

Protection and overall performance rely both on the characteristics of the material(s) from which the equipment is made and on the design and manufacture of the equipment. As a result, the only definitive tests for both the properties and the approval of protective clothing systems should be tests at the full system level (Figure 6-1).

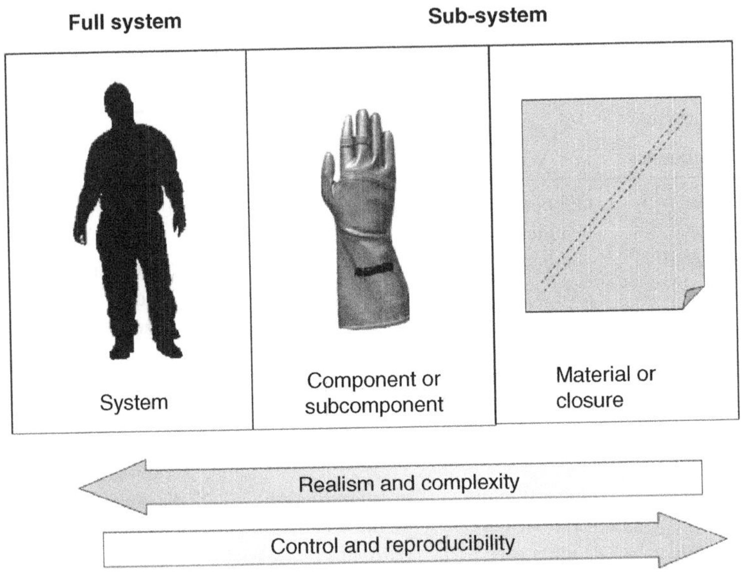

FIGURE 6-1 Test scales.

Nevertheless, such tests are elaborate and time consuming. In particular, because of the hazards involved in such large-scale testing, it is very rare for a very hazardous agent test to be performed at the system or even the component level, and when it is, a short list of agents only will be evaluated.

It is important to identify when substitution of a simpler test is possible or preferable, and whether such a test requires further validation before it can be useful. Such simpler tests are referred to here as subsystem tests because they generally involve the use of a smaller piece of the protective system: for example, material or "swatch" tests, or tests on a smaller component or subcomponent. Such an approach often makes control of variables easier, although the disadvantage is that achieving realistic conditions may be difficult. Subsystem tests may use a variety of test platforms on which the protective concept is mounted. Human testing is preferred when it can be performed safely, reliably, and with adequate population distribution. The most realistic alternative test platforms are the most human-like in size and form, taking into account, for a particular test, those particular human properties that may be important, such as skin surface characteristics or sweating.

In general, then, while a full system test on a human being is theoretically the test of choice, for it to be used in practice, a validated test on the system level must be available for the performance property under investigation, and system testing must have a significant added value over subsystem testing. In most cases, this value is highly evident in the qualification phase, as it is rare that system-level performance can be fully predicted by subsystem tests. In other phases, subsystem tests can be considered for substitution.

Material-Level Tests. Swatch- or material-level tests generally involve the use of a piece of material a few centimeters in diameter. Materials are tested as systems when multiple layers with different protective capabilities are of interest. The material system is positioned in a test rig that can reproducibly generate some set of preconditioning conditions (environment, challenge agent, wear or use); the same rig may be used for the subsequent performance evaluation part of the test, or alternatively, the material system may be moved to a different apparatus for these evaluations. Most properties that can be are evaluated first at the material level, although interpretation of results should always be performed with caution.

Durability tests are relatively easily performed and interpreted, where a material is stressed in a variety of reproducible manners (stretching, flexing, abrading, etc.) and its resistance to puncture, fracture, or other forms of stress-related breakage are subsequently evaluated. It is relatively easy to develop realistic values for various parameters and to devise test rigs that will accomplish the necessary stresses. To achieve realistic conditions of use, such tests may need to be performed under extreme environmental conditions; for example, cold may induce phase changes in polymers that significantly lower their resistance to stress-related effects during use.

Agent testing on a material is much more difficult to relate to full system performance, although it is an important stage in the qualification of materials. Where protective performance against CBRN agents is being evaluated, it is important that the test rig be leaktight so that no agent can reach the detection system that did not first pass through the material. The shape and size of the test rig can have significant effects on the outcome even when all other conditions are held equal, and if so, the rationale for a particular configuration must be understood, and standardized if necessary. The more realistic the shape and size of the material item being tested, the more realistic the results will be. Issues such as the amount of space above or beneath the material, airflow patterns, and edge effects can all be important.

In most cases, tight control and reproducibility of exposure and challenge conditions are made easier by the use of material tests, making them particularly appropriate for standardized testing. There are, however, numerous exceptions; for example, generation and measurement of controlled uniform concentrations of challenge vapors and aerosols are often more difficult on a bench scale than in a large exposure chamber.

Attempting to control these conditions using material tests is sometimes counterproductive. This can be because the act of measuring affects the outcome, because detectors cannot be scaled down appropriately, or because the property is too system dependent. Examples of possible issues are:

- Many detector sampling systems that actively sample the air can affect the outcome of the measurement by forcing air where it would not otherwise go.
- Depletion of permeated agent by the detector may drive greater permeation than would have happened in reality.
- Larger-scale effects of movement and ventilation are impossible to reproduce.

One example of an issue that is difficult to control and reproduce realistically is the overall airflow pattern of the system. Depending on the air permeability of the material and design of the system, airflow patterns can be highly varied beneath an item of PPE. To achieve reproducible results for a small-scale test, airflow needs to be tightly controlled. This necessarily implies that it will be nothing like the highly variable flow patterns observed in reality. Test rig parameters that can affect the outcomes are numerous; sometimes, correlations can be drawn with larger-scale testing, but these can be unreliable; hence, screening using such approaches should be performed with caution, as even relative performance trends are not always reliable.

Component or Subsystem Tests. Examples of components and subsystems that can profitably be evaluated include smaller PPE items, such as gloves, boots, or respirators, which can be evaluated on simulated hands, feet, or headforms, respectively. Clothing materials can be evaluated as "sleeves" of material mounted on a cylinder or simulated limb platform. Closures can sometimes be screened in this manner also. The test platforms can be designed to simulate limited human characteristics.

There are numerous benefits conferred by this scale of test, as more realistic test conditions are achievable to allow better correlation with full system results; at the same time, variables can be controlled and isolated through test design. At this level, integration issues between different components and closure concepts can also be examined in a preliminary manner. When a test chamber is necessary, if a full system test chamber is already available, it is generally easily adapted to the use of subsystem tests; otherwise, appropriate chambers can be constructed for laboratory use. The test platform is often more easily instrumented than in a material-scale test, because of the increased size. The physical modeling that can be achieved at this scale of testing is particularly beneficial when combined with theoretical or numerical modeling.

Another possible advantage of a subsystem test over a full system test lies in the area of anthropometry. The effects of changing anthropometry may be observed in a controlled manner by varying the dimensions of the test platform within the range of human anthropometry without requiring the construction of full mannequins or the use of human beings. At the system level, anthropometry is a particular difficult challenge to obtain in a controlled manner or to reproduce realistically, as few if any mannequins are available in different anthropometries; therefore, the selection of one particular set of anthropometric measurements necessarily biases the results with respect to PPE sizing and fitting.

A robotic torso with a single movable head and arms connected with a breathing machine has been developed for the test of inward leakage of PAPR systems (for which the facial anthropometry issues are somewhat less critical to the protective outcome, due to the positive pressure) [361]; the head can nod and tip backward as well as turning side to side, and the arms can pump from front to back.

Various headform test platforms for the evaluation of respirators (some still in developmental stages) have been developed on behalf of or by the British Ministry of Defence, the Canadian Department of National Defence [422], and NIOSH, as discussed in an ISO standard [149]. Various versions have incorporated motion, breathing, sweating, and human-like anthropometry. ISO has proposed a standard

16900-5, *Respiratory Protective Devices: Methods of Test and Test Equipment; Part 5: Test Tools* to describe standard headform and torso test platforms.

Full System Tests. Such tests are performed using a full PPE system on either a person or a mannequin platform. Where possible, the use of humans is usually preferred, as all of the true performance variables are evaluated, and various forms of additional feedback are often provided that aid in the correct performance of the test (a person will usually tell you when closures are under- or overtightened) as well as in other assessments (they will also let you know when it was uncomfortable or impractical!). Some human performance parameters are also simply too complex to simulate well using a mannequin: for example, evaluating overall human performance at extreme temperatures while wearing PPE.

Nevertheless, some of these advantages are simultaneously disadvantages, as it is difficult to obtain a large enough group of human participants to evaluate correctly all of the possible variables in the human population, and reproducibility and repeatability become difficult. Also, it is apparent that some types of evaluations cannot be performed on humans for ethical and safety reasons; these include exposure to hazardous environments and lengthy or repeated tests with controlled movements. In this case, mannequins that reproduce the necessary aspects of human performance must be used. Identifying what the necessary human aspects are can be difficult before the measurement is actually performed, and implementing multiple types of human characteristics can be challenging. As a result, mannequins are generally designed to have selected specific features, whereas others may be omitted.

Examples of human-like features that may be included include soft sealing surfaces, motion, breathing, and sweating. Additional features that may be incorporated to advantage in mannequin platforms include embedded sensors and easily decontaminable surfaces. A full-scale robotic test platform has been developed for the Canadian Department of National Defence for PPE testing against airborne hazards [423]. Other somewhat less human-like versions have been developed by various governments and test agencies to test against live agent, flame, and thermal performance.

6.3.3 Item To Be Tested

The item to be tested may consist of a subcomponent, such as a closure or layer of material, an entire item, such as a glove or a respirator, or the entire protective system. The hierarchy of tests to be performed should be determined based on a number of factors. As noted above, an appropriate test at the system level is preferred, and if it has been performed, a lower-level test is not necessary. However, it is usual that much of the testing will be performed at the subsystem level.

As discussed previously, the location in the life cycle of the item affects how tests should be prioritized, and the closer to the point of being fielded an item is, the more crucial it is that tests be performed under realistic conditions, whereas well before and after fielding, tests may simply be used for screening purposes. The plan for testing throughout the life cycle of the equipment should be laid out early. If a particular test is relevant for life-cycle management after an item has been procured, for comparison, it is important to have generated data using that test well prior to procurement preferably over as long a period of use or storage as possible. If tests at the system level are to be performed, prior testing at the subsystem level should lead up to such tests and support their interpretation. If particular critical tests may be difficult for an item or system to pass, they should be performed earlier rather than later so that issues can be addressed, and such critical tests are bound to be item-specific based on the particular design and technologies incorporated.

Systems. PPE systems and ensembles may be differentiated primarily on their purpose and the user group that employs them. They may be general CBRN protective systems, or include specialty protection such as B, C, R, explosive, fire, thermal, ballistic, or immersion. Obviously, when a system has any particular capability, whether it be a protective or other function, relevant tests must be performed. Although such tests may be performed initially on the subsystem level, generally the capabilities will be confirmed at the system level wherever possible or practical.

At the system level, while other evaluations may be performed in addition, the important parameters that are typically assessed include various human factors and system integrity tests:

- User comfort, including resistance to thermal burden and climatic extremes
- System functionality to permit or enhance performance for all required user tasks, and assess integration with other equipment, such as vehicles and weapons
- Ease of donning, doffing, and decontamination
- System integrity to penetration by airborne agents, and durability and resistance to penetration or damage by other hazards, such as water spray and fire
- System integration among components
- Capability to fit the user population adequately (in terms of all the factors above)

Components and Subcomponents. The components that make up a system may include the following items of PPE:

- *Dermal protection:* gloves, boots, socks, overgarments, stand-alone garments, undergarments, hoods
- *Respiratory protection:* consists of separable subcomponents, such as facepiece, air-purifying elements, and air supply devices, such as tanks, blowers, and hoses
- *Other personal protective equipment or items:* uniforms, helmets, ballistic protection, explosion protection, cooling systems, immersion protection, straps/belts/carriers/holsters, and backpacks

Tests that are particularly relevant to components are those that test the quality of construction when materials and design are translated into production-quality items.

These include dealing with such issues as successful manufacture of seams, layers, coated materials, interfaces, and connections: for example, when sewing up multiple layers, inserting a closure, or coating a polymer as it is formed into a glove or boot. Integrity, resistance to agent, and durability, particularly, may be affected by imperfections introduced during production. Reproducibility of sizing should also be examined at this level.

Closures, Seals, Connections, and Attachments. As referred to here, closures are that part of PPE that must adjust or open and close in normal use: for example, to fit properly and to put on and remove clothing. Various styles of closures exist, and their role in CBRN PPE is always at least in part to provide integrity to the system against penetration by an agent when they are closed. This does not necessarily mean that they will be air- or liquid-tight, depending on the level of protection required, since cooling by convection through closures may be required for user comfort. They may be single- or multiple-use. Examples include zippers, seams, hook and loop fasteners, elastic polymer seals and bands, drawstrings, and adhesive strips. The respirator face seal itself is not usually adjustable (i.e., it does not constitute a closure); instead, the harness performs this function.

Seals, connections, and attachments are required to mate subcomponents together, whether permanently during manufacture or by the user. Examples include seams between various materials within a component, formed by sewing or polymer seaming techniques, and screw threads or bayonet mounts between separable subcomponents of a respirator. Typically, closures, seals, and attachments will need to be manufactured into some form of subcomponent or component before they can be evaluated effectively. Facets of their performance, such as integration and ease of putting on and removal, may require evaluation at the system level, while preliminary evaluations or investigations of other performance characteristics may be performed using subcomponents or components.

Materials. The many types of materials suitable for incorporation into CBRN PPE, and their appropriateness for particular applications, have been discussed in Section 4.5. Materials may perform different functions at different locations within PPE, and that PPE may be intended for different user groups. Therefore, there is no single set of evaluations that will prove a given material adequate for use, but as always, evaluations must be customized for the application.

The types of questions that must be asked when deciding what types of evaluations are relevant include all of those generally relating to the concept of operations: that is, what types of individual and combined hazards and environments the material might be exposed to as a result of its intended use. Materials intended for single-use concepts will have different durability requirements than those intended for multiple-use concepts. Consider also where the material might be used within the PPE and how that might affect the likelihood that it will experience a particular set of conditions: for example, materials located on the soles of the boots will experience significantly different conditions than those on the top of the head, meaning that different durability tests might be relevant. Materials from different components may also need to be tested in layers where that is their intended use.

Resistance to various aspects of handling and use in a CBRN environment is usually tested initially at the material level; such tests include durability (resistance to abrasion, puncture, flexing, breaking, fatigue, etc.) and resistance to penetration or permeation by various agents, particularly after various preconditioning treatments that could degrade subsequent performance. Such preconditioning may include aging materials through various aspects of an entire life cycle of handling, including wear, laundering, sweat, and encounter with various contaminants, such as common solvents, moisture, or decontamination processes. We note that it is rare that the entire sequence of such events would be simulated in the laboratory; typically, only one or two preconditioning treatments might be performed. However, field tests on systems may replicate realistic preconditioning, following which the system can be broken down into subcomponents (e.g., material swatches) for a variety of subsequent performance evaluations.

Intrinsic material properties can also have an effect on the type of test to be performed: air-permeable materials may need to be tested differently from air-impermeable materials. Some few materials have been so well characterized that the necessary information does not need to be gathered: for example, certain glasses and metals. Materials may need to be evaluated for their ability to take certain additional treatments, such as repellency or camouflage patterns, and may need to be reevaluated for performance once treated, or for the durability of the treatment.

Packaging. The durability of packaging and its ability to withstand or hold out various possible challenges, such as extreme temperature and humidity, rough handling, lengthy storage periods, and environmental contaminants, are typically evaluated. The susceptibility of the item inside the packaging may determine the evaluations required.

6.3.4 User-Related Parameters

Some of the test parameters will be set based on the user's characteristics: in particular, certain parameters that reflect human physiology, such as work and breathing rate, ambient temperature, and humidity. The concept of the use and role of the intended user, and the resulting performance duration requirements, determine the work rate from which test parameters are derived, as well as the duration of the test. The high humidity and warmth characteristic of the environment next to the skin at a high work rate should also be considered, as well as movement. Flow-rate parameters (which simulate breathing) can have a profound effect on the performance of RPDs. While many APEs are tested with unidirectional flow at constant flow rates characteristic of possible peak flows at high end work rates, this is a relatively unrealistic test condition, but the difficulty in establishing standardized methods that include more realistic breathing parameters has prevented their widespread adoption.

6.3.5 Challenge CBRN Agent

As discussed previously, selection of the particular challenge CBRN agents to use for testing is a difficult task. No matter what approach is employed, there are always significant limitations in the outcome when the hazardous properties of tens of thousands of substances are being simulated by a few selected substances. There have been many attempts to create lists of hazardous materials to use for testing, and each has strengths and weaknesses. Many of the general drawbacks of standard test methods, described in Section 6.2.3, apply when trying to use standard lists. Each list has been developed for a particular reason, and a particular set of assumptions has been made to shorten it to a manageable number. It is rare that these assumptions are documented, and even rarer that the same assumptions are relevant to the next application of interest, or equally valid when new protective concepts are being evaluated. Nevertheless, use of standard lists is valuable in that it permits comparison of performance behavior among different protective concepts evaluated at different times or in different laboratories.

Chemical Agents. When selecting chemical challenge agents, certain physicochemical properties of the chemical are important. Toxicity is a primary consideration: Is the chemical itself a known toxic hazard, or is a suitable nontoxic simulant available? The properties that govern a chemical agent's ability to pose a hazard to the protected individual depend on the given mechanism of protection (blocking, filtering, adsorption, reaction, or some combination of these) as well as the detailed design and chemical makeup of the protective system itself. Chemical structure, solubility, and reactivity (particularly with respect to important generic categories such as acid/base, oxidation/reduction) are relevant. Properties that govern potential concentration, persistency, and physical form (liquid, vapor, or aerosol) must also be considered, such as surface tension, volatility, polarity, and vapor density.

Lists of hazardous chemical agents for testing have been developed by, for example, ASTM [424], NIOSH [425], NFPA [409,410], CSA/CGSB [3], ISO [426], and NATO; typically, when produced by military organizations, such lists are sensitive or classified and hence are not discussed in detail here. ASTM F1186 covers a standard classification system for chemicals that may aid in the generation of lists [427]. The ASTM F1001 list [424] (see Table 6-2) is based on choosing high-production-volume common industrial chemicals that represent various chemical classes; they are not

TABLE 6-2 ASTM F1001 List of Chemicals

Acetone	Diethylamine	Methyl chloride gas
Acetonitrile	*N,N*-dimethylformamide	Nitrobenzene
Anhydrous ammonia gas	Ethyl acetate	Sodium hydroxide, 50% aqueous
1,3-Butadiene gas	Ethylene oxide gas	Sulfuric acid (concentrated)
Carbon disulfide	*n*-Hexane	Tetrachloroethylene
Chlorine gas	Hydrogen chloride gas	Tetrahydrofuran
Dichloromethane	Methanol	Toluene

TABLE 6-3 NIOSH CBRN APR Standard List of Chemicals for Testing the Air-Purifying Element

Ammonia	Hydrogen cyanide	Phosgene
Cyanogen chloride	Hydrogen sulfide	Phosphine
Cyclohexane	Nitrogen dioxide	Sulfur dioxide
Formaldehyde		

necessarily chosen in particular for their hazardous nature, although many are toxic or corrosive. Such a list may be particularly helpful when the ability of a material to resist physical degradation by a challenge agent type is being evaluated (material integrity tests). NFPA 1991 [409] uses the entire list for generic representation of material resistance to permeation and visor seam resistance to permeation, and all of the liquids on the list for penetration resistance of suit seams. To this list are added phosgene, cyanogen chloride, dimethylsulfate, hydrogen cyanide, sulfur mustard, and sarin for permeation tests on materials and seams. For NFPA 1994 [410], the list of liquid chemicals includes sulfur mustard, soman, acrolein, acrylonitrile, and dimethylsulfate, and ammonia and chlorine are used as test gases.

The NIOSH list for CBRN air-purifying respirator purifying elements [425] (see Table 6-3) is based in part on a similar approach but takes into account to some degree different classes of chemicals that are particularly difficult for active carbon systems to remove, as well as a few of the classical CWAs that could be used in a terrorism event; two agents, sarin and sulfur mustard, that are respiratory hazards and can permeate through respirator materials are also tested. Again, not all the test agents used are extremely hazardous, and their performance is deemed to be representative of an entire class.

In any of these cases, there are difficulties with the use of these lists should a particular hazardous agent have chemical behavior that differs significantly from that of the supposedly representative test agent, a not unlikely occurrence. A more robust approach to selection of test agents that is being considred for the generation of appropriate simulants is a systematic methodology based on relevant chemical properties using a process such as multivariate data analysis [428]. In this approach a large series of physicochemical properties are collated for chemicals known to be toxic, as well as those that are less so (and hence are potentially useful as simulants). Principal component analysis is used to simplify the property space and to identify the important parameters that characterize particularly hazardous agents, and test agents can be selected to map out the entire hazardous property space systematically; simulants can easily be identified that have similar important properties to test agents but are potentially less toxic.

The military has tended to focus on a select list of chemical warfare agents that have been militarized and are highly toxic; many of these chemicals were militarized precisely because they are difficult to protect against. Thickened agents are generally included in the list because these have been weaponized. Some simulant tests are performed in which simulants are representative of particular groups of toxic compounds. Because the military hazard list under consideration is often shorter, and

TABLE 6-4 CSA/CGSB Z1610 List of Chemicals of Concern for Air-Purifying Elements

Acrolein	Fluorine	Nitrogen dioxide
Allyl alcohol	Formaldehyde	Nitrosyl chloride
Arsine	Germane	Perfluoroisobutylene
Boron trifluoride	Hydrogen cyanide	Phosgene
Carbonyl fluoride	Hydrogen fluoride	Phosphine
Chlorine	Hydrogen selenide	Phosphorus trichloride
Chlorine dioxide	Hydrogen sulfide	Selenium hexafluoride
Chlorine trifluoride	Methyl chloroformate	Silicon tetrafluoride
Cyanogen	Methyl hydrazine	Stibine
Cyanogen chloride	Methyl isocyanate	Tellurium hexafluoride
Diborane	Nickel carbonyl	Tungsten hexafluoride
Dichlorosilane	Nitric acid	
Breakdown Products		
Acetic acid	Hydrogen chloride	Phosphorus oxychloride
Ammonia	Nickel	Silane
Carbon monoxide	Nitric oxide	

these agents have been known for decades, the simulant's behavior relative to the hazard agent is likely to have been validated. Supplemental lists of toxic industrial chemical hazards of interest for testing have recently been produced by NATO based on a variety of criteria, including production volume and toxicity to the respiratory tract and skin.

The CSA/CGSB Z1610 standard [3] has attempted to approach the problem by systematically estimating the hazard posed to a particular route of entry. For protection using air-purifying respirators, this hazard is based on the toxicity of the agent combined with its ability to defeat traditional active carbon elements. This means that the test chemical list is populated by highly toxic agents, most of which are low-boiling chemicals (boiling point $< 60°C$). No simulants are used and the list is relatively comprehensive, including potential toxic breakdown products produced from reaction with carbon impregnants (Table 6-4). It is currently difficult to populate the same sort of list for dermally active chemicals, as dermal toxicity values are relatively poorly known; the standard has proposed a list including the NFPA dermal test chemicals with the addition of gaseous HCl.

For evaluation of RPD or system leakage, the general approach that has been taken is that it is not always necessary to specify any particular agent. The test agent may be a simulant that is any small aerosol or gaseous material that simulates the behavior of highly toxic vapors such as sarin and small aerosols, which are most likely to enter through small leaks in the respirator. Nevertheless, the limitations of this approach should be recognized, in that at some point the physicochemical properties of the simulant will be different from that of the agent; size, shape, charging effects, and depletion by contact with surfaces will all differ and may affect the outcome significantly in certain cases [337].

Chemical Simulants. A chemical screening program was performed in 1988 by the U.S. Army to develop a suitable list of simulants for various test formats [429]. The physicochemical parameters considered included:

- Vapor pressure
- Vapor density
- Melting point
- Surface tension
- Vapor viscosity
- Diffusion coefficient in air
- Molar refraction
- Flash point
- Oxygen index
- Heat of combustion
- Heat of formation
- Heat capacity
- Hydrolysis rate
- Hildebrandt solubility parameter
- Water solubility
- Dielectric constant
- Toxicity
- Liquid density
- Boiling point
- Molar volume
- Liquid viscosity
- Volatility
- Refractive index
- Decomposition temperature
- Autoignition temperature
- Heat of vaporization
- Heat of fusion
- Energy of vaporization
- Specific heat
- Octanol–water partition coefficient
- Hygroscopicity
- Critical temperature, pressure, volume, and density
- Dipole moment
- Chemical reactivity

Many of the chemical simulants now used (examples are given in Table 6-5) are based on the outputs of this database.

Biological Agents. As in the case of chemical agents, the properties that govern a biological agent's ability to pose a hazard to the person protected depend on the mechanism of protection. In many cases, biological agents are protected against by blocking or filtration. In this case, it is the physical properties of the biological

TABLE 6-5 Common Chemical Simulants Used in Various Types of Tests

Simulant	For (Agent, Test Type)
Dimethylmethylphosphonate (DMMP)	Liquid CWA nerve agent
Diisopropylfluorophosphate	GD nerve agent reactivity
Bis(2-ethylhexyl)phosphite	VX agent reactivity
Paraoxon	VX and GD
2-Chloroethylethyl sulfide (2-CEES)	HD
Chloroethylphenyl sulfide	HD agent reactivity
Methyl salicylate (MeS)	Persistent CW agent simulant particularly mustard; safe for human use [430]
Carbon tetrachloride	Organic vapor, for adsorbent capacity
Sulfur hexafluoride (SF_6)	Vapor leakage; safe for human use
Sodium chloride (NaCl)	Aerosol leakage; safe for human use
Oils	Penetration bench tests, aerosol leakage, penetration and deposition; safe for human use
Amorphous silica doped with fluorescein and tinopal	Aerosol leakage, penetration, and deposition; safe for human use
Polystyrene latex spheres doped with fluorescent compounds	Aerosol leakage, penetration, and deposition (size controlled)

challenge—aerosol or liquid suspension, shape, size, and charge distributions—that determine the effectiveness of protection, not the nature of the organism itself. Where use of an organism is particularly desirable in order to examine properties such as viability and reactive kill, representative organisms from various categories may be selected (e.g., bacterial, viral, spore-forming). Simulants having similar size and shape, susceptibility, and die-off rates are available in some cases.

Hazard agents are rarely used in test methods, but the list could potentially include most of the agents identified in Section 2.7.13. *Bacillus anthracis* is used in particular when test methods evaluate the capability of a biocidal approach, as it is generally considered the potential agent that is most difficult to kill.

Biological Agent Simulants. It is usual to simulate many physical properties using some form of inert aerosol or solution, particularly by using a realistic dissemination method in the case of aerosols. Nonliving simulants include the aerosols listed in Table 6-5. A commonly used bacterium is *Staphylococcus aureus* (ATCC 6538), a vegetative gram-positive organism with a mean diameter of 0.5 to 1.5 μm [431]. A sporulating simulant for *B. anthracis* is *Bacillus atrophaeus* (ATCC 9372) (also previously known as *Bacillus globigii*, BG) [432]. *B. atrophaeus* spores are somewhat smaller than those of *B. anthracis* (which is about 0.85 μm in diameter and 1.2 to 1.7 μm long) [432,433]. Because viruses cannot be cultured without a host, and are otherwise difficult to detect at the concentration needed, viruses whose presence can easily be monitored by host infection are used. *MS2 coliphage* is a common choice: It is 25 nm in diameter and its propagating host is *Escherichia coli* (ATCC 15597). *Phi-X-174 bacteriophage* is used in viral penetration testing for materials [410,434].

Radiological and Nuclear Agents. The R/N agents that are of interest are summarized in Section 2.8. The main difference between radiological challenge and nuclear challenge characteristics is in the generally accepted size range of the particulate to be used for testing. Radiological materials will be disseminated more efficiently when they are relatively small particulates around the range of 1 μm in diameter that remain suspended in air. Nuclear fallout will contain larger sizes, more like dusts. For PPE testing, the larger R/N particles are mainly of interest in terms of the hazards posed by the contaminated equipment itself—the amount of radioactive hazard posed by contaminated dusts may be quite high—such as tests for decontaminability, since removal by filtration and blocking is quite effective in comparison with the smaller aerosols, which will be more of a worst case for filtration and penetration evaluations.

Radiological and Nuclear Simulants. R/N agents may be somewhat more easily simulated than the other agents, for the simple reason that there is often no need to simulate the actual hazard itself. There are no reactive technologies, and the way that R/N agents interact with PPE and decontaminants is based entirely on physical properties that can, in theory, be simulated by other types of dry particulates that are nonradioactive. Inert aerosol simulants include those listed in Table 6-5. Larger inert particles to simulate nuclear dusts may be obtained using ISO dust standards (e.g.,

[435]). These may be relevant when considering how larger dusts may foul a piece of equipment.

If a nonradioactive simulant is to be used, one particularly important property to be simulated that may be unique to these particles is charge. One example of a simulant safe for human use that mimics R behavior and can be used for certain types of evaluations is given in a U.S. patent application [436]. It is a combination of hydrated silica gel, a fluorescent dye, and a salt to add conductivity and therefore generate charging effects similar to those of a radiological particulate. Short-lived low-energy isotopes themselves may also be used as simulants that can be dispersed in a manner similar to that of radiological agents and are easily detected. The chemical properties of radioactive materials, such as their solubility into materials, can be simulated by nonradioactive analogs.

On the other hand, safe laboratory practice for handling of radiological materials is well established, and such materials are particularly well suited for easy, selective, sensitive detection, making-low hazard radioactive simulants particularly easy to use. Low concentrations of short-half-life materials of appropriate particle size can be used.

6.3.6 Amount and Physical Form of Agent

The method by which the amount of agent used for testing is selected for test methods is generally rooted somewhere in the anticipated challenge conditions. In other words, the best method for determining test conditions is to use the actual worst-case concentration and duration of exposure for a given agent, obtained from a combination of release data and modeling as outlined in general terms in Section 3.4. However, this starting point is, of course, impossible to define for every agent, and in any case, often does not translate into practical test conditions.

The usual difficulty is in deciding what the worst case actually consists of: Is higher concentration worse than higher dose? Is higher contamination density worse that bigger drop size? The answers may depend on a variety of factors that differ from application to application. Limitations and assumptions that are often made in doing so are outlined where relevant below. In general, the concentrations used by the military are based on hazard estimates for delivery from various weapons and dissemination systems, and are classified. Information can be found in a variety of NATO publications. The discussion here focuses on general principles and values that have been reported in the open literature.

Liquid Chemical. It is apparent that liquid chemical could be encountered in any drop size and contamination density, depending on the generation method and how close a person approaches the release point. For certain user groups, operational limitations may be used to reduce the magnitude of liquid hazard that they would be permitted to encounter (e.g., not approaching within 100 m of the release point or within the hot zone, or not entering an area with visually obvious liquid contamination).

Modeling of the evaporation rate of liquids has been used to estimate the length of time some CWAs (and other liquids) would persist in the liquid phase [69,437].

This also puts limits on the likelihood of encounter of a significant liquid hazard for certain user groups that are at a distance from the release point; for example, first receivers remote from a release event may be unlikely to encounter certain types of liquid agent because decontamination (at a minimum, removal of contaminated clothing from exposed persons) and evaporation will have combined to eliminate it before casualties are received.

Modeling has usually not been used to generate data on drop-size distributions, although it is possible to use for contamination densities. Since for many materials, the likelihood of permeation increases significantly as the drop size increases, particularly when pressure is applied, drop size is an important test parameter. It is generally impractical to devise a test method that uses some form of realistic drop-size distribution. Generally, one or more drops of a single size are applied manually or using relatively simple automation procedures. Drop sizes and contamination densities that have been used in civilian test methods range from 1 μL drops applied at 5 to 10 g·m^{-2} [410] through bulk liquid [409]. Dilution of liquid hazard agent into other solvents is not recommended, as this can alter permeation behavior significantly.

Vapor Chemical. It is generally somewhat easier to characterize vapor releases; there is an ultimate concentration limit based on the volatility. Indoor release characteristics can be described further by the amount released vs. the volume within which it is contained, and ventilation conditions. Concentrations generally decrease quite rapidly with time as the agent is diluted, also putting limits on duration of exposure from any individual release event. Vapor may continue to be evolved from a persistent liquid hazard for some time, but this is unlikely to be a worst-case scenario. Models as described in Section 3.4.2 are used routinely, and simple models can yield useful data, because the protective performance of most materials and systems is linear with respect to vapor challenge concentration and dose until saturation of the materials is reached.

Certain types of operational restrictions are often employed that will also reduce the likelihood of encounter with higher concentrations or longer durations, limiting the necessary range of test conditions. One example is the use of guidance such as the *Emergency Response Guidebook* [66], which designates the worst-case sizes of initial isolation and protective action zones (Figure 3-2) based on the agent released, its amount, and the environmental conditions. By remaining outside the initial isolation zone, responders can be assured that the dose used is unlikely to exceed a certain value no matter how long they remain in the area. For example, the CSA/CGSB standard [3] suggests that a maximum possible vapor dosage of 40,000 mg·min·m^{-3} is achievable outside the hot zone for a single-release event; this limit assumes that the hot-zone perimeter is located correctly based on various indicators, and is potentially hundreds of meters from the release point.

Another type of operational limitation is placed by NIOSH and is used in part to set test dosages for CBRN air-purifying respirators. In the United States, it is not permissible to use an air-purifying respirator in an IDLH environment; this principle was originally developed for the industrial workplace so that in case of respirator failure, serious injury would not occur and escape is always possible. It is, however, a somewhat impractical approach to apply in a CBRN event, as it is rare that a person would, in fact, know whether they were in an IDLH environment, and in fact most of the area of the event where responders would be expected to operate would be above IDLH levels. Assuming that use of an APR is more likely to be set by distance from the release rather than knowledge of the concentration and chemical, the concentration at a given location would be based on the volatility of the chemical and not on its toxicity, therefore, selected test dosages would be set more consistently based on volatility and possible release amount.

Notwithstanding, many of the test dosages for toxic industrial chemicals (Table 6-3) used in the NIOSH APR standard [425] are based in some way on the IDLH value for the test chemical. This and other factors lead to a number of inconsistencies as to how this drives protective design. First, test chemicals that are not, in fact, toxic are used as simulants for an entire category of agent (e.g., cyclohexane as a simulant for hydrocarbon vapors); protecting against them at *their* relatively high IDLH concentration is irrelevant and means that a canister must build in a substantial amount of protection against a nonhazard agent when less might be sufficient to protect against the actual hazard agents. In fact, when toxic agents are actually used to set the criteria, test concentrations are lower, and the total capacity to protect against them is potentially *less*.

Further, the selection of test concentration has also been applied inconsistently in that the test concentrations used for some CWAs are orders of magnitude above IDLH (in the APR standard [425] as well as in the corresponding NFPA 1994 class 3 standard [410] for DPE for use with CBRN APR). Taken as a whole, this approach could benefit from a systematic reworking.

Aerosol Chemical. Deriving test concentrations for aerosol chemicals is particularly difficult. Hazard aerosols may be generated using a large number of different methods, depending on the starting hazard material, for example, sprayed liquids, volatilized and recondensed solids, or inert dry chemical aerosols coated with a chemical agent. Their size distribution and persistency will vary depending on a large number of factors, and there is often a cogenerated vapor concentration. This wide variety of possible exposure conditions is difficult to simulate in a well-controlled test format. Partly as a result of the general difficulty in selecting and controlling chemical aerosol challenge conditions, the test concentrations and size distributions selected for chemical aerosols are often not based on known hazards, and performance against chemical aerosols is sometimes not considered explicitly at all, particularly when other airborne hazards, such as vapors or biological aerosols, are being tested. Nevertheless, when selecting test concentrations for aerosols, careful consideration should be made of the actual physical form of the aerosol of concern and of the possible volume of the original container, release method, and how the aerosol's behavior might evolve over time after release, in order to obtain the most realistic possible performance behavior of the system being tested. This will probably mean that more than one type of aerosol test agent and format will be required.

Aerosol Biological. There is little information in the open literature based on which it would be possible to develop realistic test concentrations of biological agents. Anthrax, although not the most hazardous possible agent, is worst-case in some ways, for reasons to do with its persistency and relative ease of manufacture. Simple example calculations can be performed; for example, release of 1 g of anthrax-containing aerosol uniformly into the ventilation system of a large building, assuming losses of 10%, would result in doses of over 60 minutes on the order of 10^{-2} mg·min·m^{-3}, with local concentration values actually highly dependent on deposition and loss rate, efficiency of dispersion and air circulation, time after dispersion, and other factors. Modeling and measurements have been performed of a 1 to 3 g anthrax letter scenario [438] for both opening and sorting, yielding total doses on the order of 10^5 spores received over perhaps 10 minutes, with a few hundred counts per minute of aggregated spore-containing particles sampled using active sampling (a few hundred agent-containing particles per liter of air). Models assuming a release of about 2 g of *B. anthracis* in a 50×2000 m^3 building result in airborne concentrations as high as 10^8 spores per cubic meter near the release point (first floor) but four orders of magnitude lower at the most remote point in the building (the fiftieth floor) [439,440].

Biological concentrations are more often expressed in more easily measurable units, such as colony-forming units (CFU, a single-particle agglomerate of organisms or spores usually of undefined size); a typical range might be 10^4 to 10^5 CFU per liter of air [441]. This type of quantitation is of marginal use for a reproducible test with interlaboratory comparability unless the particle size distribution or some other estimate of the number of organisms in the agglomerate is also available [442].

Contact Biological. Such challenges are usually intended to simulate bloodborne contamination, and thus use a solution of organism in simulated body fluids. One example of an organism is the *phi-X-174 bacteriophage* virus, which is easily quantified in a laboratory situation [434]. Such hazards are actually likely to be of far lower magnitude than those in a BWA release, but this type of protection is often included regardless: first, as an indicator of the integrity of the system, and second, so that PPE can be deemed to provide routine protection against bloodborne hazards in the community. A dry aerosol surface deposition device has been demonstrated that loads *B. atrophaeus* spores and *S. aureus* organisms in the range 10^4 to 10^5 CFU·cm^{-2} [443,444].

Aerosol Radiological. The inhalation hazard from radiological aerosols has been described by NATO [73] in terms of the dose delivered in sieverts over a duration of 15 minutes in the vicinity of the release point indoors, or 100 m downwind of the release outdoors. Two of the worst-case hazard scenarios (in terms of biological impact to an unprotected person) are indoor release of $^{90}Sr(NO_3)_2$, a beta emitter, or $^{238}PuO_2$, an alpha emitter, either of which can be derived from a radioisotope thermoelectric generator that can yield as much as 1.1×10^{16} Bq of particles of about 2 μm aerodynamic diameter with specific activity on the order of 5×10^{11}

$Bq.g^{-1}$. Modeled doses of 60,000 and 30,000 $mg{\cdot}min{\cdot}m^{-3}$ are delivered, yielding committed effective doses over 50 years of 3×10^4 and 3×10^6 Sv, respectively. Outdoor release reduces the dose by as much as a factor of 1000. The hazard from a radiological aerosol attack in a city has been analyzed in some detail by Andersson et al. [19]. For a 1000-TBq ^{90}Sr source in a ceramic matrix, the surface deposition for horizontal outdoor surfaces outside the immediate blast zone was estimated to be 0.1 to 10 $MBq{\cdot}m^{-2}$. The maximum airborne concentration in the range of a few blocks downwind from the release was in the range 10^8 to 5×10^9 $Bq{\cdot}s{\cdot}m^{-3}$ at ground level, depending on the details of the model, dropping by a couple of orders of magnitude at distances of a few kilometers downwind.

Nuclear. Information on the level and types of hazard from a nuclear accident has been provided by NATO [73]. The International Nuclear Event Scale of the International Atomic Energy Agency can be used to describe the initial magnitude of the event to be modeled [445].

6.3.7 Test Environments

Testing can occur in a variety of environments, which may range from tightly controlled laboratory conditions through to field conditions. Having already specified the item to be tested, the parameter(s) to be evaluated, and the type, form, and amount of agent, as discussed in Section 6.3.6, the test environment must be specified to be compatible additionally with the test scale and platform as well as the remaining relevant environmental conditions, such as temperature, relative humidity, and wind speed. Even more extreme environments, such as rain or immersion, may also be simulated.

Laboratory. Benchtop testing may be performed with minimal environmental control or in highly controlled circumstances, depending on the type of test.

Full-Scale Chamber. Chamber conditions may be used to permit full-scale testing in a more controlled environment than the field.

Field. Field testing is realistic and may be preferred when multiple variables, such as user comfort, equipment integration, and wear, are being evaluated. Field testing may be performed in a moderately controlled environment (e.g., if realistic conditions of use are indoors), but often will include an outdoor component. It is customary to combine field testing with some form of training activity, and although this approach has many benefits, full control of the test conditions is not one of them. Possible environmental conditions may be bounded by the location and time of year, but often the tester will have to accept the weather as it occurred, whether or not sufficiently representative of the expected range of exposure. Release of hazard agents in the field is usually not permissible, and even some simulants may be considered too hazardous for outdoor release.

6.3.8 Preconditioning

The CBRN PPE is often expected to protect even after it has been worn under normal conditions of use for some time prior to actual exposure. Preconditioning is performed to achieve some worst-case or realistic conditioning, similar to that which might occur in normal use, capable of degrading materials or systems, after which, some form of performance evaluation follows. The duration of preconditioning is important in most cases (e.g., breathing through a canister at a certain rate prior to exposure, or a simulated wear test), while in other cases it is simply necessary to achieve an equilibrium state with the preconditioning environment prior to the test.

Mechanical and Physical Effects. The effect of wear can be simulated or generated realistically. Mechanical issues that may affect performance include abrasion, puncture, tear, and stretch. There are various approaches, which can include, for example, repeatedly flexing the sole of a boot to simulate walking, stretching, or abrading a material. Clothing could be placed on a moving mannequin or the entire system worn in chamber trials, obstacle courses, or full-scale field trials. Laundering may also affect system performance characteristics by fatiguing or shrinking material (as well as by adsorption of laundry chemicals onto adsorbents). Variables include temperature, washing chemicals (and hardness of water), loading, number and intensity of wash and rinse cycles, type of washing machine, drying method and temperature, and ironing conditions.

Environmental Effects. One important form of preconditioning involves the temperature and relative humidity. For example, a material that is to be worn next to the skin may be preconditioned at skin temperature and high relative humidity. It may not be sufficient simply to carry out the performance test under those conditions, as it could take some time for equilibrium characteristics of the material to be achieved. Longer exposures to extreme temperatures may simulate storage and aging of systems. Dirt and environmental exposures such as seawater and rain are also relevant; these types of exposures can be achieved in field trials or by pretreating materials. Exposure to other chemicals that might be encountered in the workplace is discussed below.

Chemical Exposure. Various types of exposure to chemical contaminants can affect performance. Examples of chemicals and various environmental contaminants that may be encountered in routine use prior to CBRN exposure include solvents, fuels and oils, insect repellent, firefighting foam, and cleaning agents, including laundry detergents. Systems are expected to work for some period of time while and after being exposed to decontaminating solutions, particularly for persons whose task it is to perform decontamination of equipment or of other people, and thus preconditioning with a decontamination solution is important. Direct protective performance may be affected; for example, many chemical contaminants can poison active carbon or affect polymer structure such that it is more permeable. Other important properties, such as optical clarity of visors or tensile strength of materials, could also be affected.

Body fluids may contaminate PPE, particularly clothing. When contamination is external, resistance to both the body fluids themselves, which could contain infectious agents, and to degradation of the materials is important. Contamination can also be internal; for example, when soldiers are expected to spend many hours in PPE, body wastes and sweat are probable contaminants, and blood is likely in a hostile CBRN environment.

Parameters for pretreatment with liquids in a controlled manner using swatches of material are:

- Side of material to be exposed
- Angle of material to the horizontal
- Amount of liquid to be applied per surface area (contamination density)
- Application method (dipping, submersing, pouring, spraying, application of separate droplets)
- Application pressure
- Duration of exposure
- Removal of contamination (decanting, blotting, wiping with tissue, evaporation)
- Time and environmental conditions until further evaluation (temperature and air movement can both enhance evaporation)

Examining this list of possible preconditioning scenarios, it is apparent that there are a huge number of variables and possible sets of conditions. Without knowing something about the characteristic performance of the type of material or system being investigated, it is difficult to know how to prioritize the evaluation conditions. Therefore, for new concepts it is important to do some prescreening to try to establish their particular weaknesses that would lead to worst-case exposures. As already noted, preconditioning may be performed at the material, component, or full PPE level. Material-level preconditioning is most reproducible, but the manner in which the pretreatment is applied may, in fact, be quite unrealistic in terms of how "worst-case" it actually is. Component preconditioning is of benefit, as wear issues such as manufacturing durability of seams and seals can be included and can be made incrementally more realistic; however, taking the case of simulating human motion as an example, the full range of motion and realistic encounter with abrading surfaces is unlikely to be achieved. Laundering combined with controlled wear trials is a much more reproducible way to obtain worn material with or without sweat contamination. Field trials offer the most realistic approach, particularly when it comes to environmental contaminants, but are labor and cost intensive with limited control and reproducibility.

6.3.9 Representative Sampling

Test samples that are subsamples of an item or a component (e.g., a swatch of material) should be taken in such a way that they are representative of the entire quantity of material or quantity of clothing items they are meant to represent. This may depend on the objective of the test: Is it to qualify a single lot of material for use or to investigate the effect of storage under particular extreme conditions, in which case a smaller subset would be sampled? Alternatively, is it to determine reproducibility of material production or effect of storage under a variety of conditions, in which case many different lots or storage conditions would need to be sampled? In addition, different phases in the life cycle of the clothing item, and the variability shown by different test methods, can require different:

- Sampling frequencies
- Number of items, lots, sizes, storage locations sampled
- Number of sampling locations within an item
- Numbers of samples per sampling location

The largest number of test samples possible taken from the largest relevant variety of sources of sample is desirable.

For sampling of cloth materials, typically smaller swatches or samples are cut out of larger rolls or items. If the protective material to be tested consists of more than one layer, the material "system" should be layered as worn and the swatch cut from the layered system. Care should be taken to avoid cutting the swatch too near the edge of a roll of material as well as to avoid obvious areas of imperfection related to the fabrication or manufacture of the material that would not be used in manufacture. Swatches of material should be cut to a size such that when test cells are assembled, a tight seal between cell parts is formed and no leakages along edges can occur.

For sampling of polymeric materials such as the elastomers used in respirators or eyepiece materials, they may be cut from a molded item, although this can make them difficult to mount into test cells when the item is not itself flat; alternatively, the polymer might be cast into sheets specifically for the purposes of testing during the design and qualification phases. The thickness of the polymeric sample should in general be characteristic of the thinnest dimension used in the protective item. When sampling is performed within a system in simulated use, multiple locations should still be monitored based on knowledge of the system's design.

6.4 DESIGNING METHODS AND SETTING CRITERIA

The performance criteria for a protective system can essentially be divided into two categories: those that relate to its ability to reliably protect the person within, and those that relate to the ability of the person to continue to perform the necessary tasks while remaining protected.

6.4.1 Setting Priorities

As has already been discussed, design of PPE is an exercise in trading off the risks associated with underprotection with those associated with overprotection. As part of setting criteria, it must be appreciated that use of the most extreme performance requirements will always limit performance requirements in some other area. Therefore, the approaches listed in Chapter 3 are essential to ensure that realistic and comprehensive protective and human capabilities are achieved.

In the next sections we discuss in general how testing is approached for CBRN protective systems. Those evaluations associated with protection against hazards other than CBRN, such as ballistic or fire, often included in CBRN protective systems, are not discussed in comparable detail, although their relevance should not be neglected.

6.4.2 Degradation of Performance

In addition to understanding performance under optimum (new) conditions, it is important to remember what specific parameters can degrade performance when designing evaluation methods for a system, item, or material, as discussed earlier. Examples of factors that should be considered include:

- Aging in storage and with wear
- Climatic considerations: temperature, water, and wind
- Dirt and contaminants
- Exposure to multiple hazards
- Compromise of integrity
- User factors
- Use with other items

6.4.3 CBRN Protective Performance Measures Using Items or Systems

Ultimately, all protection measures are related to the prevention of excessive injury to the wearer. In the case of exposure to CBR agents, providing protection involves reducing the dose reaching the wearer through relevant routes of entry. Evaluations of dose reaching the wearer are quite difficult to achieve realistically except in system-level evaluations, where the test can be designed to measure dose by sampling within the PPE over the entire exposure in a test chamber or field test.

The most relevant methods for determining the protective performance of a system are those of the simulated workplace protection style. Simulated workplace protection factors (SWPFs) are measured by simulating the hazard environment and workplace conditions of use for persons wearing full PPE systems, and take into account such important variables as user anthropometry, equipment fit, and type of seal under realistic conditions of use (challenge environment, activity routine). SWPF evaluations cannot be performed using human subjects in a toxic environment. Instead, they are performed using simulated toxic materials as challenge agents and thus are termed *man-in-simulant tests* (MISTs). Alternatively, toxic challenges can be used with mannequin-based platforms; such tests are often best viewed as a validation of the fidelity of the simulant, as current mannequin platforms are limited in their ability to simulate human characteristics.

MIST evaluations should be performed using representative protective ensembles. Other items of equipment that can affect performance (e.g., helmets and ballistic protection) should also be worn, as these can either increase protection by providing an additional barrier, or decrease protection by interfering with the quality of the seal between components or to the skin.

MIST (and mannequin-based) methodologies include:

- **Respiratory tests:** evaluations of respiratory and eye protection against airborne chemical, biological, and radiological aerosols and chemical vapors
- **Dermal aerosol tests:** evaluations of cutaneous protection against airborne chemical, biological, and radiological aerosols
- **Dermal vapor tests:** evaluations of cutaneous protection against persistent chemical agent vapors
- **Dermal liquid tests:** evaluations of cutaneous protection against persistent chemical agent liquids

The best-designed tests sample multiple locations within the PPE in order to take into account differing leakage into the system through seals and closures. In this case, the performance criteria are relatively easily described in theory, as they are based on a combination of exposure conditions and allowable dose. Tests that are merely comparative—say, the leakage into a new item of PPE must be less than that of the one it is replacing—are poorly designed and should be considered for replacement where reasonable, since it is not clear in this case whether either system is over- or underprotecting. All tests should be possible to perform on an unprotected wearer for comparison, yielding a quantitative protective performance measure.

It is usual for the exposure conditions to encompass the reasonable worst-case exposure to a particular class of highly hazardous agent (typically, using a simulant for that class of agent). This may require evaluation at several different exposure conditions, as the effect of certain parameters (such as wind speed) on system-level performance may be unknown before the test. Liquid testing is particularly problematic, as it is difficult to generate liquid contamination in a realistic manner, and drop size and the method of contact with the drop have a considerable impact on the result.

The allowable dose is usually that dose which results in negligible effects based on human toxicological studies. Skin as a route of entry may be negligible, or if it is significant, body region variability of permeation rates becomes an important factor [82,83]. Dose may be cumulative (in the case of systemically acting agents) or the agent may act locally only. In either case, the allowable dose must be known at many different locations in order to get an accurate picture of potential dermal toxicity.

Skin or the PPE may also act as a reservoir for agent that could enter the body via other routes of entry after removing the PPE (e.g., biological agent), in which case it is the potential inhalation dose that should be measured after removal and decontamination procedures have been performed. It is thus important to understand the nature of the hazard before designing a test procedure. For a PPE item as opposed to an entire system, ultimately it will need to be incorporated into a system for testing, since integration of the closure with other items is still an important feature of the item. Nevertheless, prescreening is usual; the best tests are simply scaled-down versions of the entire system test in which the item is placed on a nearly anthropomorphic test platform, such as a breathing headform for a respirator. In this case, exposure conditions and dose measurement can usually be similar to those used in a whole-system test.

Test Environment. Protection performance testing of full systems is almost always performed in full-scale test chambers [446], often in a dedicated building. The chamber will usually have controlled temperature, relative humidity, wind, and challenge test agent. Control of these parameters while maintaining uniformity throughout can be difficult; temperature and relative humidity may be particularly difficult to achieve at the extremes, depending on the ambient conditions. Recirculation and cleaning of air before exhausting is usually required to some degree. Slightly smaller chambers can be used for test items.

Airborne challenge test agent can be generated continuously or in a single burst that is allowed to decay over time. Continuous generation is often preferred to maintain a constant concentration over time; a system test protocol will usually involve an activity routine that varies with time, and hence the challenge should not, in order to incorporate the effect of activities on the result both reproducibly and uniformly. For aerosols, a more likely realistic scenario involves release of an aerosol over a relatively short duration of time, followed by dispersion, agglomeration and fallout. To simulate these factors, a single release followed by a period of time during which a rapidly repeating activity routine is performed would be a better compromise.

Liquid droplets of test agent are usually placed with a syringe on the outside of the protective system at the beginning of the test, although in theory they could be reapplied over the course of the test if rapid application techniques are available. This permits control of the location of the challenge and ensures that dosimeters are located correctly. If the dosimeter covers the entire body (e.g., underclothing impregnated with a colorimetric detection system capable of detecting liquid penetration), a contamination transfer test format could be used with the contamination located instead on the environment within the chamber.

The duration of the test should be at least long enough to replicate a variety of realistic changes in parameters; activity routines should be at least 20 to 30 minutes in duration in order to permit a number of different activities to be performed, and for human test subjects to be breathing hard and sweating. The duration of individual activities should be short compared to any change in concentration over the course of the test. Much longer test durations in a wider variety of temperature conditions can potentially be achieved with mannequins to observe the possibility of closure or material failures with protracted use, although field trials will yield a wider variety of failure modes.

Dosage and Protection Performance Metrics. Depending on the physical form of the agent and the design of the protective system, there are various ways in which

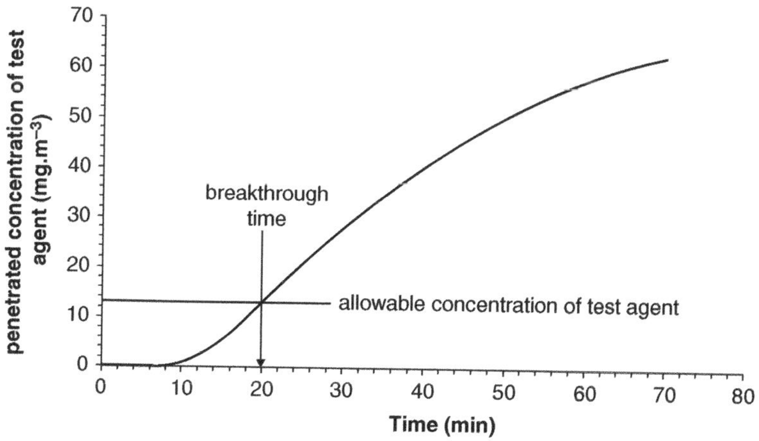

FIGURE 6-2 Breakthrough curve for gas testing of an air-purifying element such as an active-carbon bed in a canister.

dose can reach the wearer, and each is measured differently. Vapors and liquids can reach the wearer by a combination of penetration and permeation. In the case of permeation, characterization parameters are breakthrough time and permeation rate (the slope of measured permeated mass vs. time), as well as total permeated dose. Penetration is usually characterized by percent penetration or inward leakage (the concentration within or behind the protective item divided by the concentration outside, expressed as a percentage), or even simply by the qualitative observation of penetration; however, penetration can also have a breakthrough time, as for example, in Figure 6-2, where the breakthrough of a gas through a canister is plotted as a function of time. Most correctly, the breakthrough time would be set as that time at which an allowable dose would be reached by the wearer (where the dose at a given time is the area under the curve), but it is frequently the case that test methods specify an allowable concentration instead, past which toxic effects have been observed by unprotected persons over a certain specified time period.

In many cases, the allowable concentration or dosage of toxic materials has not been determined for skin. Under these circumstances, for industrial protective clothing, zero allowable dose has frequently been the default value, potentially resulting in considerable overprotection for many toxic industrial chemicals. This default assumption is being considered for revision [447].

The overall vapor protection performance of a system is generally characterized by a *protection factor*, PF:

$$PF = \frac{Ct_{\text{outside}}}{Ct_{\text{inside}}} \qquad (6\text{-}1)$$

where Ct is the vapor dosage either outside the system or inside, above the skin of the wearer, at a given location within the system. Penetration through materials and closures, as well as permeation through materials, will all contribute to the result. Note that the PF will vary from location to location, depending on the location of leakage into the system relative to the measurement.

Vapor dose above the skin is best measured using dosimeters placed on the skin that simulate the characteristics of how agent is taken up by skin. As described by equation (6-2), a dosimeter with *uptake rate* u having an active sample area A takes up mass m of challenge agent proportional to the dose Ct to which it is exposed. For accurate dosimetry, u should be linear at values that are low relative to its overall capacity and similar to that of skin, which is on the order of 0.1 to 10 cm·min^{-1} for chemical warfare agents:

$$m = uACt \tag{6-2}$$

Aerosols can penetrate systems, remaining airborne, and their concentrations or dosages can thus be used as protective performance measures in exactly the same manner as vapors. Note, however, that in the case of aerosols (unlike vapors), mass and number of particles are no longer related in a constant manner, as particulates are virtually never generated in a single size and always aggregate as they are dispersed and circulate in the test system.

Different measurement methods may bias toward one of counting particles, measuring particle diameter or measuring total mass, and will thus yield different results for amount penetrated; this is exacerbated by the fact that the removal of particulates by filtration and deposition mechanisms will vary in efficiency for different particle sizes (an extensive discussion of this effect for microorganisms is given in [448]), meaning that the challenge particle size distribution will almost certainly differ from that ultimately measured within the protective system. Hence, for the most generally useful information, amount penetrated or PF values should, strictly speaking, be expressed for a given small particle size range within which these variations are negligible. Therefore, a given protective system will yield multiple PF values (differing by location, and depending on the airborne hazard—vapor, or a given size range of aerosol particle).

Alternatively, when aerosols deposit onto surfaces, the deposition (mass deposited per unit area) is the raw parameter that is considered. The *deposition velocity* v_d is described as

$$v_d = \frac{F}{C} \tag{6-3}$$

where F is the flux (amount per unit area per unit time) of the material passing through a given area A and C is its concentration above the surface. This equation can be rearranged to be identical in form to equation (6-2):

$$m = v_d ACt \tag{6-4}$$

where A is the surface area sampled. Chemical aerosols may also generate evolved vapors, in which case the vapor concentration can be measured in addition. For bioaerosols, tests should be based on the viability of the organism under realistic exposure conditions.

System-Level Test Agent Sampling Approaches. In a system-level test, the test agent's concentration and movement will be modulated by realistic processes typical of natural airflow patterns both interior and exterior to the protective system. Exterior airflow patterns will probably be turbulent as opposed to laminar and may affect the material in any direction; interior airflow patterns will depend on bellowing effects, material air permeability, the spacing of the material from the body, and the shape of the body. Hence, if used, mannequin platforms should be as similar as possible to human beings in all possible regards.

Test agent must be sampled in a manner that will not affect the outcome once it has penetrated the system. To measure correctly dose penetrated or concentration at a given location, test agent that is removed for the purposes of measurement must be a small fraction of that present, or should be sampled in a realistic manner. *Active sampling* (using forced airflow into a collection system or detector, Figure 6-3) is convenient, as it permits online sampling, but it can disturb the outcomes unless the sampling flow rates are very low compared with the natural flow conditions beneath the system; it may also require penetration of the protective layers by cables or hoses. *Passive sampling* (Figure 6-4) in which the test agent diffuses into a capture medium is preferred, but analysis is usually more labor intensive, as the mass accumulated by the passive dosimeter is analyzed off-line, and only total accumulated dose is obtained. Depending on the test agent form and how it can be recovered from the sampler, skin itself may be used as the sampler, or alternatively, some other absorbing or adsorbing detection system is used.

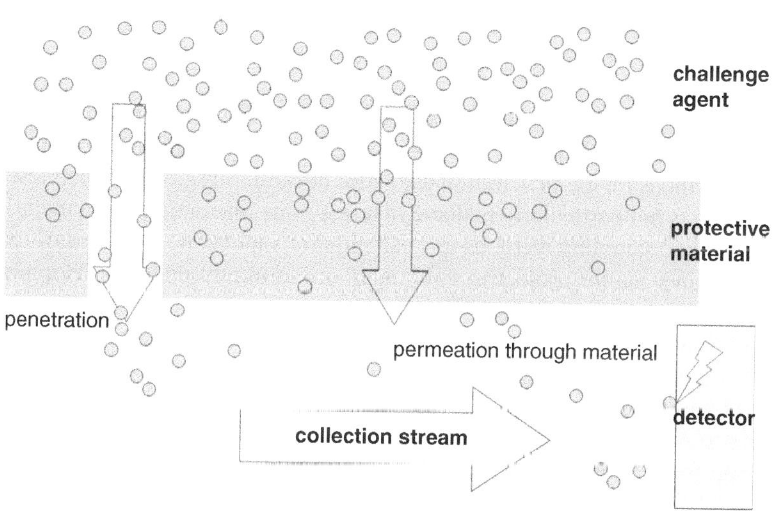

FIGURE 6-3 Active sampling arrangement to measure mass permeating through a system or material using a collection fluid stream.

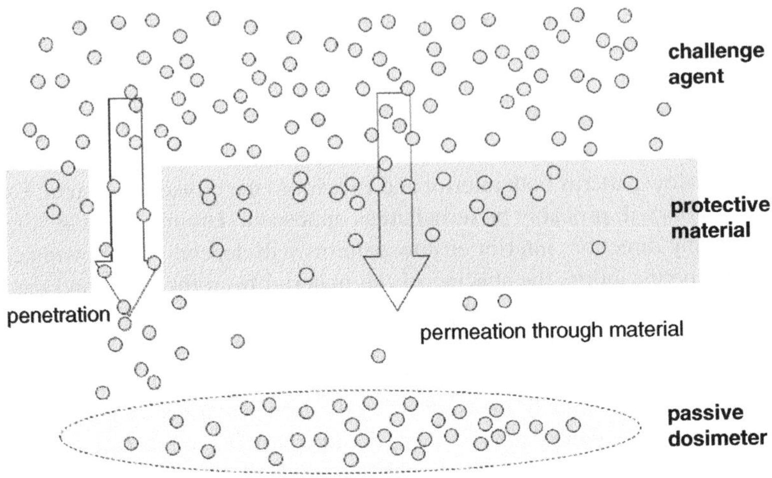

FIGURE 6-4 General arrangement to measure cumulative dose permeating through a system or material using a passive dosimeter.

For respirators, aerosols are the most common test agent to measure the protection factor against airborne substances. For this purpose, active sampling is possible and convenient because the air exchange rates within the system due to breathing and additional air supplies are very high (many liters per minute in and out). In addition, respirators often have drinking tubes that permit access to the nose cup without damaging the respirator's integrity. The airflow patterns are dominated by these factors, and sampling flow rates of a few hundred $mL{\cdot}min^{-1}$ into the detector are considered negligible. If there is no drinking tube, the interior of the respirator is sampled by attaching a sampling tube that penetrates the facepiece or visor. Either number count or mass-based sampling systems are used, and the results are available in real time.

For liquid and vapor exposures of full protective systems, sampling must be performed in situ, as the PPE cannot be removed before measuring. Existing online active sampling methods for gas penetration use sulfur hexafluoride gas as an inert test agent [409], which is not particularly realistic in terms of its interactions with materials or systems. Online sampling methods that can monitor vapor concentrations of realistic CWA simulants within protective equipment are currently under development. In the meantime, passive dosimeters that absorb test agent approximately the way that skin absorbs semipersistent or persistent CWAs are used (see Figure 6-4); one such dosimeter, the passive adsorbent dosimeter, is designed for monitoring of penetrated dosage of vapor-phase methyl salicylate in the vapor MIST and is described in the appendix to ASTM F2588 [449].

Inert aerosol exposures at the system level can be of two types in principle, dry aerosol and wet aerosol. Dry aerosol sampling of the skin is usually performed by skin washing: for example by demarcating an area of skin and swabbing to recover the aerosol into a liquid collection medium. Other options are recovery by vacuuming, tape stripping, or collection onto a surrogate surface such as a patch sampler [450,451].

With a liquid (or dry but vapor-evolving) aerosol, vapor concentration should also be measured if the test is simulating an agent with a measurable vapor pressure; however, this is technically challenging when using vapor dosimeters, as their response or uptake rate to any aerosol that deposits on their surface is difficult to calibrate and could be different from their response to the vapor itself.

For bioaerosols, swabbing should be done in an appropriate medium that will maintain viability but quench any decontaminants or reactive substances that might be recovered at the same time as the organism. If desired, the aggregates deposited can be broken up prior to viability testing to get a total organism or spore count. Subsequent viability testing is usually based on growing the organisms in an appropriate culture medium and counting the number of colony-forming units (CFU). For passive protective systems in which no decontaminants or reactive substances are present, an alternative approach for bioaerosol skin sampling involves using a contact (or RODAC) plate that contains a growth medium and pressing the medium's surface against the skin. In this case, only CFU deposited on the skin are measured (as opposed to individual organism numbers). Recovery using this method is lower than with repeated swabbing, but considerable processing time can be saved, along with better viability retention for less hardy organisms. If the concentration of organism on the surface to be sampled by the contact plate is too high, the method cannot count colonies reliably, whereas in the swabbing method, serial dilutions can be performed to lower concentration as needed before plating for growth.

6.4.4 CBRN Protective Performance Measures Using Materials

In a material-level swatch test format, it is much more difficult to perform realistic exposure and measure exposure dose correctly than in a system-level test. Because of the scaled-down nature of the test, attempts to measure dose can affect the outcome of the test as the challenge agent is withdrawn from the system. The control that is possible over exposure conditions is viewed as an advantage of this type of test format, but this same control means that only a very select set of conditions is evaluated, and many realistic issues that affect performance, such as airflow around the material, movement, humidity variations across the material, and contact with skin are either poorly modeled or neglected at this stage. The test cell must be large enough that edge effects are negligible, and it must be leak-free and decontaminable. Side-by-test test cell configurations can permit higher throughput or more replicates.

Owing to a natural desire for automated time-resolved sampling, many swatch tests use some form of active sampling collection system that collects concentration data at intervals (using a detector located in the stream, or removing samples and measuring remotely) (Figure 6-3). As noted above, this method of measurement is convenient and allows for real-time measurement and higher test throughput than using a passive dosimeter, making it particularly useful for screening materials. However, it does not always permit an accurate reflection of the cumulative dose that would have been obtained under more realistic measurement conditions, as the penetrating agent must be removed from beneath the material to be measured, which tends to affect the driving forces behind the various processes. Also, only total permeated mass can be measured by this method, as the concentration of agent in the collection stream is an arbitrary function of the collection stream's volume and flow rate. The volume of the space beneath the material and the flow rate will affect the final concentration measured, partly because a larger volume will dilute the test agent as it permeates through, and partly because when the volume is smaller, the permeated or penetrated test agent is more likely to be adsorbed onto any adsorbent within the protective material system by back-diffusion.

Vapor Testing. Generation of a challenge vapor below the saturation concentration for use in a relatively small volume can be very difficult to perform reproducibly and in a controlled manner. Flash evaporation of small, metered liquid volumes into an airstream with subsequent effective mixing is the usual approach, with the resulting concentrated airstream usually diluted again into another clean airstream to achieve the concentration desired. This stream then flows over or through the material to be tested. If the results are scaled for the challenge concentration, the concentration of the test agent in principle makes no different to the outcome until it becomes so great that it saturates the adsorbent or changes the barrier properties of the material as it dissolves into it.

For air-permeable adsorbent materials, at higher flows, the test agent vapor may break directly through the material without being adsorbed. Control of incident wind conditions in any direction other than perpendicular to the material is particularly difficult. Additionally, the airstream beneath will affect the outcome. However, since incident wind perpendicular to the material is the worst case, this is not a major issue with the test format. A theoretical incident wind speed (V_w) is set using the measured pressure drop across the material, or alternatively, flow through it (F_{mat}) and its previously measured flow resistance R_{mat}. Usually, to obtain more realistic results, an additional parameter is included that takes into account a theoretical geometry in which the material is spaced at a specified distance around a cylinder (representing a limb). For example, to correct the flow for a swatch test with a perpendicular incident wind to match the case of a cylinder of radius 10 cm and a 2 to 5-mm spacing of the material from the cylinder, we use [4]

$$F_{\mathrm{mat}}(\mathrm{cm}^3 \cdot \mathrm{min}^{-1}) = 0.057 \frac{V_w(\mathrm{m} \cdot \mathrm{s}^{-1})^2}{R_{\mathrm{mat}}(\mathrm{mmH_2O}/[\mathrm{cm} \cdot \mathrm{s}^{-1}])} \qquad (6\text{-}5)$$

When testing an air-impermeable material, flow of vapor parallel to the material surface is generally used to maintain constant challenge concentration.

Liquid Testing. Liquid tests can be intended to monitor penetration, permeation, or both. Liquids can be applied as bulk or as individual droplets; either penetration of liquid through material defects or pores, or permeation of vapor, may be monitored. Permeation is measured quantitatively by any of the methods described previously for

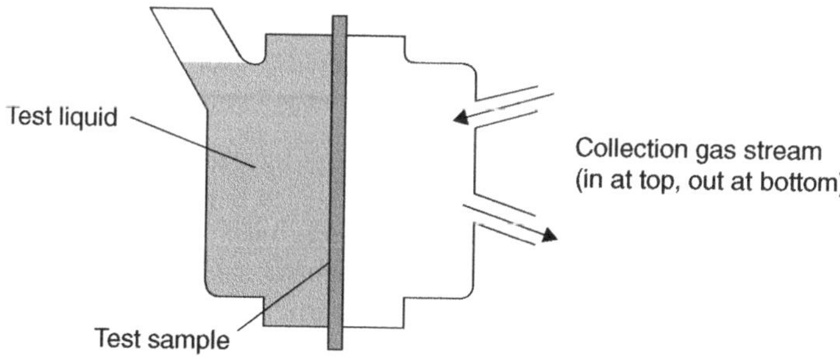

FIGURE 6-5 Schematic of a liquid or vapor permeation test cell.

system-level testing. Dosimetry or detection of penetrated liquid (liquid/liquid penetration test) is not usually quantitative. In general, methods depend on colorimetric response using some form of detector paper, and the fraction of the surface that has colored in reaction to the presence of liquid is used as the test criterion. Generally, no penetration at all would be allowed when testing for CWA penetration.

A typical ASTM F739 [452] test cell for bulk liquid application or vapor permeation with active sampling is illustrated in Figure 6-5. The test cell can be rotated 90° clockwise so that the liquid is applied on top in droplets. Similar test formats are described by NATO [4], generally used for droplet application. Small droplet application can be difficult to perform reproducibly—automated application is preferred—and deciding the drop size, drop density, and pattern for application is often somewhat arbitrary given that a realistic liquid challenge will be randomly distributed and polydisperse in size. Issues such as edge effects in the test cell can be problematic, where different results may be obtained depending on the exact pattern of the application even when all else is kept equal. Placing liquid drops on seams and seals is often used to test their integrity.

Airflow perpendicular to the material is generally used only for air-permeable materials. Airflow parallel to the material surface over the top of liquid droplets is often present, primarily to evaporate the droplet in a realistic manner. For example, small droplets of sulfur mustard may evaporate in a few hours, whereas depending on the concept of the use of the item (i.e., how long it would be worn after exposure), a test will continue for longer to observe the material breakthrough. If there is no airflow and the sample is covered, the test is termed *occluded*, simulating the presence of a liquid drop that is covered by other items.

Pressure placed on the material during or after application of the liquid simulates contact between the body and other objects after liquid encounter, or a drop falling onto the PPE. This pressure can be applied by the use of a specified weight, by an automated piston, or by allowing a drop to fall from a specified height. For air permeable materials, pressure almost always causes increased penetration. For air-impermeable materials, pressure is likely to spread the droplet, which can in fact increase protection against the permeation since a larger area will be covered by the drop, which may be less likely to permeate and could evaporate faster. Hence, the worst-case test may be different in these two cases. Allowing a drop to fall from a height will result in an initial higher pressure where the drop contacted, but the drop will probably spread as a result of splash. Therefore, the cell must be large enough to capture the entire drop realistically as it spreads.

Chemical, Radiological, and Inert Aerosol Testing. Inert aerosols are often used as a test of the filtering capacity of materials and canisters. Control and measurement of aerosol characteristics can be challenging; particle size distributions can vary as a function of time and generation method, and more than one type of analyzer may be required to assess the full distribution. Different analyzers also measure or report different apparent diameters [453,454] as well as having different losses at different particle sizes [455]. Many aerosol generation methods result in charge being deposited onto the aerosol, which can affect the filtration effectiveness of the test item, and therefore charge neutralization may be required.

A schematic for test apparatus for measurement of the filtration efficiency of fabric is shown in Figure 6-6. The apparatus is based, in part, on ASTM Standard F1215-89 (now withdrawn) for determining the filtration efficiency of flat-sheet filter media.

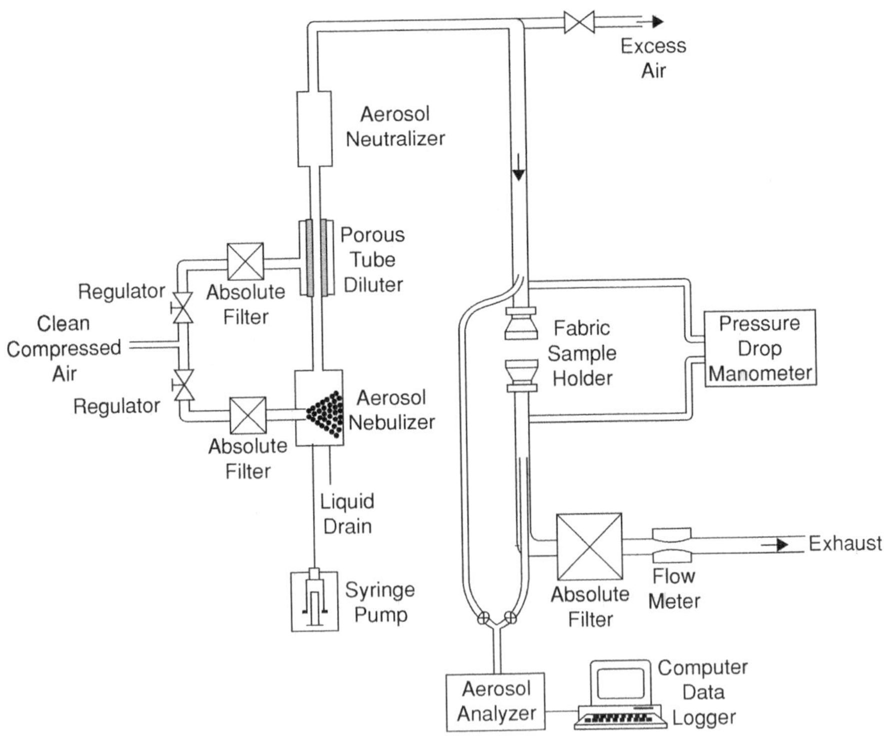

FIGURE 6-6 Inert aerosol fabric penetration test apparatus used at Research Triangle Institute. (Reproduced by permission of, and © RTI International, date unknown.)

The apparatus is constructed of polished stainless steel (providing a smooth static-free duct), is temperature and humidity controlled for operation over a wide range of conditions, and is typically operated with a constant 25-Pa pressure drop across the swatch. The test aerosol consists of oleic acid, whose concentration is diluted as needed and then neutralized before passing through the fabric sample. The flow rate through the fabric is controlled to the appropriate pressure drop (permitting comparison across fabrics of different air permeabilities). The aerosol analyzer measures particle concentration over a variety of particle sizes (0.3 to 5 μm optical diameter), and the penetration P for each particle size bin can be determined as the concentration downstream divided by the concentration upstream:

$$P = \frac{C_{\text{downstream}}}{C_{\text{upstream}}} \tag{6-6}$$

The penetration velocity v is calculated as

$$v = P \times v_{\text{face}} \quad \text{(at 25 Pa)} \tag{6-7}$$

where v_{face} is the measured face velocity of the airstream at the pressure drop given. Figure 6-7 shows the penetration P and penetration velocity v for an example material.

Biological Aerosol Penetration and General Viability Testing. For evaluations using a viable biological aerosol, the approach is similar to that described above. Typically, a Collison nebulizer [456] or a sparging nebulizer [457] is used to generate the

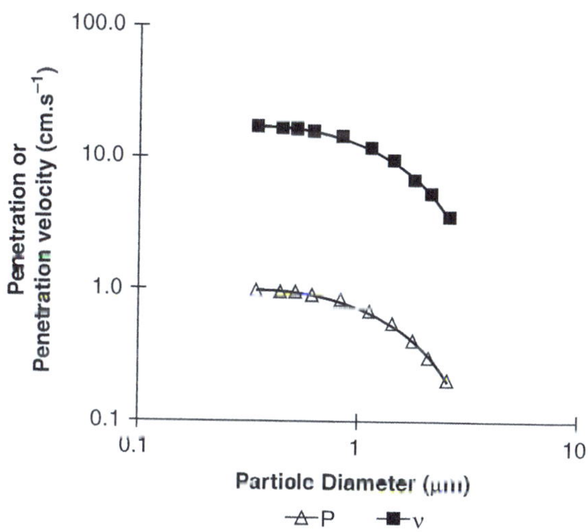

FIGURE 6-7 Penetration and penetration velocity plotted against particle diameter for a nonfiltering fabric, obtained using the apparatus described in Figure 6-6.

aerosol, which can then be diluted and dried. Charge neutralization is not performed, as this would affect the viability of the biological material. The airstream containing the bioaerosol can be split if desired so that control and experimental fabric samples or filters are measured simultaneously. Collection of the aerosol after passing through the sample is performed using, for example, a multistage Andersen impactor [458] containing a suitable growth substrate such as a petri dish. The viability of the biological material collected is determined by culturing and counting. In this case, only the viability of that portion of the aerosol that is not captured by the fabric is determined. The same experiment can be used to measure filtration efficiency relative to a nonfiltering control when samples are not biocidal.

For experiments in which microbicidal activity of a material is being determined, control of the flow rate, temperature, and relative humidity are all important, as the reaction with the biocide is likely to be highly dependent on contact time and conditions. Viability testing is often performed using a contact geometry, in which a biocidal material is placed in a medium or against a culture plate, with either the material or the medium having been dosed with organisms. The apparent time dependency and efficiency of the biocidal action are often determined in this manner. This type of approach lacks fidelity for a number of reasons. First, the nature of the contact between the organism and the biocide, as well as the environmental conditions, are unrealistic compared with an exposure in air. Second, in some cases the biocide can leach out into the solution, which is again unrealistic. Alternatively, the organism in contact with the material can be removed by swabbing or extraction into solution after the contact time period. Whichever method is used, during the growth period, any biocide that might have transferred to the growth medium must be quenched by adding an appropriate reagent, whose presence must not itself interfere with growth; for example, biocidal quaternary ammonium salts are neutralized by letheen broth [459].

6.4.5 Human Performance Measures Using Items or Systems

Human performance measures are extremely difficult to obtain in a standardized and realistic manner. As always, the preference should be to use human test subjects to obtain the outcomes. However, the variability among humans is immense, even given only a single variable, and obtaining representative performance using a test subpopulation is challenging when taking into consideration the sheer number of possible variables that can affect both performance and the validity of test outcomes. The physiology of any given test subject can vary from day to day or even within a given day, and for some tests, having been tested previously can affect the outcome of the next test by, for example, changing the physical or emotional state or level of training.

The potentially relevant variables (probably not exhaustive) include:

- Anthropometry and weight
- Fitness and health
- Gender

- Age
- Role or occupation and relevant training
- Motivation, emotional state, or state of stress
- State of rest or tiredness
- Activity immediately prior to test that can elevate metabolism, and work rate during the test
- Time of the day, month, and season of the test
- Nutrition and hydration
- Environmental conditions of the test and of the test subject's environment prior to the test

The relevance of each depends on the operational capability or parameter being evaluated.

It is apparent from examining this list that the most realistic outcomes will be obtained using a test population that is drawn from the actual user group while being as large and representative as possible. A previous understanding of the impact of some of the difficult-to-control parameters that can result in variable results for a given person is important. The person's previous status may be understood or controlled to the extent possible prior to the test by controlling their state of rest or nutrition or hydration status, understanding their possible acclimatization to test environmental conditions, providing uniform training prior to performing the test, and other factors.

Realistic evaluations of operational capability while wearing the full system are essential (although not always required during procurement) and are often the last stage of a development or acquisition program. Mannequin- or headform-type platforms for testing items and systems are often a desirable alternative, particularly when ranking is being performed, due to the ability to perform controlled repeated testing. Mannequin platforms are rarely available in multiple sizes or to represent anything other than the "average male," which probably makes tests performed using these types of platforms even further from reality than when assessing CBRN protective performance. However, the ability to introduce human-like features in mannequins is improving: with sweating and skin temperature control now possible, for example. The ability to perform scans of humans, as well as the capability for rapid prototyping and production of headforms and molds, means that the anthropometric fidelity of headforms and mannequins should improve over the next few years. There is a paucity of data relating meaningful operational performance to outcomes of controlled nonhuman platform tests, and hence it is likely that some combination of test designs crossing the full range of platforms will be the best approach for some years to come.

6.4.6 Human Performance Measures Using Materials

The measures of human performance that can be evaluated at the material level are relatively few. Certain material-level physical parameters can be related to human performance, such as hand properties, which can be related to comfort, and thermal and moisture vapor transmission, which can be related to thermal burden. For example, a Canadian standard [3] lists the values for evaporative resistance of materials and how they may be related to working time in a poorly ventilated PPE system, as limited by the thermal burden under warm conditions. However, truly meaningful absolute criteria are rare for these types of methods, for all the reasons mentioned in Section 6.4.5, combined with the additional variables introduced by system design. The thermal burden can be affected more by the ventilation of a system than by the materials of which it is made, and subjective comfort rankings are often based more on one particular "most uncomfortable" feature of a system than on a combination of factors that can be individually optimized.

6.5 SOURCES OF METHODS

Each section that follows starts with a brief discussion and then gives a table of selected standards that contain relevant information on test methods and conditions; an exhaustive discussion of the contents or relative merits of each standard or test method is not generally included. Essentially all of the standards that specify performance requirements for CBRN PPE of any sort either contain relevant test methods unique to the standard or reference to other standards; these are listed explicitly in the following sections only where unique or particularly informative information is given. Standards that list requirements at the material level are also given in this section. A more comprehensive list of item- or system-level standards is provided in Section 7.3.

Throughout this section, tables of standard test methods are given. The acronyms for the organizations issuing the standards are as follows:

- AATCC: American Association of Textile Chemists and Colorists (www.aatcc.org)*
- ANSI: American National Standards Institute (www.ansi.org)
- ASTM: American Society for Testing and Materials (www.astm.org)
- BSI: British Standards Institution (www.bsigroup.com)
- CGSB: Canadian General Standards Board (www.tpsgc-pwgsc.gc.ca/ongc-cgsb/index-eng.html)*
- CSA: Canadian Standards Association (www.csagroup.org)
- EN: European Norm (Committee for European Norms; www.cen.eu)
- ISO: International Organization for Standardization (www.iso.ch)
- JIS: Japanese Industrial Standard (Japan Standards Association; www.js.or.jp)
- NATO: North Atlantic Treaty Organization (www.nato.int)
- NFPA: National Fire Protection Association (United States; www.nfpa.org)

* AATCC and CGSB have large stables of textile test methods, but only a few particularly relevant ones are mentioned here.

- NIJ: National Institute of Justice (United States; www.nij.gov)
- NIOSH: National Institute of Occupational Safety and Health (United States; www.cdc.gov/niosh)

Note that these lists are not exhaustive but are intended to be representative of many methods commonly used and are almost all available for purchase or download online at the sites indicated above.

6.6 PRECONDITIONING AND PRETREATING

A number of factors can affect the outcome of a test, and depending on the nature of the test and the test platform, some preconditioning or pretreating of the test samples may be relevant. A list of relevant standards is given in Table 6-6.

6.7 PHYSICAL PROPERTIES AND SURVIVABILITY

6.7.1 Dimensions and Weight

A list of relevant standards is given in Table 6-7.

6.7.2 Electrostatics

Static electrification is primarily a problem related to the possible degradation of performance of electronic equipment used by the wearer when it receives a static discharge, after charge has accumulated on the PPE. Selected test methods and requirements are outlined in Table 6-8.

6.7.3 Environmental Stressors

In this section we describe tests that examine the performance of PPE during or after it has been exposed to extreme environmental conditions, such as sunlight, rain, and low temperature (Table 6-9). Some relevant preconditioning information is also given in Section 6.6.

6.7.4 Physical Stressors

An overall summary on wear tests for textiles is given in ASTM D3181 10, *Standard Guide for Conducting Wear Tests on Textiles*, and for mechanical properties of plastics in ISO 6721-1:2011, *Plastics: Determination of Dynamic Mechanical Properties; Part 1: General Principles.* Various standard tests for physical stressors such as abrasion, wear, flex, puncture, and tear are given in Tables 6-10 through 6-12. Other durability tests are addressed in the next section.

6.7.5 Other Durability

Other types of durability tests for materials and RPDs are given in Tables 6-13 and 6-14, respectively.

TABLE 6-6 Preconditioning and Pretreatment Standards

Standard	Title	Comments
ASTM D1776-08e1	Standard Practice for Conditioning and Testing Textiles	Preconditioning
CAN/CGSB 4.2 No. 58 (2007)	Textile Test Methods: Dimensional Change in Domestic Laundering of Textiles	Textile—laundering
CAN/CGSB 4.2 No. 2-M88 (2001)	Conditioning Textile Materials for Testing	Preconditioning
EN 1811:2011	Reference Test Method for Nickel Release	Simulated sweat
EN 13274-5:2001	Respiratory Protective Devices—Methods of Test; Part 5: Climatic Conditions	Preconditioning
ISO 3175-1:2010	Textiles: Professional Care, Drycleaning and Wetcleaning of Fabrics and Garments; Part 1: Assessment of Performance After Cleaning and Finishing	Cleaning and finishing
ISO 3175-2:2010	Textiles: Professional Care, Drycleaning and Wetcleaning of Fabrics and Garments; Part 2: Procedure for Testing Performance When Cleaning and Finishing Using Tetrachloroethene	Dry cleaning
ISO 6330:2000	Textiles: Domestic Washing and Drying Procedures for Textile Testing	Textile—laundering
NATO AEP-38 (2011)	Operational Requirements, Technical Specifications, and Evaluation Criteria for CBRN Protective Clothing	Preconditioning, pretreatments with battlefield contaminants
NIOSH CET-APRS-STP-CBRN-0311 (v1.2 2005)	Laboratory Durability Conditioning Process for Environmental, Transportation, and Rough Handling Use Conditions on Chemical, Biological, Radiological, and Nuclear (CBRN) Respiratory Protective Devices (RPD); Standard Conditioning Procedure (SCP)	Preconditioning for durability; transportation and storage tests
NIOSH CET-APRS-STP-CBRN-0411 (v1.1 2005)	Laboratory Durability Conditioning Process for Environmental, Transportation, and Rough Handling Use Conditions on Chemical, Biological, Radiological, and Nuclear (CBRN) (Air-Purifying or Self-Contained) Escape Respirator	Preconditioning for durability; transportation and storage tests

TABLE 6-7 Dimensional and Weight Test Methods

Standard	Title	Comments
AATCC 96-2009	Dimensional Changes in Commercial Laundering of Woven and Knitted Fabrics Except Wool	Fabric
AATCC 135-2010	Dimensional Changes of Fabrics After Home Laundering	Fabric
AATCC 150-2010	Dimensional Changes of Garments After Home Laundering	Garments
AATCC 158-2005	Dimensional Changes on Drycleaning in Perchloroethylene: Machine Method	
AATCC 187-2009	Dimensional Changes of Fabrics: Accelerated	Fabric
ASTM D1460-86(2010)	Standard Test Method for Rubber Property: Change in Length During Liquid Immersion	Rubber
ASTM D3767-03(2008)	Standard Practice for Rubber: Measurement of Dimensions	Rubber
ASTM D3773 / D3773M-10	Standard Test Methods for Length of Woven Fabric	Woven textile
ASTM D3774-96(2008)e1	Standard Test Methods for Width of Textile Fabric	Textile
ASTM D3775-08	Standard Test Method for Warp (End) and Filling (Pick) Count of Woven Fabrics	Woven fabric
ASTM D3776/ D3776M-09ae1	Standard Test Methods for Mass per Unit Area (Weight) of Fabric	Fabric
ASTM D3882-08	Standard Test Method for Bow and Skew in Woven and Knitted Fabrics	Woven and knitted fabric
ASTM D6132-08	Standard Test Method for Nondestructive Measurement of Dry Film Thickness of Applied Organic Coatings Using an Ultrasonic Gage	Films
BS EN 12127:1998	Textiles—Fabrics: Determination of Mass per Unit Area Using Small Samples	Fabric
CAN/CGSB 4.2 No. 58-2007	Textile Test Methods: Dimensional Change in Domestic Laundering of Textiles	Textile
ISO 3759:2011	Textiles, Preparation, Marking and Measuring Specimens and Garments in Tests for Determining Dimensional Change	Textile
ISO 3801:1977	Woven Fabrics: Determination of Mass per Unit Length and Mass Per Unit Area	Woven fabric
ISO 5077:2007	Textiles: Determination of Dimensional Change in Washing and Drying	Textile
ISO 5084:1996	Textiles: Determination of Thickness of Textiles and Textile Products	Textile

(continued)

TABLE 6-7 (*Continued*)

Standard	Title	Comments
ISO 7771:1985	Textiles: Determination of Dimensional Changes of Fabrics Induced by Cold-Water Immersion	Textile
ISO 9073-1:1989	Textiles: Test Methods for Nonwovens; Part 1: Determination of Mass per Unit Area	Nonwoven fabric
ISO 9073-2:1995	Textiles: Test Methods for Nonwovens; Part 2: Determination of Thickness	Nonwoven
ISO 9073-7:1995	Textiles: Test Methods for Nonwovens; Part 7: Determination of Bending Length	Nonwoven

TABLE 6-8 Tests for Electrostatic Properties

Standard	Title	Comments
ASTM D4470-97(2010)	Standard Test Method for Static Electrification	
EN 1149-1:1996	Protective Clothing: Electrostatic Properties; Surface Resistivity (Test Methods and Requirements)	
EN 1149-2:1997	Protective Clothing: Electrostatic Properties; Part 2: Test Method for Measurement of the Electrical Resistance Through a Material (Vertical Resistance)	
EN 1149-3:2004	Protective Clothing: Electrostatic Properties; Part 3: Test Methods for Measurement of Charge Decay	
EN 1149-5:2008	Protective Clothing: Electrostatic Properties; Material Performance and Design Requirements	

TABLE 6-9 Tests for Environmental Stressors

Standard	Title	Comments
AATCC 111-2009	Weather Resistance of Textiles: Exposure to Daylight and Weather	Daylight, weather
AATCC 169-2009	Weather Resistance of Textiles: Xenon Lamp Exposure	UV, light
AATCC 186-2009	Weather Resistance: UV Light and Moisture Exposure	UV, light, moisture
AATCC 192-2009	Weather Resistance of Textiles: Sunshine-Arc Lamp Exposure with and Without Wetting	Light, wet or dry
ISO 105: various dates, various subparts	Textiles: Tests for Colour Fastness	Color fastness to light, rubbing, perspiration, accelerated aging, wet scrubbing, organic solvents, water, seawater, chlorinated water, perspiration, spotting (acid, alkali, water), hot water, milling, dry heat, etc.
ISO 4675:1990	Rubber- or Plastics-Coated Fabrics: Low-Temperature Bend Test	Coated fabrics, low temperature
NIOSH RCT-ASR-STP-0118 (v1.1 2005)	Determination of Low-Temperature Operations: Minimum Temperature per Applicant, Open-Circuit, Self-Contained Breathing Apparatus	SCBA, low temperature
NIOSH RCT-ASR-STP-0143 (v1.1 2005)	Determination of Low-Temperature Operation: Minimum per Manufacturer, Closed-Circuit, Self-Contained Breathing Apparatus	Closed-circuit SCBA, low temperature

TABLE 6-10 Abrasion Tests

Standard	Title	Comments
ASTM D1044-08	Standard Test Method for Resistance of Transparent Plastics to Surface Abrasion	Visors, etc.
ASTM D1630-06	Standard Test Method for Rubber Property: Abrasion Resistance (Footwear Abrader)	Rubber
ASTM D2228-04(2009)	Standard Test Method for Rubber Property: Relative Abrasion Resistance by the Pico Abrader Method	Rubber
ASTM D3884-09	Standard Guide for Abrasion Resistance of Textile Fabrics (Rotary Platform, Double-Head Method)	Fabric
ASTM D3885-07a	Standard Test Method for Abrasion Resistance of Textile Fabrics (Flexing and Abrasion Method)	Fabric
ASTM D3886-99(2007)e1	Standard Test Method for Abrasion Resistance of Textile Fabrics (Inflated Diaphragm Apparatus)	Fabric
ASTM D3939-11	Standard Test Method for Snagging Resistance of Fabrics (Mace)	Fabric
ASTM D4157-10	Standard Test Method for Abrasion Resistance of Textile Fabrics (Oscillatory Cylinder Method)	Fabric
ASTM D4158-08	Standard Guide for Abrasion Resistance of Textile Fabrics (Uniform Abrasion)	Fabric
ASTM D4966-10	Standard Test Method for Abrasion Resistance of Textile Fabrics (Martindale Abrasion Tester Method)	Fabric
ASTM D4970/D4970M-10e1	Standard Test Method for Pilling Resistance and Other Related Surface Changes of Textile Fabrics: Martindale Tester	Fabric
ASTM D6770-07 (2011)	Standard Test Method for Abrasion Resistance of Textile Webbing (Hex Bar Method)	Webbing
ASTM F735-06	Standard Test Method for Abrasion Resistance of Transparent Plastics and Coatings Using the Oscillating Sand Method	Visors, etc.
EN 530:2010	Abrasion Resistance of Protective Clothing Material: Test Methods	
ISO 12947-1:1998, -2:1998, -3:1998, -4:2000	Determination of the Abrasion Resistance of Fabrics by the Martindale Method	Fabric
ISO 12945-1:2000 and -2:2000	Determination of Fabric Propensity to Surface Fuzzing and to Pilling	Fabric

TABLE 6-11 Break, Stretch, Tear, Flex, Bend, Strength, and Durability Tests: Materials

Standard	Title	Comments
ASTM D747-10	Test Method for Apparent Bending Modulus of Plastics by Means of a Cantilever Beam	Plastics
ASTM D1424-09	Standard Test Method for Tearing Strength of Fabrics by Falling-Pendulum Type (Elmendorf) Apparatus	Fabric
ASTM D2136-02	Standard Method for Coated Fabrics: Low-Temperature Bend Test	Coated fabric
ASTM D2261-07ae1	Standard Test Method for Tearing Strength of Fabrics by the Tongue (Single Rip) Procedure (Constant-Rate-of-Extension Tensile Testing Machine)	Fabric
ASTM D5034-09	Standard Test Method for Breaking Strength and Elongation of Textile Fabrics (Grab Test)	Textile fabric
ASTM D5035-11	Standard Test Method for Breaking Force and Elongation of Textile Fabrics (Strip Method)	Fabric
ASTM D5587-08	Standard Test Method for Tearing Strength of Fabrics by Trapezoid Procedure	Fabric
ASTM D6614-07	Standard Test Method for Stretch Properties of Textile Fabrics: CRE Method	Textile fabric
ASTM D6775-02(2007)	Standard Test Method for Breaking Strength and Elongation of Textile Webbing, Tape and Braided Material	Webbing
ASTM F392 / F392M-11	Standard Test Method for Flex Durability of Flexible Barrier Materials	Polymerics
ISO 1421:1998	Rubber- or Plastics-Coated Fabrics: Determination of Tensile Strength and Elongation at Break	Coated fabrics
ISO 4674-1:2003	Rubber- or Plastics-Coated Fabrics: Determination of Tear Resistance; Part 1: Constant Rate of Tear Methods	Coated fabrics
ISO 4674-2:1998	Rubber- or Plastics-Coated Fabrics: Determination of Tear Resistance; Part 2: Ballistic Pendulum Method	Coated fabrics
ISO 7854:1995	Rubber- or Plastics-Coated Fabrics: Determination of Resistance to Damage by Flexing	Rubber- or plastics-coated fabric
ISO 5978:1990	Rubber- or Plastics-Coated Fabrics: Determination of Blocking Resistance	Rubber- or plastics-coated fabric

(continued)

TABLE 6-11 (*Continued*)

Standard	Title	Comments
ISO 9073-3:1989	Textiles: Test Methods for Nonwovens; Part 3: Determination of Tensile Strength and Elongation	Nonwovens
ISO 9073-4:1997	Textiles: Test Methods for Nonwovens; Part 4: Determination of Tear Resistance	Nonwovens
ISO 9073-18:2007	Textiles: Test Methods for Nonwovens; Part 18: Determination of Breaking Strength and Elongation of Nonwoven Materials Using the Grab Tensile Test	Nonwovens
ISO 13934-1:1999	Textiles: Tensile Properties of Fabrics; Part 1: Determination of Maximum Force and Elongation at Maximum Force Using the Strip Method	Fabrics
ISO 13934-2:1999	Textiles: Tensile Properties of Fabrics; Part 2: Determination of Maximum Force Using the Grab Method	Fabrics
ISO 13935-1:1999	Textiles: Seam Tensile Properties of Fabrics and Made-Up Textile Articles; Part 1: Determination of Maximum Force to Seam Rupture Using the Strip Method	Seams in fabrics
ISO 13935-2:1999	Textiles: Seam Tensile Properties of Fabrics and Made-Up Textile Articles; Part 2: Determination of Maximum Force to Seam Rupture Using the Grab Method	Seams in fabrics
ISO 13937-1:2000	Textiles: Tear Properties of Fabrics; Part 1: Determination of Tear Force Using Ballistic Pendulum Method (Elmendorf)	Fabrics
ISO 13937-2:2000	Textiles: Tear Properties of Fabrics; Part 2: Determination of Tear Force of Trouser-Shaped Test Specimens (Single Tear Method)	Fabrics
ISO 13937-3:2000	Textiles: Tear Properties of Fabrics; Part 3: Determination of Tear Force of Wing-Shaped Test Specimens (Single Tear Method)	Fabrics
ISO 13937-4:2000	Textiles: Tear properties of fabrics; Part 4: Determination of Tear Force of Tongue-Shaped Test Specimens (Double Tear Test)	Fabrics

TABLE 6-12 Burst, Puncture, Cut, and Ballistic Tests

Standard	Title	Comments
ASTM D2582-09	Test Method for Puncture-Propagation Tear Resistance of Plastic Film and Thin Sheeting	Puncture, film
ASTM D6797-07	Standard Test Method for Bursting Strength of Fabrics Constant-Rate-of-Extension (CRE) Ball Burst Test	Bursting, fabric
ASTM D3786 / D3786M-09	Standard Test Method for Bursting Strength of Textile Fabrics: Diaphragm Bursting Strength Tester Method	Bursting, fabric
ASTM D3787-07	Standard Test Method for Bursting Strength of Textiles: Constant-Rate-of-Traverse (CRT) Ball Burst Test	Bursting, textile
ASTM F1342-05	Test Method for Protective Clothing Material Resistance to Puncture	Puncture, material
ASTM F1790-05	Test Method for Measuring Cut Resistance of Materials Used in Protective Clothing	Cut, material
ASTM F1790-04	Standard Test Method for Measuring Cut Resistance of Materials Used in Protective Clothing	Cut, material
EN 863:1995	Protective Clothing: Mechanical Properties—Test Method: Puncture Resistance	Puncture, material
ISO 13995:2000	Protective Clothing: Mechanical Properties—Test Method for the Determination of the Resistance to Puncture and Dynamic Tearing of Materials	Puncture, dynamic tearing, material
ISO 13997:1999	Protective Clothing: Mechanical Properties—Determination of Resistance to Cutting by Sharp Objects	Cut, material
ISO 13996:1999	Protective Clothing: Mechanical Properties—Determination of Resistance to Puncture	Puncture, material
EN 863:1995	Protective Clothing: Mechanical Properties—Test Method: Puncture Resistance	Puncture, material
Euro-Asian Interstate Council for Standardization, Metrology and Certification (EASC): Russia-GOST R50744-95	Ballistic Standard	Ballistic, Russian body armor

(continued)

TABLE 6-12 (*Continued*)

Standard	Title	Comments
ISO 13938-1:1999	Textiles: Bursting Properties of Fabrics; Part 1: Nonwoven Materials	Burst, nonwoven
pr ISO 14876-2-2002	Protective Clothing: Body Armour; Part 2: Bullet Resistance	Bullet, body armor; draft standard
pr ISO 14876-3-2002	Protective Clothing: Body Armour; Part 3: Knife Stab Resistance	Stab, body armor; draft standard
pr ISO 14876-4-2002	Protective Clothing: Body Armour; Part 4: Needle and Spike Stab Resistance; Requiremnets and Test Methods	Needle/spike, body armor; draft standard
NATO STANAG 2920 Ed. 2	Ballistic Test Method for Personal Armour Materials and Combat Clothing	Test method for personal armor used by NATO forces
NIJ Standard-0101.06	Ballistic Resistance of Body Armor	U.S. bullet-resistant vests used by law enforcement
UK Home Office Police Scientific Development Branch (2003).	Standard for Ballistic Resistance and Knife and Spike Resistance	UK police body armor
Underwriters' Laboratory UL 752 (2006)	Ballistic Standards	Body armor
U.S. Army MIL-STD-662F (1997)	V50 Ballistic Test for Armor	Bullet-resistant vests used by U.S. Army

TABLE 6-13 Other Durability Tests: Materials

Standard	Title	Comments
ASTM D751-06	Standard Test Methods for Coated Fabrics	Fabrics, adhesion/ delamination
ASTM D 3359-09e2	Standard Test Methods for Measuring Adhesion by Tape Test	Laminates, visors, coatings
ISO 5978:1990	Rubber- or Plastics-Coated Fabrics: Determination of Blocking Resistance	Coated fabrics, sticking

TABLE 6-14 Other Durability Tests: RPDs

Standard	Title	Comments
ISO (proposed) 16900-6	Respiratory Protective Devices: Methods of Test and Test Equipment; Mechanical Resistance; Strength of Components	In preparation
NIOSH RCT-ASR-STP-0100 (v1.1 2005)	Determination of Strength of Hoses and Couplings: Type C and CE, Supplied-Air Respirators	Hoses and couplings
NIOSH RCT-ASR-STP-0101 (v1.1 2005)	Determination of Tightness of Hoses and Couplings: Type C and CE, Supplied-Air Respirators	Hoses and couplings
NIOSH RCT-ASR-STP-0104 (v1.1 2005)	Determination of Air-Regulating Valve 100,000 Cycles Performance: Demand and Pressure-Demand, Type C and CE, Supplied-Air Respirators	Regulator valves

6.7.6 Survivability Against Heat and Flame

The documents ASTM D4723-07e2, *Standard Classification Index of and Descriptions of Textile Flammability Test Methods*, and EN 1103:2005, *Textiles: Fabrics for Apparel—Detailed Procedure to Determine the Burning Behavior*, provide some overview of relevant test approaches and procedures. Relevant vocabulary can be found in ISO 4880:1997, *Burning Behaviour of Textiles and Textile Products: Vocabulary*. Table 6-15 lists relevant standard heat and flame resistance test methods. Note that test methods designed to evaluate the thermal performance of materials and systems as it pertains to the human physiological state (as opposed to those designed to evaluate the survivability of the materials themselves) are given in Section 6.9.6.

6.8 CBRN PERFORMANCE

6.8.1 Generic Integrity and Protection Factor Methods

This category of test includes the use of challenge aerosol, vapor, or liquids as a generic simulant for the protective performance against either airborne or liquid hazards. Respiratory SWPF methods use either fine particulate [3,460,461] or vapor (SF_6) (see Table 6-16), while dermal SWPF/integrity methods use vapor (SF_6), pressure integrity, or liquid sprays or splashes (see various ISO methods in Table 6-17). A variety of less generic tests that are targeted to vapor CWAs, liquid CWAs, and aerosol chemicals are described in later sections.

6.8.2 Chemical Vapor and Liquid Protection

In this section we outline those tests that are performed specifically with, or targeted to, protective performance against vapor- and liquid-phase chemicals, as listed in Tables 6-18 through 6-20. Aerosol protection assessments are described in Sections 6.8.1 and 6.8.3.

TABLE 6-15 Heat and Flame Resistance Test Methods

Standard	Title	Comments
ASTM D1230-10	Standard Test Method for Flammability of Apparel Textiles	
ASTM D6413-08	Standard Test Method for Flame Resistance of Textiles (Vertical Test)	
ASTM D7138-08	Standard Test Method to Determine Melting Temperature of Synthetic Fibers	
ASTM D7140-07	Standard Test Method to Measure Heat Transfer Through Textile Thermal Barrier Materials	
ASTM D7571-10	Standard Specification for Retained Sewn Seam Strength After Exposures to Hot Air and Open Flame	
ASTM F1358-08	Test Method for Effects of Flame Impingement on Materials Used in Protective Clothing Not Designated Primarily for Flame Resistance	
ASTM F1930-00 (2008)	Standard Test Method for Evaluation of Flame Resistant Clothing for Protection Against Flash Fire Simulations Using an Instrumented Manikin	
EN 136:1998 (2003) Sections 8.5 and 8.6	Respiratory Protective Devices: Full Face Masks—Requirements, Testing, Marking; Flammability and Resistance to Thermal Radiation	
EN 367:1992	Protective Clothing: Protection Against Heat and Fire—Method of Determining Heat Transmission on Exposure to Flame	
EN 13274-4:2001	Respiratory Protective Devices: Methods of Test; Part 4: Flame Tests	
ISO 6940:2004	Textile Fabrics: Burning Behaviour—Determination of Ease of Ignition of Vertically Oriented Specimens	
ISO 6941:2003	Textile Fabrics: Burning Behaviour—Measurement of Flame Spread Properties of Vertically Oriented Specimens	
ISO 6942:2002	Protective Clothing: Protection Against Heat and Fire—Method of Test: Evaluation of Materials and Material Assemblies When Exposed to a Source of Radiant Heat	
ISO 10047:1993	Textiles: Determination of Surface Burning Time of Fabrics	

TABLE 6-15 (*Continued*)

Standard	Title	Comments
ISO 12127-1:2007	Clothing for Protection Against Heat and Flame: Determination of Contact Heat Transmission Through Protective Clothing or Constituent Materials; Part 1: Test Method Using Contact Heat Produced by Heating Cylinder	
ISO 13506:2008	Protective Clothing Against Heat and Flame: Test Method for Complete Garments—Prediction of Burn Injury Using an Instrumented Manikin	
ISO 14116:2008	Protective Clothing: Protection Against Heat and Flame—Limited Flame Spread Materials, Material Assemblies and Clothing	
ISO 15025:2000	Protective Clothing: Protection Against Heat and Flame—Method of Test for Limited Flame Spread	
ISO (proposed) 16900-10	Respiratory Protective Devices: Methods of Test and Test Equipment; Part 10: Resistance to Heat, Ignition and Flame	In preparation
ISO 17493:2000	Clothing and Equipment for Protection Against Heat: Test Method for Convective Heat Resistance Using a Hot Air Circulating Oven	
NATO AEP-38 (2011) F.11	Operational Requirements, Technical Specifications and Evaluation Criteria for CBRN Protective Clothing	Nuclear heat flash test

6.8.3 Particulate and Biological: Penetration and Filtration Methods

U.S. Particulate Filtration Classification Standards. In the United States, NIOSH certifies respirator particulate filter media for nonpowered air-purifying respirators to the following classes [465]: N-, R-, and P-series, with three levels of filter efficiency (95%, 99%, and 99.97%) in each class. All filter tests employ the nominally most penetrating aerosol size, 0.3 μm aerodynamic mass median diameter. The N-series is tested against a mildly degrading aerosol of sodium chloride (NaCl), while the R- and P-series filters are tested against a highly degrading liquid aerosol of dioctylphthalate (DOP) oil. N and R concepts are intended to be limited use (e.g., single shift), whereas the P-series has no implied time limit for performance. The common term *HEPA filter* (high-efficiency particulate air filter) refers to any filter that removes 99.97% of airborne particles measuring 0.3 μm [466].

TABLE 6-16 Integrity, Inward Leakage, Protection Factor Tests: RPDs

Standard	Title	Comments
CSA/CGSB Z1610-11: B.2-B.8 and C.5.4	Protection of First Responders from Chemical, Biological, Radiological, and Nuclear (CBRN) Events	
EN 136:1998 (2003) Section 8.16	Respiratory Protective Devices: Full Face Masks—Requirements, Testing, Marking; Inward Leakage	SF_6 or NaCl
ISO/DIS 16900-1 (draft)	Respiratory Protective Devices: Methods of Test and Test Equipment; Part 1: Determination of Inward Leakage	Draft standard
JIS T 8159:2006	Leakage Rate Testing Method for Respiratory Protective Devices	
NIOSH TEB-CBRN-APR-STP-0352 (v1.0 2008)	Determination of Laboratory Respirator Protection Level (LRPL) Values for CBRN Self-Contained Breathing Apparatus (SCBA) Facepieces or CBRN Air Purifying Respirator (APR)	
NIOSH TEB-CBRN-APR-STP-0452 (v2.0 2008)	Determination of Laboratory Respirator Protection Level (LRPL) Values for CBRN Air-Purifying Escape Respirator (APER)	
NIOSH TEB-CBRN-APR-STP-0552 (v1.0 2008)	Determination of Laboratory Respirator Protection Level (LRPL) Values for CBRN Tight-Fitting Powered Air Purifying Respirator (PAPR)	
NIOSH TEB-CBRN-APR-STP-0553 (v1.0 2008)	Determination of Laboratory Respiratory Protection Level (LRPL) Values for CBRN Loose-Fitting Powered Air Purifying Respirator (PAPR)	

TABLE 6-17 Generic Penetration and Integrity Tests: Materials and DPE/PPE Systems

Standard	Title	Comments
AATCC 118-2007	Oil Repellency: Hydrocarbon Resistance Test	Textiles
AATCC 193-2007	Aqueous Liquid Repellency: Water/Alcohol Solution Resistance Test	
ASTM D5151-06	Test Method for Detection of Holes in Medical Gloves	
ASTM F1052-09	Test Method for Pressure Testing Vapor Protective Suits	
ASTM F1296-08	Standard Guide for Evaluating Chemical Protective Clothing	
ASTM F1359-07	Standard Test Method for Liquid Penetration Resistance of Protective Clothing or Protective Ensembles Under a Shower Spray While on a Mannequin	
CSA/CGSB Z1610-11: B.9 and C.5.3	Protection of First Responders from Chemical, Biological, Radiological, and Nuclear (CBRN) Events	
EN 14786:2006	Protective Clothing: Determination of Resistance to Penetration by Sprayed Liquid Chemicals, Emulsions and Dispersions—Atomizer Test	
EN 464:1994	Protective Clothing: Protection Against Liquid and Gaseous Chemicals, Including Aerosols and Solid Particles—Test Method: Determination of Leak-tightness of Gas-tight Suits (Internal Pressure Test)	
ISO 13982-2:2004	Protective Clothing for Use Against Solid Particulates; Part 2: Test Method of Determination of Inward Leakage of Aerosols of Fine Particles into Suits	
ISO 17491-3:2008	Protective Clothing: Test Methods for Clothing Providing Protection Against Chemicals; Part 3: Determination of Resistance to Penetration by a Jet of Liquid (Jet Test)	
ISO 17491-4:2008	Protective Clothing: Test Methods for Clothing Providing Protection Against Chemicals; Part 4: Determination of Resistance to Penetration by a Spray of Liquid (Spray Test)	

(*continued*)

TABLE 6-17 (*Continued*)

Standard	Title	Comments
ISO 811:1981	Textile Fabrics: Determination of Resistance to Water Penetration—Hydrostatic Pressure Test	Textiles
ISO 4920:1981	Textiles: Determination of Resistance to Surface Wetting (Spray Test) of Fabrics	Textiles
ISO 9073-11:2002	Textiles: Test Methods for Nonwovens; Part 11: Run-off	Textiles
ISO 9865:1991	Textiles: Determination of Water Repellency of Fabrics by the Bundesmann Rain-Shower Test	Textiles
ISO 23232:2009	Textiles: Aqueous Liquid Repellency—Water/Alcohol Solution Resistance Test	Textiles
ISO/DIS 17491-1	Protective Clothing: Test Methods for Clothing Providing Protection Against Chemicals; Part 1: Determination of Resistance to Outward Leakage of Gases (Internal Pressure Test)	Draft standard
ISO/DIS 17491-2	Protective Clothing: Test Methods for Clothing Providing Protection Against Chemicals; Part 2: Determination of Resistance to Inward Leakage of Aerosols and Gases (Inward Leakage Test)	Draft standard
ISO/NP 17491-5	Protective Clothing: Test Methods for Clothing Providing Protection Against Chemicals; Part 5: Determination of Resistance to Penetration by a Spray of Liquid (Manikin Spray Test)	Draft standard
NFPA 1991-05	Standard on Vapor-Protective Ensembles for Hazardous Materials Emergencies	SF_6 test
NFPA 1994-07: 8.4 and 8.5	Standard on Protective Ensembles for First Responders to CBRN Terrorism Incidents	

TABLE 6-18 APE Gas and Vapor Capacity and Breakthrough Tests (Chemical)

Standard	Title	Comments
CSA/CGSB Z1610-11: B.7.11	Protection of First Responders from Chemical, Biological, Radiological, and Nuclear (CBRN) Events	
EN 12083:1998	Respiratory Protective Devices: Filters with Breathing Hoses (Non-mask Mounted Filters); Particle Filters, Gas Filters, and Combined Filters; Requirements, Testing, Marking	
EN 12083:1998/AC: 2000	Respiratory Protective Devices: Filters with Breathing Hoses (Non-mask Mounted Filters); Particle Filters, Gas Filters, and Combined Filters; Requirements, Testing, Marking	
EN 14387:2004 + A1: 2008	Respiratory Protective Devices: Gas Filter(s) and Combined Filter(s); Requirements, Testing, Marking	
ISO 16900-4:2011	Respiratory Protective Devices—Methods of Test and Test Equipment; Part 4: Determination of Gas Filter Capacity and Migration, Desorption and Carbon Monoxide Dynamic Testing	
NIOSH TEB-APR-STP-0033B (v2.3 2008)	Determination of Ammonia Service-Life Test, Air Purifying Respirators with Canisters	
NIOSH TEB-APR-STP-0033D (v2.1 2007)	Determination of Ammonia Service-Life Test, Tight-Fitting Powered Air Purifying Respirators with Gas Mask Canister(s)	
NIOSH RCT-APR-STP-0034 (v1.1 2005)	Determination of Carbon Monoxide Service Life	
NIOSH RCT-APR-STP-0035 (v1.1 2005)	Determination of Chlorine Service Life	
NIOSH RCT-APR-STP-0036 (v1.1 2005)	Determination of Chlorine Dioxide Service Life	
NIOSH RCT-APR-STP-0037 (v1.1 2005)	Determination of α-Chloroacetophenone (CN) Service Life	
NIOSH RCT-APR STP-0038 (v1.1 2005)	Determination of Ethylene Oxide Service Life	
NIOSH TEB-APR-STP-0039B (v2.1 2009)	Determination of Formaldehyde Service-Life Test, Air Purifying Respirators with Canisters	

(*continued*)

TABLE 6-18 (*Continued*)

Standard	Title	Comments
NIOSH TEB-APR-STP-0039C (v2.1 2009)	Determination of Formaldehyde Service-Life Test, Powered Air Purifying Respirators with Cartridges	
NIOSH RCT-APR-STP-0040 (v1.1 2005)	Determination of Hydrogen Chloride Service Life	
NIOSH RCT-APR-STP-0041 (v1.1 2005)	Determination of Hydrogen Cyanide Service Life	
NIOSH RCT-APR-STP-0042 (v1.1 2005)	Determination of Hydrogen Fluoride Service Life	
NIOSH TEB-APR-STP-0043B (v2.1 2009)	Determination of Hydrogen Sulfide Service-Life Test, Air Purifying Respirators with Canisters	
NIOSH TEB-APR-STP-0045B (v2.0 2008)	Determination of Methylamine Service-Life Test, Air Purifying Respirators with Canisters	
NIOSH TEB-APR-STP-0045D (v2.0 2008)	Determination of Methylamine Service-Life Test, Tight-Fitting Powered Air Purifying Respirators with Gas Mask Canister(s)	
NIOSH TEB-APR-STP-0046B (v2.3 2008)	Determination of Organic Vapor (Carbon Tetrachloride) Service-Life Test, Air Purifying Respirators with Canisters	
NIOSH TEB-APR-STP-0046D (v2.0 2006)	Determination of Organic Vapor (Carbon Tetrachloride) Service-Life Test, Tight-Fitting Powered Air Purifying Respirators with Gas Mask Canister(s)	
NIOSH RCT-APR-STP-0047 (v1.1 2005)	Determination of Phosphine Service Life	
NIOSH TEB-APR-STP-0048B (v2.0 2008)	Determination of Sulfur Dioxide Service-Life Test, Air Purifying Respirators with Canisters	
NIOSH TEB-APR-STP-0048D (v2.0 2008)	Determination of Sulfur Dioxide Service-Life Test, Tight-Fitting Powered Air Purifying Respirators with Gas Mask Canisters	
NIOSH RCT-APR-STP-0050 (v1.1 2005)	Determination of *O*-Chlorobenzylidene Malononitrile (CS) Service Life	
NIOSH RCT-APR-STP-0062 (v1.1 2005)	Determination of Nitrogen Dioxide Service Life	

TABLE 6-19 Protection Against Liquid and Vapor-Phase Chemicals: Material- and Item-Level Tests

Standard	Title	Comments
ASTM F739-07	Standard Test Method for Resistance of Protective Clothing Materials to Permeation by Liquids or Gases Under Conditions of Continuous Contact	
ASTM F903-03 (2004)	Standard Test Method for Resistance of Materials Used in Protective Clothing to Penetration by Liquids	
ASTM F 1194-99	Standard Guide for Documenting the Results of Chemical Permeation Testing of Materials Used in Protective Clothing	
ASTM F1383-99a	Standard Test Method for Resistance of Protective Clothing Materials to Permeation by Liquids or Gases Under Conditions of Intermittent Contact	
ASTM F1407-99a	Standard Test Method for Resistance of Chemical Protective Clothing Materials to Liquid Permeation-Permeation Cup Method	
ASTM F2130-01	Standard Test Method for Measuring Repellency, Retention, and Penetration of Liquid Pesticide Formulation through Protective Clothing Materials	
CSA/CGSB Z1610-11: C.5.3	Protection of First Responders from Chemical, Biological, Radiological, and Nuclear (CBRN) Events	Contains modifications on the NFPA 1994 method
EN 374-3:1994	Protective Gloves Against Chemicals and Micro-organisms—Part 3: Determination of Resistance to Permeation by Chemicals	
EN 14325:2004	Protective Clothing Against Chemicals: Test Methods and Performance Classification of Chemical Protective Clothing Materials, Seams, Joins and Assemblages	
ISO 6529:2001	Protective Clothing: Protection Against Chemicals: Determination of Resistance of Protective Clothing Materials to Permeation by Liquids and Gases	
ISO 6530:2005	Protective Clothing. Protection Against Liquid Chemicals—Determination of Resistance of Materials to Penetration by Liquids	

(*continued*)

TABLE 6-19 (*Continued*)

Standard	Title	Comments
ISO 13994:1998	Clothing for Protection Against Liquid Chemicals: Determination of the Resistance of Protective Clothing Materials to Penetration by Liquids Under Pressure	
ISO 22608:2004	Protective Clothing: Protection Against Liquid Chemicals—Measurement of Repellency, Retention, and Penetration of Liquid Pesticide Formulations Through Protective Clothing Materials	
ISO/CD 6529	Protective Clothing: Protection Against Chemicals—Determination of Resistance of Protective Clothing Materials to Permeation by Liquids and Gases	Draft standard
NATO AEP-38 (2011) F.4	Operational Requirements, Technical Specifications and Evaluation Criteria for CBRN Protective Clothing	CWA and TIC vapor swatch test
NATO AEP-38 (2011) F.7	Operational Requirements, Technical Specifications and Evaluation Criteria for CBRN Protective Clothing	CWA liquid swatch test—diffusive flow
NATO AEP-38 (2011) F.8	Operational Requirements, Technical Specifications and Evaluation Criteria for CBRN Protective Clothing	CWA liquid swatch test—incident wind
NIOSH RCT-ASR-STP-0103 (v1.1 2005)	Determination of Gasoline Permeation of Hoses and Couplings: Type C and CE, Supplied-Air Respirators	
NFPA 1994-07: 8.7	Standard on Protective Ensembles for First Responders to CBRN Terrorism Incidents	Contains modifications on the ASTM F739 swatch test method

TABLE 6-20 Protection Against Liquid- and Vapor-Phase Chemicals: Item- and System-Level Tests

Standard	Title	Comments
ASTM F2588-11	Man-in-Simulant Test Method	Whole-system vapor test; see [462], which describes the chamber/ methods in some detail
NATO AEP-38 (2011) F.2	Operational Requirements, Technical Specifications and Evaluation Criteria for CBRN Protective Clothing	Whole-system test—vapor; includes ASTM F2588 methods but also mannequin-based testing
NATO AEP-38 (2011) F.3	Operational Requirements, Technical Specifications and Evaluation Criteria for CBRN Protective Clothing	Whole system test—liquid; see also [463,464] for method description and toxicological information
NIOSH CET-APRS-STP-CBRN-0451 (v1.1 2005)	Determination of Chemical Agent Permeation and Penetration Resistance Performance Against Sulfur Mustard (HD) Liquid and Vapor of the Chemical, Biological, Radiological, and Nuclear (CBRN) Air-Purifying Escape Respirator Standard Test Procedures (STP)	
NIOSH CET-APRS-STP-CBRN-0450 (v1.1 2005)	Determination of Chemical Agent Permeation and Penetration Resistance Performance Against Sarin (GB) Vapor of Chemical, Biological, Radiological, and Nuclear (CBRN) Air-Purifying Escape Respirator	
NIOSH NPPTL-STP-CBRN-PAPR-0550 (v0.0 2006)	Determination of CBRN Powered Air Purifying Respirator (PAPR) Performance During Dynamic Testing Against the Chemical Agent Vapor Sarin (GB) Chemical Biological, Radiological, and Nuclear (CBRN) Standard Testing Procedure (STP)	

(continued)

TABLE 6-20 (*Continued*)

Standard	Title	Comments
NIOSH NPPTL-STP-CBRN-PAPR-0551 (v0.0 2006)	Determination of CBRN, Powered Air Purifying Respirator (PAPR) Performance During Dynamic Testing Against Chemical Agent Distilled Sulfur Mustard (HD) Vapor and Distilled Sulfur Mustard (HD) Liquid Chemical, Biological, Radiological, and Nuclear (CBRN) Standard Testing Procedure (STP)	
NIOSH RCT-CBRN-STP-0200 (v1.1 2005), -0201 (v1.1 2005)	Determination of Open Circuit, Self-Contained Breathing Apparatus (SCBA) Performance During Dynamic Testing Against Chemical Agents of Sarin (GB) Vapor and Distilled Sulfur Mustard (HD) Vapor and Liquid	
NIOSH RCT-CBRN-APR-STP-0350 (v0.1 2005)	Determination of Full Facepiece, Tight-Fitting, Negative-Pressure, Air Purifying Respirator (APR) Performance During Dynamic Testing Against the Chemical Agent Vapor Sarin (GB)	
NIOSH RCT-CBRN-APR-STP-0351 (v0.1 2005)	Determination of Full-Facepiece, Tight-Fitting, Negative-Pressure, Air Purifying Respirator (APR) Performance During Dynamic Testing Against Chemical Agent Distilled Sulfur Mustard (HD) Vapor And Liquid CBRN	
U.S. Army (2002) TOP 10-2-022	Chemical Vapor and Aerosol System Level Testing of Chemical/Biological Protective Suits	

European Particulate Filtration Classification Standards. As illustrated in Table 6-21, in Europe, the terms *EPA*, *HEPA*, and *ULPA filter* are used. The higher-efficiency filters are intended primarily for building applications, and the (presumed) most penetrating particle size used for testing may not be the same in each group. Various pertinent test methods exist [467–471].

General Comments. Relevant aerosol and particulate penetration methods for respirator filters are given in Table 6-22 and for materials in Table 6-23. Associated methods that examine the capability to resist clogging are included.

System-Level-Tests. One example of an aerosol whole-system test is performed at Research Triangle Institute, North Carolina (Figure 6-8). In this test, a person performing activities is exposed to a fluorescent aerosol consisting of a fluorescently tagged silica dust whose aerodynamic diameter is about 3 μm, at varying wind speeds. The deposition of the challenge aerosol onto skin is monitored qualitatively

TABLE 6-21 Filtration Efficiency Classes According to EN 1822-1:2009

Group	Filter Class	Integral Value of Filtration Efficiency (%)[a]
EPA	E10	85
	E11	95
	E12	99.5
HEPA	H13	99.95
	H14	99.995
ULPA	U15	99.9995
	U16	99.99995
	U17	99.999995

Source: [467]

[a]Local penetration values at specific leakage points may be permitted to be higher.

using black-light visualization and quantitatively by skin swabbing and fluorescent analysis.

Exposure can be performed using a biological aerosol; in this case, spores of *Bacillus atrophaeus* can be used as a simulant for *Bacillus anthracis*, and skin can again be sampled by swabbing followed by culturing and counting the spores deposited (after agglomerates have been broken up). Instead of swabbing, the spores can be collected as agglomerates by transferring from skin by direct contact to an agar plate instead, although recovery is somewhat less effective. It is somewhat unclear which recovery method is to be preferred, as the infectivity of the spores is probably not linearly related to either the number of spore agglomerates or the total number of spores but to a combination of the two. Recent studies have indicated that particle number and surface area are more important determinants of harmful health effects than is mass; this would indicate that filtration test methods should focus on these metrics rather than mass-based penetration [473,474].

Regardless of the nature of the particulate challenge, by performing removal following a prescribed protocol that includes the use of various rooms or areas with controlled airflow, and using various monitoring procedures for the resuspended aerosol combined with extraction or swabbing of the clothing, the efficacy of capture of the aerosol by clothing and of decontamination procedures can also be monitored [472]. Note, however, that the chemical aerosol is not a good simulant for reactive decontaminants and will not monitor that component of the decontamination process. The simulant *B. atrophaeus* is a reasonable simulant for evaluating biocidal action of decontaminants against *B. anthracis*.

6.8.4 Biological: Methods Assessing Resistance to Microbial Growth

ASTM E2756-10, *Standard Terminology Relating to Antimicrobial and Antiviral Agents*, contains useful descriptions distinguishing among different types of reagents and their activity. Methods that assess resistance to microbial growth are given in Table 6-24. Many of these methods are included primarily because they can have

TABLE 6-22 Aerosol and Biological Penetration and Filtration: RPDs and Filters

Standard	Title	Comments
EN 143:2000	Respiratory Protective Devices: Particle Filters; Requirements, Testing, Marking	Limitations with regard to its ability to characterize electret filters correctly
EN 13274-8:2002	Respiratory Protective Devices: Methods of Test; Part 8: Determination of Dolomite Dust Clogging	
EN 13274-7:2008	Respiratory Protective Devices: Methods of Test; Part 7: Determination of Particle Filter Penetration	
ISO/DIS 16900-3	Respiratory Protective Devices: Methods of Test and Test Equipments; Part 3: Determination of Particle Filter Penetration	Draft standard
NIOSH TEB-APR-STP-0001 (v2.0 2009)	Determination of Particulate Filter Penetration (PAPR)	
NIOSH TEB-APR-STP-0051 (v2.0 2007)	Determination of Particulate Filter Efficiency Level for P100 Series Filters Against Liquid Particulates for Non-powered, Air Purifying Respirators	
NIOSH TEB-APR-STP-0052 (v2.0 2007)	Determination of Particulate Filter Efficiency Level for P99 Series Filters Against Liquid Particulates for Non-powered, Air Purifying Respirators	
NIOSH TEB-APR-STP-0053 (v2.0 2007)	Determination of Particulate Filter Efficiency Level for P95 Series Filters Against Liquid Particulates for Non-powered, Air Purifying Respirators	
NIOSH TEB-APR-STP-0054 (v2.0 2007)	Determination of Particulate Filter Efficiency Level for R100 Series Filters Against Liquid Particulates for Non-powered, Air Purifying Respirators	
NIOSH TEB-APR-STP-0055 (v2.0 2007)	Determination of Particulate Filter Efficiency Level for R99 Series Filters Against Liquid Particulates for Non-powered, Air Purifying Respirators	
NIOSH TEB-APR-STP-0056 (v2.0 2007)	Determination of Particulate Filter Efficiency Level for R95 Series Filters Against Liquid Particulates for Non-powered, Air Purifying Respirators	
NIOSH TEB-APR-STP-0057 (v2.0 2007)	Determination of Particulate Filter Efficiency Level for N100 Series Filters Against Solid Particulates for Non-powered, Air Purifying Respirators	
NIOSH TEB-APR-STP-0058 (v2.0 2007)	Determination of Particulate Filter Efficiency Level for N99 Series Filters Against Solid Particulates for Non-powered, Air Purifying Respirators	
NIOSH TEB-APR-STP-0059 (v2.0 2007)	Determination of Particulate Filter Efficiency Level for N100 Series Filters Against Solid Particulates for Non-powered, Air Purifying Respirators	

TABLE 6-23 Aerosol and Biological Penetration and Filtration: Materials and DPE

Standard	Title	Comments
ASTM F1670-03	Standard Test Method for Resistance of Materials Used in Protective Clothing to Penetration by Synthetic Blood	
ASTM F1671-03	Standard Test Method for Resistance of Materials Used in Protective Clothing to Penetration by Blood-borne Pathogens Using Phi-X174 Bacteriophage Penetration as a Test System	
ASTM Fl819-04	Standard Test Method for Resistance of Materials Used in Protective Clothing to Penetration by Synthetic Blood Using a Mechanical Pressure Technique	
ASTM F1862-00a	Standard Test Method for Resistance of Medical Face Masks to Penetration by Synthetic Blood (Horizontal Projection of Fixed Volume at a Known Velocity)	
ASTM F2053-00 (2006)	Standard Guide for Documenting the Results of Airborne Particle Penetration Testing of Protective Clothing Materials	
ASTM F2100-04	Standard Specification for Performance of Materials Used in Medical Face Masks	
ASTM F2101-01	Standard Test Method for Evaluating the Bacterial Filtration Efficiency (BFE) of Medical Face Mask Materials, Using a Biological Aerosol of *Staphylococcus aureus*	Intended to be for "clothing" and not respirators, and stops at 99.9% efficiency
ASTM F2299-03	Standard Test Method for Determining the Initial Efficiency of Materials Used in Medical Face Masks to Penetration by Particulates Using Latex Spheres	Intended to be for "clothing" and not respirators, and stops at 99.9% efficiency
ISO 16603:2004	Clothing for Protection Against Contact with Blood and Body Fluids: Determination of the Resistance of Protective Clothing Materials to Penetration by Blood and Body Fluids—Test Method Using Synthetic Blood	Material test
ISO 16604:2004	Clothing for Protection Against Contact with Blood and Body Fluids: Determination of Resistance of Protective Clothing Materials to Penetration by Blood-borne Pathogens Test Method Using Phi-X 174 Bacteriophage	Material test

(continued)

TABLE 6-23 *(Continued)*

Standard	Title	Comments
ISO 22612:2005	Clothing for Protection Against Infectious Agents: Test Method for Resistance to Dry Microbial Penetration	Material test
NATO AEP-38 (2011) F.9	Operational Requirements, Technical Specifications and Evaluation Criteria for CBRN Protective Clothing	Aerosol swatch test
NATO AEP-38 (2011) F.10	Operational Requirements, Technical Specifications and Evaluation Criteria for CBRN Protective Clothing	Reaerosolization; see also [472] for a description of a possible test method, and see description below
NATO AEP-38 (2011) F.3	Operational Requirements, Technical Specifications and Evaluation Criteria for CBRN Protective Clothing	Whole-system test—aerosol—see descriptions below
U.S. Army (2002) Test Operating Procedure (TOP) 8-2-501	Permeation and Penetration Testing of Air Permeable, Semipermeable, and Impermeable Materials with Chemical Agents or Simulants (Swatch Testing)	Swatch test—aerosol
U.S. Army (2002) TOP 10-2-022	Chemical Vapor and Aerosol System Level Testing of Chemical/Biological Protective Suits	Whole-system test—aerosol

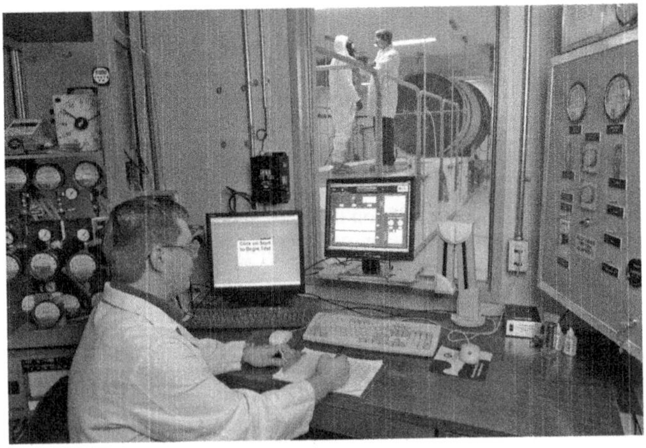

FIGURE 6-8 System-level aerosol test facility; control room in the foreground and aerosol chamber viewed through the window. (Reproduced with permission of Research Triangle Institute, © J.W. Crawford/RTI International 2008.)

TABLE 6-24 Methods Assessing Resistance to Biological Growth

Standard	Title	Comments
AATCC 30-2004	Antifungal Activity, Assessment on Textile Materials: Mildew and Rot Resistance of Textiles	
AATCC 100-2004	Antibacterial Finishes on Textile Materials: Assessment of	
AATCC 147-2004	Antibacterial Activity of Fabrics, Assessment of Textile Materials: Parallel Streak Method	
ASTM E1052-96(2002)	Standard Test Method for Efficacy of Antimicrobial Agents Against Viruses in Suspension	
ASTM E1053-11	Standard Test Method to Assess Virucidal Activity of Chemicals Intended for Disinfection of Inanimate, Nonporous Environmental Surfaces	
ASTM E1054-08	Standard Test Methods for Evaluation of Inactivators of Antimicrobial Agents	
ASTM E1428-99(2009)	Standard Test Method for Evaluating the Performance of Antimicrobials in or on Polymeric Solids Against Staining by *Streptoverticillium reticulum* (A Pink Stain Organism)	
ASTM E1482-04	Standard Test Method for Neutralization of Virucidal Agents in Virucidal Efficacy Evaluations	
ASTM E1882-10	Standard Test Method for Evaluation of Antimicrobial Formulations by the Agar Patch Technique	
ASTM E1891-10a	Standard Guide for Determination of a Survival Curve for Antimicrobial Agents Against Selected Micro-organisms and Calculation of a *D* Value and Concentration Coefficient	
ASTM E2111-05	Standard Quantitative Carrier Test Method to Evaluate the Bactericidal, Fungicidal, Mycobactericidal, and Sporicidal Potencies of Liquid Chemical Microbicides	
ASTM E2149-10	Standard Test Method for Determining the Antimicrobial Activity of Immobilized Antimicrobial Agents Under Dynamic Contact Conditions	
ASTM E2180-07	Standard Test Method for Determining the Activity of Incorporated Antimicrobial Agent(s) In Polymeric or Hydrophobic Materials	
ASTM E2315-03(2008)	Standard Guide for Assessment of Antimicrobial Activity Using a Time Kill Procedure	

(*continued*)

TABLE 6-24 (*Continued*)

Standard	Title	Comments
ASTM E2414-05	Standard Test Method for Quantitative Sporicidal Three Step Method (TSM) to Determine Sporicidal Efficacy of Liquids, Liquid Sprays, and Vapor or Gases on Contaminated Carrier Surfaces	
ASTM E2614-08	Standard Guide for Evaluation of Cleanroom Disinfectants	
ASTM E2720-10	Standard Test Method for Evaluation of Effectiveness of Decontamination Procedures for Air Permeable Materials when Challenged with Biological Aerosols Containing Human Pathogenic Viruses	
ASTM E2721-10	Standard Test Method for Evaluation of Effectiveness of Decontamination Procedures for Surfaces When Challenged with Droplets Containing Human Pathogenic Viruses	
ASTM E2722-09	Standard Test Method for Using Seeded Agar for the Screening Assessment of Antimicrobial Activity in Fabric and Air Filter Media	
ASTM E2756-10	Standard Terminology Relating to Antimicrobial and Antiviral Agents	
ASTM E2783-10	Standard Test Method for Assessment of Antimicrobial Activity for Water Miscible Compounds Using a Time Kill Procedure	
ASTM E2800-11	Standard Practice for Characterization of *Bacillus* Spore Suspensions for Reference Materials	
ASTM G21-09	Standard Practice for Determining Resistance of Synthetic Polymeric Materials to Fungi	
ASTM WK27438	New Specification for Antimicrobial Medical Gloves	Draft standard
ASTM WK30532	New Test Method for Microbial Contact Transfer: Skin-to-Surface Transfer	Draft standard
ASTM WK4789	Test Method for Quantitative Sporicidal Three-Step Method (TSM) to Determine Sporicidal Efficacy of Liquids and Vapor or Gases on Contaminated Carrier Surfaces	Draft standard
ASTM WK6355	Decontamination Factor Testing of *Bacillus* Using Coupon Method	Draft standard
CGSB-2.161-97 (1997)	Assessment of Efficacy of Antimicrobial Agents for Use on Environmental Surfaces and Medical Devices	
CGSB-4.2/28.2-M91 (1997)	Resistance to Micro-organisms: Surface-Growing Fungus Test—Pure Culture	

useful standard challenge and growth assessment methods, even though they are not intended for evaluation of antimicrobial surfaces and materials.

6.9 HUMAN FACTORS

6.9.1 Comfort, Fit, and Function

Dexterity, Tactility, and Other Hand and Foot Functionality Issues. Relevant issues to the use of hands include muscle activity, dexterity, touch sensitivity, finger pinch, and forearm torque strength [162]. Tactility can be investigated by tests such as those that measure the tactile responses of the thumb and fingers using small objects such as monofilaments that are pointed at the pad of the finger, and using increasing target forces until a response is felt [475].

The Purdue pegboard test [476] and the ASTM modified version [477] primarily investigate very fine finger dexterity, although tactility and finger pinch also come into play in handling the pegs. Various discomfort effects (both increase and reduction) or effects on strength are not assessed. The method, in use for decades, comes with a vast quantity of reference data on expected performance levels. A glove found inadequate in this test may prove to be adequate for performing some other task requiring slightly less fine motor control (e.g., assembly or disassembly of a weapon) [154]. ASTM F489-96 [478] looks at the grip of materials used for footwear soles and heels.

Fabric Hand Properties. Because hand properties are currently a fairly subjective concept, it has been difficult to develop objective methodologies, or at least to set targets or requirements using such methods that relate to desirable subjective properties. The Kawabata methodology, based on a series of different instrumental measurements, has been in use for a number of years [479] and attempts to capture, for textiles, the characteristic tensile, bending, shearing, and compression behaviors, along with surface friction and roughness, and fabric construction (thickness, weight). It is used to reflect the subjectively described properties of stiffness, antidrape stiffness, crispness, fullness and softness, smoothness, and "total hand value."

A much more qualitative series of characteristics (the handfeel spectrum descriptive analysis) can be used to rank fabrics. One study included properties such as graininess, grittiness, fuzziness, volume, thickness, stretch, friction on the hand, friction between layers, springiness and depression depth, compression resilience intensity and rate, force to gather and to compress, stiffness, and noise intensity and pitch [480] to predict the comfort of military fabrics. Independently, the comfort affect labeled magnitude (CALM) scale was developed to describe comfort for a given fabric (ranking from "greatest imaginable comfort" through "neither comfortable nor uncomfortable" to "greatest imaginable discomfort"). The correlation between the various properties and comfort was assessed. It is noteworthy that an increase in any of the properties noted above, to a large enough value, correlates with discomfort. The main subjective components contributing most to discomfort in this study were found

TABLE 6-25 Standard Methods for Fabric Hand and Material Stiffness

Standard	Title	Comments
AATCC 5-2006	Fabric Hand: Guidelines for the Subjective Evaluation of	
ASTM D1388-08	Standard Test Method for Stiffness of Fabrics	
ASTM D4032-08	Standard Test Method for Stiffness of Fabric by the Circular Bend Procedure	
ASTM D1388-08	Standard Test Method for Stiffness of Fabrics	
ASTM D6828-02(2007)	Standard Test Method for Stiffness of Fabric by Blade/Slot Procedure	

to be surface texture/depth, volume, and noise, while when Kawabata data were included, shear was a noteworthy positive attribute, and compression and friction were negative. However, it must be emphasized that "comfort" is strongly subjectively associated with the intended use: for example, a fleecy material is very comfortable in the winter but would potentially be viewed as quite uncomfortable in a ball gown. This type of superimposed judgment for a specified use is stronger the more informed the judge is of what "works" for them in that context. In this study, the materials were being examined in the context of a uniform to be worn as a relatively invisible garment, in contact with the skin, for a neutral-to-warm thermal environment. Different results would undoubtedly have been obtained for a different application.

The fabric extraction technique attempts to capture many of these properties in a single measurement and consists of a device that pulls a swatch of fabric through an orifice in a controlled manner, generating a force vs. displacement curve. The properties of this curve can be correlated with hand properties, in particular when calibrated against one or more reference fabrics [481].

Standard material level tests of stiffness are given in Table 6-25. Any material should not cause allergic contact dermatitis (skin irritation), as verified by, for example, the modified Draize test described by Marzulli and Maibach [482]. The Oeko-Tex Standard 100 can also be used to test materials for emission of various hazardous chemicals [483].

Moisture Accumulation. One of the more irritating aspects of wearing PPE involves the accumulation of sweat. There are few tests specific to this issue; however, AATCC 79-2010 [484] deals with absorbency of textiles and could be used for wicking materials.

System Utility and Functionality. System utility is best evaluated using performance measures customized for the user group. All of the activities that could be performed by the group should be evaluated in field trials. To provide some guidance and assure minimum levels of performance, standard methods for demonstrating the utility of items and systems are given in Table 6-26. U.S. Pentagon Report A274274 [99]

TABLE 6-26 Standard Methods for Utility and Functionality

Standard	Title	Comments
ASTM F1154-99a(2004)	Standard Practices for Qualitatively Evaluating the Comfort, Fit, Function, and Integrity of Chemical-Protective Suit Ensembles	
CSA/CGSB Z1610-11: C.3	Protection of First Responders from Chemical, Biological, Radiological, and Nuclear (CBRN) Events	Several user- and work-rate-related recommended standard activity routines
EN 13274-2:2001	Respiratory Protective Devices: Methods of Test; Part 2: Practical Performance Tests	
ISO (proposed) 16900-7	Respiratory Protective Devices: Methods of Test and Test Equipment; Part 7: Practical Performance Tests	In preparation
ISO/TR 16982:2002	Ergonomics of Human–System Interaction: Usability Methods Supporting Human-Centred Design	
ISO/TS 18152:2010	Ergonomics of Human–system Interaction: Specification for the Process Assessment of Human-System Issues	
NIJ Standard-0117.00: 6.2, 6.3, 6.8 through 6.18 (draft)	Bomb Suit Standard for Public Safety	Draft standard; a variety of dynamic tests for putting on and removing PPE and for mobility and range of motion using people

outlines some additional activity routines useful for evaluating ergonomics of law enforcement PPE.

6.9.2 Field of View

Little information has been published that can relate a measured field of view with performance capabilities. Each different field can be evaluated independently by using the area within that field (e.g., lateral, downward, upward, for each of eye; see Figure 6-9 for examples), or the overlapping (binocular) or full (ambinocular) field for both eyes can be determined. The details of how the area is calculated and compared with the unobstructed field are slightly different for each method.

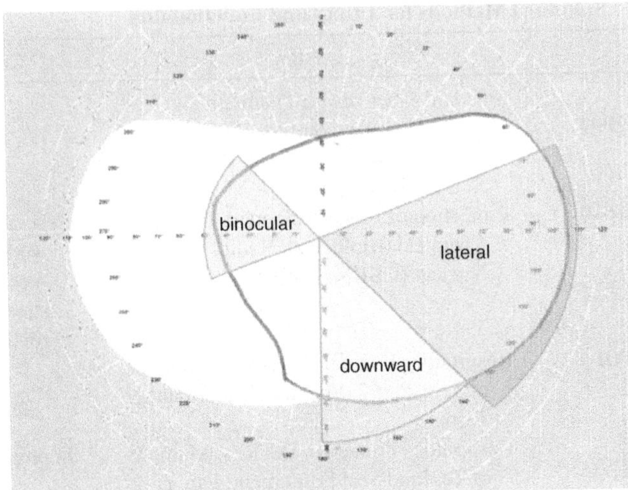

FIGURE 6-9 Angular areas and unobstructed field of view for the left eye (mask facing the viewer); only that part of the measured field falling within the solid line that demarcates the assumed unobstructed field of view for the eye is counted.

Standard methods for field of view are given in Table 6-27. There are two classes of methods, those that use an apertometer to gauge the field of view of the respirator itself in a static manner, and those that use human test subjects. There are significant advantages and disadvantages to each approach. The static headform used with an apertometer has the advantage that it can give results that are highly repeatable and comparable between respirators; however, the mounting apparatus must be suitably

TABLE 6-27 Field of View Methods

Standard	Title	Comments
EN 136:1998 (2003) Section 8.18	Respiratory Protective devices: Full Face Masks—Requirements, Testing, Marking; Field of Vision	Includes reference unobstructed field of view data
ISO CD/16900-11	Methods of Test and Test Equipment; Part 11: Determination of Field of Vision	Headform/ apertometer method: draft standard
NIJ Standard-0117.00: 6.4, 6.5, and 6.6 (Draft)	Bomb Suit Standard for Public Safety	Both static and dynamic tests using humans; draft standard
NIOSH CET-APRS-STP-CBRN-0312 (v1.1 2005)	Determination of Field of View for Full-Facepiece Chemical, Biological, Radiological, Nuclear (CBRN) Respiratory Protective Devices (RPD)	Headform/ apertometer method

FIGURE 6-10 Apertometer that uses an adjustable respirator mounting apparatus instead of a headform.

flexible, noting that the headforms recommended by the standards may not be particularly forgiving with regard to properly mounting different designs and sizes of respirator. An adjustable mounting apparatus may be a better solution than a headform to increase the versatility of the method (an example of an apertometer based on this design is illustrated in Figure 6-10). Disadvantages of the headform/apertometer method are (1) that there are limitations on possible sizes to be tested when using a single size-medium headform (this issue is overcome when using an adjustable mounting apparatus), and (2) that eye, head, and body movement are not evaluated, nor is possible further obstruction by ancillary equipment such as helmets easily included.

Human testing is much more irreproducible because individuals have different head shapes and eye sockets, so the unobstructed field of view is highly variable from person to person; this is further complicated by the different sizes of respirator worn. The best solution is likely to evaluate initially using the apertometer method with a headform, followed by functionality trials, using many people with varying anthropometries and respirator sizes, that explicitly incorporate the field of view evaluations by ensuring inclusion of activities that require good peripheral vision.

6.9.3 Visual Acuity

Dynamic tests for clarity, distortion, and haze are given in U.S. Pentagon Report A521824 [485]. A standard headform test platform has been described [486] for evaluating the formation and propagation of mist in respirators, goggles, and other headwear under various climatic conditions (−40°C to +50°C, RH up to 100%);

TABLE 6-28 Visual Acuity Methods

Standard	Title	Comments
ASTM D1003-00	Standard Test Method for Haze and Luminous Transmittance of Transparent Plastics	
ANSI/ISEA Z87.1-2010	American National Standard for Occupational and Educational Personal Eye and Face Protection Devices	
National Institute of Justice, NIJ Standard-0117.00: 6.19 through 6.25 (draft)	Bomb Suit Standard for Public Safety	Methods for fogging, distortion, and resistance to abrasion; draft standard
NIOSH CET-APRS-STP-CBRN-0314 (v1.1 2005)	Determination of Lens Fogging on Full-Facepiece Chemical Biological Radiological Nuclear (CBRN) Air Purifying Respirator	
NIOSH CET-APRS-STP-CBRN-0316 (v1.1 2005)	Determination of Haze, Luminous Transmittance, and Abrasion Resistance Properties of the Primary Lens System Material for Full-Facepiece Respiratory Protective Devices (RPD)	
US MIL-DTL-43511D, Section 4.4.5.(1990)	Distortion	

the heated headform is capable of mimicking breathing and sweating across the full human range and includes a facility to humidify exhaled breath. Standard methods for determination of factors that can reduce visual acuity, such as distortion, haze, and fogging, are outlined in Table 6-28.

6.9.4 Communications

The modified rhyme test [487] is used routinely for military respirators and evaluates a listener's ability to comprehend single words, providing an indication of the quality of speech transmission. Evaluations should be done in the presence of background white or pink noise in the range 40 to 80 dBA [3,333] and resulting signal-to-noise ratios in the range 2 : 1 to 1 : 1. An automated objective test system for speech intelligibility [488] (using the Speech Perception in Noise (SPIN) test [489]) was developed to assess the impact of head-borne personal protective equipment on speech intelligibility and transmission. The system comprised talker and listener headforms, speech recordings, and speech recognition software. A recording of sentences from the SPIN test was transmitted from the speaker in the talker headform, to microphones

TABLE 6-29 Methods for Determination of Effectiveness of Communication

Standard	Title	Comments
ANSI S3.5-1997 (2007)	Methods for the Calculation of the Speech Intelligibility Index	
ANSI/ASA S12.6-2008	Methods for Measuring the Real-Ear Attenuation of Hearing Protectors	
NIOSH RCT-APR-STP-0030 (v1.1 2005)	Determination of Noise Level Test, Powered Air Purifying Respirator with Hoods or Helmets	
NIOSH RCT-ASR-STP-0111 (v1.1 2005)	Determination of Air Velocity and Noise Levels: Sound Level, Type C and CE, Supplied-Air Respirators	
NIOSH RCT-ASR-STP-0114 (v1.1 2005)	Determination of Sound-Level Measurement: Escape, Open-Circuit, Self-Contained Breathing Apparatus Using Hoods or Helmets	
NIOSH TEB-CBRN-APR-STP-0313 (v2.0 2007)	Determination of Communication Performance Test for Speech Conveyance and Intelligibility of Chemical Biological Radiological and Nuclear (CBRN) Full-Facepiece Air Purifying Respirator	

in the ears of the listener headform and speech recognition software recorded the speech received. The responses were scored by hand.

ANSI has a standard Speech Intelligibility Index [490], and software is available for computing this index [491]. A list of relevant standards on effectiveness of communication is given in Table 6-29.

6.9.5 Air Quality and Supply

Maintaining the quality of breathing air in a respirator requires a combination of ensuring that the incoming air is clean and that the exhaled air does not return to the respirator, where it can cause a buildup of exhaled gases and fogging. Maintaining adequacy of flow and pressure are also important in any system with a compressed or powered air supply, as is minimizing breathing resistance in APRs. Various relevant test methods are given in Tables 6-30 and 6-31.

6.9.6 Thermal Performance

Table 6-32 outlines various relevant international standards covering the measurement of thermal resistance of clothing. They employ the same measurement technique, but they differ in mannequin size, test conditions, formula for calculating the insulation, and the parameters for reporting the test results.

TABLE 6-30 Methods for Determination of the Air Quality and Supply of RPDs

Standard	Title	Comments
EN 13274-6:2001	Respiratory Protective Devices: Methods of Test; Part 6: Determination of Carbon Dioxide Content of the Inhalation Air	
ISO/CD 16900-9	Respiratory Protective Devices: Methods of Test and Test Equipment; Part 9: Carbon Dioxide Content of the Inhaled Air (Dead Space)	Draft standard
ISO 8573-2 through -9 (1999 through 2007)	Compressed Air: Test Methods for . . . (various contaminants)	Tests for humidity, gaseous contaminants, oil, oil aerosol, liquid water, organic solvent, solid particles, viable microbiological
NFPA 1989-2008	Standard on Breathing Air Quality for Emergency Services Respiratory Protection	Other test methods and requirements
NIOSH RCT-APR-STP-0012 (v1.1 2005)	Determination of Air Flow for Powered Air Purifying Respirators	
NIOSH RCT-APR-STP-0063 and -0064 (v1.1 2005)	Determination of Facepiece Carbon-Dioxide and Oxygen Concentration Levels: Tight Fitting, Powered Air Purifying Respirators with the Blower Unit Running (Off)	
NIOSH RCT-ASR-STP-0105 (v1.1 2005)	Determination of Air Flow: Continuous Flow, Type C and CE, Supplied-Air Respirators	
NIOSH RCT-ASR-STP-0105A (v1.1 2005)	Determination of Air Flow: Demand and Pressure-Demand, Type C and CE, Supplied-Air Respirators	
NIOSH RCT-ASR-STP-0115 (v1.1 2005)	Determination of Rated Service Time: Constant-Flow, Escape, Open-Circuit, Self-Contained Breathing Apparatus	
NIOSH RCT-ASR-STP-0117 (v1.1 2005)	Determination of Positive Pressure: Closed-Circuit, Pressure-Demand, Self-Contained Breathing Apparatus	
NIOSH RCT-ASR-STP-0120 (v1.1 2005)	Determination of Positive Pressure: Open-Circuit, Pressure-Demand, Self-Contained Breathing Apparatus	
NIOSH RCT-ASR-STP-0121 (v1.1 2005)	Determination of Rated Service Time: Open-Circuit, Demand and Pressure-Demand, Self-Contained Breathing Apparatus	

TABLE 6-30 (*Continued*)

Standard	Title	Comments
NIOSH RCT-ASR-STP-0121A (v1.1 2005)	Determination of Rated Service Time: Closed-Circuit, Demand and Pressure-Demand, Self-Contained Breathing Apparatus	
NIOSH RCT-ASR-STP-0123 (v1.1 2005)	Determination of Gas Flow Measurements: Open-Circuit, Demand and Pressure-Demand, Self-Contained Breathing Apparatus	
NIOSH RCT-ASR-STP-0124 (v1.1 2005)	Determination of Remaining Service-Life Indicator: Open-Circuit, Demand and Pressure-Demand, Self-Contained Breathing Apparatus	
NIOSH RCT-ASR-STP-0136 (v1.1 2005)	Determination of Demand Gas Flow: Closed-Circuit, Demand and Pressure-Demand, Self-Contained Breathing Apparatus	
NIOSH RCT-ASR-STP 0137 (v1.1 2005)	Determination of Continuous Gas Flow on Constant Flow with Demand Flow: Closed-Circuit, Self-Contained Breathing Apparatus	
NIOSH RCT-ASR-STP-0139 (v1.1 2005)	Determination of Facepiece Carbon Dioxide Concentrations: Self-Contained Breathing Apparatus	

ISO 9920 provides formulas for estimating thermal insulation and evaporative resistance of clothing. The international standards for the assessment of thermal comfort or thermal strain of a clothed body include ISO 7730, ISO/TR 11079, and ISO 7933, for evaluating moderate thermal, cold, and hot environments, respectively. Their applications to thermal properties of clothing, principles, and limitations vary. ISO 15265 [492] deals with the assessment process for management of thermal stress in the working environment, while ISO 15743 [493] does the same for cold environments.

ASTM F2370 and F2371 are the only standards that address the measurement of evaporative resistance of clothing using a mannequin, and F2371 addresses methods for evaluation of body cooling systems. These standards specify the configuration of sweating thermal mannequin, test protocol, and test conditions; various sweating thermal mannequins exist [494–497]. It has been recommended that the isothermal method should be used, with a mass loss calculation, in the application of ASTM F2370 [498].

TABLE 6-31 Methods for Determination of Resistance to Breathing and Flow

Standard	Title	Comments
ASTM F778-88 (2007)	Standard Methods for Gas Flow Resistance Testing of Filtration Media	Flat specimens of filtration media
EN 136:1998 (2003) Section 8.15	Respiratory Protective Devices: Full Face Masks—Requirements, Testing, Marking; Breathing Resistance	
EN 13274-3:2001	Respiratory Protective Devices: Methods of test; Part 3: Determination of Breathing Resistance	
ISO 16900-2:2009	Respiratory Protective Devices: Methods of Test and Test Equipment; Part 2: Determination of Breathing Resistance	
ISO (proposed) 16900-8	Respiratory Protective Devices: Methods of Test and Test Equipment; Part 8: Determination of Airflow	In preparation
ISO CD 16900-12	Respiratory Protective Devices: Methods of Test and Test Equipment; Part 12: Determination of Volume Averaged Work of Breathing	Draft standard
NIOSH RCT-APR-STP-0003 (v1.1 2005)	Determination of Exhalation Resistance	
NIOSH RCT-APR-STP-0065 (v1.1 2005)	Determination of Air Flow Resistance: Breath-Responsive, Powered Air Purifying Respirators	
NIOSH RCT-ASR-STP-0106 (v1.1 2005)	Determination of Inhalation Air Flow Resistance: Pressure-Demand, Type C and CE, Supplied-Air Respirators	
NIOSH RCT-ASR-STP-0107 (v1.1 2005)	Determination of Exhalation Air Flow Resistance: Pressure-Demand, Type C and CE, Supplied-Air Respirators	
NIOSH RCT-ASR-STP-0108 (v1.1 2005)	Determination of Inhalation Air Flow Resistance: Demand, Type C and CE, Supplied-Air Respirators	
NIOSH RCT-ASR-STP-0109 (v1.1 2005)	Determination of Exhalation Air Flow Resistance: Demand, Type C and CE, Supplied-Air Respirators	
NIOSH RCT-ASR-STP-0113 (v1.1 2005)	Determination of Air Flow Resistance: Continuous-Flow, Type C and CE, Supplied-Air Respirators	
NIOSH RCT-ASR-STP-0116 (v1.1 2005)	Determination of Air Flow Resistance: Continuous-Flow, Escape, Open-Circuit, Self-Contained Breathing Apparatus with Hoods	
NIOSH RCT-ASR-STP-0122 (v1.1 2005)	Determination of Exhalation Breathing Resistance: Open-Circuit, Demand and Pressure-Demand, Self-Contained Breathing Apparatus	
NIOSH RCT-ASR-STP-0132 (v1.1 2005)	Determination of Inhalation Breathing Resistance: Open-Circuit, Demand, Self-Contained Breathing Apparatus	
NIOSH RCT-ASR-STP-0133 (v1.1 2005)	Determination of Exhalation Breathing Resistance: Open-Circuit, Pressure-Demand, Self-Contained Breathing Apparatus Using Two Second Stage Regulators	
NIOSH RCT-ASR-STP-0135 (v1.1 2005)	Determination of Inhalation and Exhalation Breathing Resistance: Closed-Circuit, Demand and Pressure-Demand, Self-Contained Breathing Apparatus	
NIOSH TEB-APR-STP-0007 (v2.0 2009)	Determination of Inhalation Resistance Test, Air Purifying Respirators, Standard Test Procedure	

TABLE 6-32 Methods for Determination of Thermal Performance of Materials and Systems: Air Permeability, Thermal Transmission, Thermal Insulation, Evaporative Resistance, and Heat Generation and Removal

Standard	Title	Comments
ASTM D737-04(2008)e2	Standard Test Method for Air Permeability of Textile Fabrics	Air-permeable textiles
ASTM D1518-11	Standard Test Method for Thermal Resistance of Batting Systems Using a Hot Plate	Conduction, convection, and radiation for dry specimens of textile fabrics, battings, and other materials
ASTM D7024-04	Standard Test Method for Steady State and Dynamic Thermal Performance of Textile Materials	Used for the determination of the temperature-regulating factor; conduction of dry specimens
ASTM F1291-10	Standard Test Method for Measuring the Thermal Insulation of Clothing Using a Heated Manikin	Clothing
ASTM F1868-02	Standard Test Method for Thermal and Evaporative Resistance of Clothing Materials Using a Sweating Hot Plate	Thermal resistance and evaporative resistance provided by a fabric, batting, or other type of material
ASTM F2370-10	Standard Test Method for Measuring the Evaporative Resistance of Clothing Using a Sweating Manikin	
ASTM F2371-10	Standard Test Method for Measuring the Heat Removal Rate of Personal Cooling Systems Using a Sweating Heated Manikin	Garments based on various cooling technologies can be evaluated fairly and objectively by taking into account convective and evaporative heat
EN 342:2004	Protective Clothing: Ensembles and Garments for Protection Against Cold	Clothing
EN 702:1994	Protective Clothing: Protection Against Heat and Flame—Test method: Determination of the Contact Heat Transmission Through Protective Clothing or Its Materials	Clothing or materials

(*continued*)

TABLE 6-32 (*Continued*)

Standard	Title	Comments
ISO 7730:2005	Ergonomics of the Thermal Environment: Analytical Determination and Interpretation of Thermal Comfort Using Calculation of the PMV[a] and PPD[b] Indices and Local Thermal Comfort Criteria	
ISO 7933:2004	Ergonomics of the Thermal Environment: Analytical Determination and Interpretation of Heat Stress Using Calculation of the Predicted Heat Strain	
ISO 8996:2004	Ergonomics of the Thermal Environment: Determination of Metabolic Rate	
ISO 9237:1995	Textiles: Determination of the Permeability of Fabrics to Air	Air-permeable textiles
ISO 9886:2004	Ergonomics: Evaluation of Thermal Strain by Physiological Measurements	
ISO 9920:2007	Ergonomics of the Thermal Environment: Estimation of Thermal Insulation and Water Vapor Resistance of a Clothing Ensemble	
ISO 10551:1995	Ergonomics of the Thermal Environment: Assessment of the Influence of the Thermal Environment Using Subjective Judgement Scales	
ISO 11079:2007	Ergonomics of the Thermal Environment: Determination and Interpretation of Cold Stress When Using Required Clothing Insulation (IREQ) and Local Cooling Effects	
ISO 11092:1993	Textiles: Physiological effects—Measurement of Thermal and Water–Vapor Resistance Under Steady-State Conditions (Sweating Guarded-Hotplate Test)	Fabrics, films, coatings, foams, and leather, including multilayer assemblies
ISO 11399:1995	Ergonomics of the Thermal Environment: Principles and Application of Relevant International Standards	
ISO 15831:2004	Clothing: Physiological Effects—Measurement of Thermal Insulation by Means of a Thermal Manikin	

[a]Predicted mean vote.
[b]Predicted percent dissatisfied.

7 Selection and Use of PPE

Our intent in this chapter is not primarily to provide a user-level selection guide. Rather, it is to provide information on various standards that can assist the informed user in selecting based on all of the information in the preceding chapters. After a discussion of detailed requirements development, the expected levels of performance from relevant styles of PPE are described, and existing PPE performance standards relevant to CBRN protection are given.

7.1 OPERATIONAL REQUIREMENTS

7.1.1 Examples of CBRN Work Environments

It is important to itemize all possible work environments for the specific user group to assure adequate selection. A training, laboratory, or demilitarization environment is always different from a genuine CBRN event, with the hazards being much more controlled in the former case. High-hazard events include on-target attack, the immediate vicinity of a breached container, and any indoor release. Outdoor environments can be lower in CBRN hazards, due to less containment of the agent near ground level and lower resulting concentrations; however, it may be more difficult to leave the hazard environment totally because of larger contaminated areas, meaning a longer duration of stay. Other hazards, such as flame and explosion, are likely to be reduced when outdoors. Most people might be expected to be able to work both indoors (shorter duration and higher concentration agent) and outdoors (longer duration and lower concentration agent), and both sets of conditions should be considered. Outdoor operations include the possibility of a number of uncontrolled environmental parameters affecting both the hazards and the protective system, such as heat, wind, cold, precipitation, and extreme physical and mechanical stressors.

Some examples of work roles and their associated environments include:

- Demilitarization: various roles and locations
- CBRN warfare (out-of-area operations):
  - Combat: outdoor

Personal Protective Equipment for Chemical, Biological, and Radiological Hazards: Design, Evaluation, and Selection, First Edition. Eva F. Gudgin Dickson.

  - Within collective protection (e.g., vehicle/airframe/ship, shelter, field hospital: indoor)
  - Open vehicle/ship/airframe: outdoor, high wind
  - Decontamination line: usually outdoor but partially within collective protection
  - Sampling and identification: indoor or outdoor
  - Urban operations: indoor or outdoor
  - Explosive ordnance disposal: indoor or outdoor
  - Maritime operations: may include amphibious operations outdoor
- CBRN response (domestic):
  - Sampling and identification: indoor or outdoor
  - Explosive ordnance disposal: indoor or outdoor
  - Rescue or rapid intervention: indoor or outdoor
  - Perimeter control: most often outdoor
  - Medical treatment: indoor or outdoor
  - Decontamination line: outdoor
  - Tactical operations: outdoor or indoor
- Laboratory work with a subset of CWA, TIC, R, B, simulants: indoor
- Training with a subset of CWA, TIC, R, B (often low-hazard materials), simulants: indoor or outdoor

7.1.2 Requirements Setting and Specifications

When setting the requirements for CBRN PPE prior to selection, the considerations in Chapters 1 and 3 must be thoroughly reviewed. All requirements should be documented by referring to a specific task, role, standard, or document that specifies the source of the requirement. Be careful not to translate requirements into specifications too early in the process: A requirement should indicate a general capability to be delivered, which may be broken down into more specific issues. A fictitious example of some specific statements and how they might be further expanded or documented follows:

- The clothing system shall not significantly increase the signature of the wearer. (*The rationale is fairly obvious; nevertheless, one should include more information, such as: to maintain stealth and reduce the likelihood of being targeted—by being explicit, the various subrequirements become more obvious.*)
  - The silhouette of the wearer shall not increase significantly compared to comparable operational clothing. (*It is necessary somewhere in the document to itemize the environments in which this item will be used to better develop the test conditions.*)

This particular subrequirement will then be translated into a general specification that describes acceptable performance; for example:

- The clothing system shall not increase the silhouette of the wearer by more than 20% compared to comparable operational clothing. (*Why was 20% chosen? Refer to documented scientific studies or consensus process that developed this value.*)

The general specification can be made more specific by referring to a test method and setting criteria; for example:

- A test panel of 10 individuals of various sizes shall be evaluated using the shadow method and the ratio of the silhouette when wearing the system, compared with that wearing normal operational clothing, shall be less than 1.2 in all cases. (*The test method will probably require more specifics.*)

Finally, the test method itself must be described in the requirements document or by reference to another document.

If every item does not have sufficient detail or justification, revision and critical review of the requirements document becomes very difficult, both during the process itself and when the next selection cycle comes around. For example, identified omissions in requirements, specificiations, or test conditions are hard to deal with: Was an item omitted because it was considered and found to be unimportant, too hard to deal with, or because of an oversight? In summary, it is of great benefit in the overall life-cycle management process to systematically analyze and document the process by which each specification was set, as it will save time and money later.

7.2 EXPECTED LEVELS OF PERFORMANCE FROM VARIOUS STYLES OF EQUIPMENT

A summary is provided here of the expected levels of performance from a number of typical CBRN protective materials, items, and protective ensemble configurations. These are general examples of the most common configurations only, and the lists and descriptions are not intended to be exhaustive. In this section we summarize and compare the options that have been used for the purpose in general, and in the following sections we describe the particularly relevant standards as they apply to CBRN protection in more detail, organized by geographic region of origin and applicability.

7.2.1 General Comments

Protection against many CBR agents can be enhanced by the use of certain prophylactic drugs, and most protective systems perform better when decontamination is performed immediately after contact with agent.

7.2.2 Material-Level Performance

There are many reference works that discuss chemical protective clothing selection for the workplace that can also provide valuable information when specific chemicals are of interest and can provide some assistance when designing and selecting materials for generic CBRN use (good examples are the books of Forsberg and Mansdorf and Anna [499,500]). Table 7-1 summarizes the information presented in Chapter 4, organized in terms of the types of materials that can deliver desired performance.

7.2.3 Dermal Protective Equipment Performance

Table 7-2 summarizes the type of dermal protection that can be expected from various designs of DPE used for civilian response, as described in Section 5.4. Each has been intended for a different purpose, and different strengths and limitations result. For example, civilian standards may not ask for demonstration of long-duration protection in their technical specifications, on the assumption that equipment is used for a single short-response activity.

7.2.4 Respiratory Protective Device Performance

Table 7-3 summarizes the type of respiratory protection that can be obtained from various styles of RPD when there are no issues impairing their seal to the body, such as poor fit, poor integration with other PPE, or excessive activity levels that can dislodge the seal. It is to be assumed that the RPD will have been CBRN qualified against a suitable standard unless otherwise specified in the table. More information on the detailed requirements for such systems based on various national standards are given in the following sections.

The breathing resistance from generic styles of respirators has been discussed by Clayton et al. [114]; generally, any positive-pressure flow system will offer less inhalation resistance. The protective performance of the RPD will be limited by both its ability to prevent inward leakage and its ability to prevent breakthrough of chemicals. High-quality fit is essential in any CBRN application, and many standards do not ensure that this parameter in use; the CSA Z1610 standard [3] requires and provides methods to assure that the RPD demonstrate suitable performance when worn with the DPE while performing appropriate workplace activities. In general, adequate inward leakage performance, even for SCBA, should not be assumed unless suitable SWPF testing has been performed in the full configuration.* Assigned protection factors for RPDs outlined by ANSI [501], EN [502], and CSA [33] yield a fair estimation of worst-case inward leakage performance for poorly performing or poorly fitted APRs, but may overestimate worst-case performance for SCBAs that are poorly designed or poorly integrated.

*Except for encapsulating or nearly encapsulating systems, where the combination of the RPD and the encapsulation should assure performance at any activity level.

TABLE 7-1 Material Performance Summary

Desired Protective Performance	Material/Treatment Types	Impact on Burden[a]	Sections Containing More Information[b]	Tables Containing Pertinent Standard Performance Test Methods and Requirements at the Material Level
Particulate or aerosol impermeable	Barrier material, any air impermeable, most microporous (depending on particulate size)	Barriers have high thermal and physical burden; microporous may improve evaporative cooling	4.5.3, 4.5.4, 4.5.9	Tables 6-17, 6-22, and 6-23
Particulate or aerosol filtering	Microfibrous (larger particulate), nanofibrous materials	Permits evaporative/convective cooling	4.5.2, 4.5.5	Tables 6-21, 6-22, and 6-23
Vapor impermeable	Barrier material, [e.g., laminate polymer, nanocomposite, MVP materials (for some vapors)]	MVP permits evaporative cooling; nanocomposite may be thinner	4.5.3, 4.5.4, 4.5.9	Tables 6-17, and 6-19
Vapor purifying	Active carbon, other adsorbents	Permits evaporative/convective cooling	4.5.6	Table 6-19
Liquid impermeable (bulk liquid, broad-spectrum CWA/TIC)	Barrier material, [e.g., laminate polymer, nanocomposite, MVP materials (for some liquids)]		4.5.3, 4.5.4, 4.5.9	Table 6-19
Liquid impermeable (liquid drops, selected CWA/TIC)	Barrier material, laminate polymer, nanocomposite, MVP materials (for some liquids)		4.5.3, 4.5.4, 4.5.9	Table 6-19

(continued)

TABLE 7-1 *(Continued)*

Desired Protective Performance	Material/Treatment Types	Impact on Burden[a]	Sections Containing More Information[b]	Tables Containing Pertinent Standard Performance Test Methods and Requirements at the Material Level
Liquid repellent	Hydrophobic, oleophic materials; superrepellent materials		4.5.9	Table 6-19
Liquid wicking	Wicking materials	Enhances comfort when next to skin	4.5.2	Table 6-19
Chemical self-decontaminating or reactive	Impregnated active carbon, various reactive species		4.5.6, 4.5.7	Table 6-19
Biological self-decontaminating or reactive	Impregnated filters, various reactive species		4.5.2, 4.5.5, 4.5.7	Table 6-24
Moisture vapor selectively permeable	Microporous and monolithic MVP materials	Enhances evaporative cooling	4.5.4	Table 6-32
Combined CBRN and fire/flame retardancy (FR)	FR treatment on outside layer; FR materials over CBRN protective layer		4.5.9	Table 6-15
Combined CBRN and ballistic or explosive ordnance protective	Explosive ordnance disposal or ballistic protective layer over CBRN protective layer	High thermal burden, reduces evaporative or convective cooling	4.5.9	

[a]See the first occurrence of the material type for information.
[b]Sections 4.5.8 and 4.5.9 may contain specific examples of smart or next-generation materials also.

TABLE 7-2 Civilian PPE Appicable to CBRN Use: Dermal Protection Performance Summary

Desired Protective Performance	Suitable Dermal Protective Equipment Types (When Appropriately Qualified)	Performance Comments and Impact on Burden[a]	Tables/Standards Containing Relevant Standard Test Methods and Information
All-hazard, short duration	Totally encapsulating (with SCBA): EPA level A, NFPA 1991, CSA Z1610 C1, BS 8467 category A/EN 943-2	High physiological and physical burden, short-duration air supply, highly liquid protective	Tables 7-6, 7-7, 7-9, and 7-10
Most hazard (select chemicals may be excluded), moderate duration	Nearly encapsulating (with PAPR): BS 8467 category B1/B2	High physiological and physical burden	Tables 7-9 and 7-10
Bulk liquid or splash, limited vapor or aerosol	EPA level B/C (with SCBA, PAPR, APR), BS 8428	High physiological and physical burden	Tables 7-9 and 7-10
Most dermally active vapors at high concentration and moderate or low contamination density liquids, moderate/long duration	Liquid repellent, active carbon (military style) with excellent closures and reinforcement at pressure points (with SCBA, PAPR, APR): CSA Z1610 CM (moderate duration), NATO AEP-38 (moderate or long duration)	Permits convective and evaporative cooling, better hand properties than polymeric systems	Table 7-7, NATO AEP-38 [4]
Most dermally active vapors at high concentration, and moderate or low contamination density liquids, short or moderate duration	NFPA 1994 class 2 (short duration), CSA Z1610 C2 (moderate duration), BS 8467 category B1/B2	High thermal burden	Tables 7-6, 7-7, 7-9, and 7-10
Turnout gear (heat and flame)	NFPA 1971, CSA Z1610 CF	Slightly improved cooling due to high moisture vapor transmission	Tables 7-6 and 7-7
Technical rescue	NFPA 1951		Table 7-6

(continued)

TABLE 7-2 *(Continued)*

Desired Protective Performance	Suitable Dermal Protective Equipment Types (When Appropriately Qualified)	Performance Comments and Impact on Burden[a]	Tables/Standards Containing Relevant Standard Test Methods and Information
Most dermally active vapors at low concentration, and low-contamination-density liquids, short duration	NFPA 1994 class 3	Slightly improved cooling due to high moisture vapor transmission	Table 7-6
Particulate or aerosol (high hazard)	Air-impermeable or microporous materials with excellent closures (with SCBA, PAPR filtering, APR filtering):	High to moderate physiological and physical burden	Tables 7-6, 7-7, 7-9, and 7-10
	NFPA 1991, NFPA 1994 class 2, CSA Z1610 C2, BS 8467, ISO 8194		
	EN 1073-1, ventilated, ISO 8194/13982-1, ventilated	Convective cooling will be high due to ventilation; PFs in the head region 2000–50,000	Tables 7-13 and 7-16
Particulate or aerosol (moderate or low hazard)	Air-impermeable or air-permeable microfibrous materials in DPE with good closures (with PAPR filtering, APR filtering)	Air-permeable systems generally have a lower burden	NATO AEP-38 [4]
	EN 1073-2, unventilated, ISO 8194/13982-1, unventilated	Convective cooling will be inversely correlated with PF; PFs in the suit 5–500, depending on class	Tables 7-13 and 7-16
Body fluids	NFPA 1994 class 4, CSA Z1610 C4		Tables 7-6 and 7-7
	Systems demonstrating resistance to viral penetration, with good closures: NFPA 1999, NFPA 1951, NFPA 1994, CSA Z1610 C4		Tables 7-6, 7-7, and 7-14

[a]See the first occurrence of the DPE type for information.

TABLE 7-3 Civilian PPE Appicable to CBRN Use: Respiratory Protection Performance Summary

Desired Protection Performance	Suitable RPD Types (When Appropriately Qualified)	Impact on Burden[a]
All-hazard, short duration	SCBA; supplied air or breathable gas: demonstrated CWA vapor and liquid permeation resistant	Airflow provides some convective and evaporative cooling, particularly from compressed source; low work of breathing; high weight and bulk
Moderate hazard, excluding some chemicals and high concentration, high flow and breathing rates, short duration	PAPR: tight-fitting or nearly encapsulating; APR	PAPR: most blowers and hoses provide high weight and bulk, airflow provides some convective and evaporative cooling; APR: generally low weight and bulk, moderate to poor comfort, high breathing resistance
Low hazard chemical, below IDLH	PAPR; APR	Loose-fitting PAPR most comfortable
Particulate or aerosol event, short duration	SCBA, PAPR, APR, SCBA non-CBRN within encapsulating system	
Contagious outbreak (high-hazard organism) or particulate or aerosol (high-hazard) events, long duration	PAPR, APR	
Contagious outbreak (low-hazard organism) or particulate or aerosol (low-hazard) events, long duration	APR, including FFP designs; PAPR: loose-fitting	FFPs and loose fitting usually less protective and more comfortable than tight-fitting

[a] See the first occurrence of the RPD type for information.

Most civilian standards do not demonstrate long-duration protection. Any vapor-protective APE usually has a relatively short duration of protection against low-boiling chemicals and must be changed before its service life is exceeded; as discussed in CAN/CGSB/CSA Z1610-2011 [3], some low-boiling chemicals are not removed or produce toxic by-products after reacting with the carbon impregnants, meaning that no "all-hazard" APE yet exists.

There are times when the required protection performance does not result in a clear distinction between different types of RPD during the selection process, and

TABLE 7-4 Advantages and Disadvantages of Various Types of RPDs

Type of RPD	Advantages	Disadvantages
SCBA: open circuit	Independent clean breathing gas supply: essentially all-hazard protection Lower breathing resistance than negative pressure APR Positive-pressure high-flow system Cooling effect over face, and body if exhausted into DPE	Limited air supply[a] Weight and bulk of air supply and harness
SCBA: open circuit, tethered air supply	Less limitation on air supply Weight and bulk of air supply and harness removed	Limited by tether line length and snagging
SCBA: closed circuit	Independent clean breathing gas supply: essentially all hazard protection Lower breathing resistance than negative pressure APR Positive-pressure high-flow system Extended duration of operations	Extreme weight and bulk of air supply and harness Heat from chemical generation system No CBRN approved/qualified systems
Tight-fitting PAPR	Capable of providing the highest level of protection for APRs Positive-pressure high-flow system Respirator still provides protection when power is lost Lower breathing resistance than negative pressure APR Cooling effect over face	Limited by the air purification capability of the APE Cannot be used in oxygen-deficient atmospheres Limited by battery life Weight and bulk of blower and harness Blower and hoses can be fragile
Loose-fitting PAPR	Positive-pressure high-flow system Lowest breathing resistance Cooling effect over head	Limited by the air purification capability of the APE Cannot be used in oxygen-deficient atmospheres Respirator will not provide protection when power is lost Particularly easy at high work rate to bypass the air purification system Limited by battery life Weight and bulk of blower and harness Blower and hoses can be fragile

TABLE 7-4 (*Continued*)

Type of RPD	Advantages	Disadvantages
Encapsulated PAPR	Positive-pressure high-flow system Dermal protection integrated into system Low breathing resistance Cooling effect over body	Limited by the air purification capability of the APE[b] Cannot be used in oxygen-deficient atmospheres Loss of power may result in loss of respiratory protection and breathing air quality if exhaled air is rebreathed Limited by battery life Weight and bulk of blower and harness Blower and hoses can be fragile
APR	Not limited by air supply or battery life Lightweight and low profile Less logistical burden	No positive pressure or high flow to assist in providing higher levels of protection Limited by the air purification capability of the APE Highest breathing resistance Cannot be used in oxygen-deficient atmospheres Possible at high work rate to bypass the air purification system Most sensitive to errors in fitting/sizing and changes in anthropometry over time

[a]60-min tanks provide 30 to 40 min of air at higher work rates.
[b]Compared to negative-pressure APRs, the higher airflows in PAPRs generally require heavier canisters to maintain air purification capabilities.
Source: Based on [3].

other operational and human factors considerations may be included in making the selection decision. Table 7-4 addresses the advantages and disadvantages of each type.

7.3 PERFORMANCE AND SELECTION STANDARDS AND REGULATIONS

In the following sections, a general review of various existing relevant performance and selection standards for selected jurisdictions will be provided, starting with those developed directly for CBRN or biological protection. Those that could have some applicability for select CBRN protection applications but are not intended for CBRN events, are referenced in CBRN standards, or could have applicability for future reference are also mentioned. Some major features of the standard may be described in more detail. Many of these documents also include guidance on programs and procedures to ensure appropriate sizing and fitting, care and maintenance, and use. For example, CSA Standard Z.94.4 states that a respiratory protection program must address the following components:

- Roles and responsibilities of program participants
- Hazard assessment
- Selection of the appropriate respirators
- Respirator fit testing
- Training
- Use of respirators
- Cleaning, inspection, maintenance, and storage of respirators
- Health surveillance of respirator users

7.3.1 North America

Both the United States and Canada have pertinent civilian PPE standards for CBRN response.

United States. Title 29, U.S. Code of Federal Regulations (CFR), Part 1910.132, *Personal Protective Equipment*, addresses some of the generalities pertaining to PPE that in essence lead to the requirement for provision of CBRN-specific PPE. The subsequent sections of the CFR:

- Reference ANSI Z87.1-2003, *American National Standard Practice for Occupational and Educational Eye and Face Protection* (since updated in 2010)
- Describe the requirements for respiratory protection, including application of a respiratory protection program and fit testing
- Provide RPD-related vocabulary
- Describe general occupational headwear, footwear, and glove requirements

Title 42 of the CFR Part 84:

- Describes approval of respiratory protective devices
- Includes many performance requirements to which all RPDs must adhere

The NIOSH CBRN respirator standards supplement these regulations with CBRN specific requirements. The applicable CBRN standards in the United States are summarized in Tables 7-5 and 7-6.

The generic capability of the NIOSH CBRN canister to remove toxic industrial chemicals is intended to be demonstrated by its ability to remove selected chemicals as

TABLE 7-5 U.S. CBRN PPE Standards: Respiratory

Standard	Title	Comments
42 CFR 84.181 particulate filtering facepiece respirator filtration categories	Non-powered Air-Purifying Particulate Filter Efficiency Level Determination	Particulate filtration standards, non-oil proof (N) and oil-proof (P) (e.g., N95, P100)
NIOSH (2003) CBRN APR	Standard for Chemical, Biological, Radiological and Nuclear (CBRN) Full Facepiece Air-Purifying Respirator	Intended for entry into CBRN environments with (chemical) concentrations below IDLH
NIOSH (2003) CBRN APR / SCBA (Escape)	Standard for CBRN Air-Purifying Escape Respirators and Self-Contained Escape Respirators	For escape use by general working population Existing approved systems generally 15-minute service life, hood-style, APR
NIOSH (2001) CBRN SCBA	Self Contained Breathing Apparatus (SCBA) with CBRN Protection	Certification under NFPA 1981 also required
NIOSH (2006) CBRN PAPR	Standard for Chemical, Biological, Radiological, and Nuclear (CBRN) Powered Air-Purifying Respirators (PAPR)	Loose- and tight-fitting styles Intended for entry into CBRN environments with (chemical) concentrations below IDLH
NFPA 1981-2007	Standard on Open-Circuit Self-Contained Breathing Apparatus (SCBA) for Emergency Services	

representatives of various toxic material classes,* including particular toxic chemicals of concern. Because in the United States an APR can only be used in a less-than-IDLH environment, the test concentrations are in some way generally related to the IDLH concentration of the test gas. Thus, the capacity for removal of different gases may vary significantly (as described in Section 6.3.5).

Chemical agent resistance is generally assessed against selected nerve and blister CWAs in both vapor and liquid form in either a headform (RPD) or a swatch test (DPE). The vapor protection levels required in those full systems that are intended to be vapor protective without being totally encapsulating are based on their ability to protect against dermally active chemical warfare agents. The particulate protection

*Organic vapor (with vapor pressures below those of cyclohexane and carbon tetrachloride), acid gas, base gas, hydride, nitrogen oxide, formaldehyde, carbon monoxide, and particulates.

TABLE 7-6 U.S. CBRN PPE Standards: Dermal and Ensemble

Standard	Title	Description of Ensembles and Applicability
ASTM F2061-00	Standard Practice for Chemical Protective Clothing Care and Maintenance Instructions	
NFPA 1851-2008	Standard on Selection, Care, and Maintenance of Protective Ensembles for Structural Fire Fighting and Proximity Fire Fighting	Applicable to NFPA 1971 garments
NFPA 1855	Standard for Selection, Care, and Maintenance of Protective Ensembles for Technical Rescue Incidents	Draft standard; applicable to NFPA 1951 garments
NFPA 1951-2007	Standard on Protective Ensembles for Technical Rescue Operations	Technical rescue gear with CBRN protection
NFPA 1971-2007	Standard on Protective Ensembles for Structural Firefighting and Proximity Firefighting	Turnout gear with CBRN protection (escape intent)
NFPA 1991-2005	Standard on Vapor-Protective Ensembles for Hazardous Materials Emergencies	Fully encapsulating ensembles worn with CBRN SCBA
NFPA 1992-2011	Standard on Liquid Splash-Protective Ensembles and Clothing for Hazardous Materials Emergencies	May have some applicability for roles outside the hot zone; no explicit intent for CBRN events
NFPA 1994-2007	Standard on Protective Ensembles for First Responders to CBRN Terrorism Incidents	Class 2 worn with CBRN SCBA Class 3 worn with CBRN APR Class 4: particulate only hazard, respirator unspecified
NFPA 1999-2008	Standard on Protective Clothing for Emergency Medical Operations	Potentially applicable to B event or particularly contagious outbreak event
NIJ 0116.00-2010	CBRN Protective Ensemble Standard for Law Enforcement	Law enforcement response levels (LERL) as follows: LERL 1 tactical, CWA/TIC protection worn with CBRN SCBA LERL 2 tactical, CWA/fewer TIC protection worn with CBRN SCBA LERL 3 CWA/fewer TIC protection worn with CBRN APR or PAPR LERL 4 perimeter, CWA/fewer TIC protection worn with CBRN APR or PAPR, lower durability

requirement in the NFPA 1994 class 4 DPE standard is qualitative only, and demonstration of integration with the respirator is not required. The NFPA 1994, 1951, and 1971 standards require that a specified, approved respirator be worn with the system; however, the respiratory protection in this configuration is not assessed. There is limited comprehensive CBRN selection guidance to accompany these various standards. General clothing selection guidelines are provided by ASTM F1461-07, *Practice for Chemical Protective Clothing Program*.

Canada. There is a single applicable voluntary national standard, CAN/CGSB/CSA Z1610-11 *Protection of First Responders from CBRN Events* [3], that deals with all aspects of performance of full protective ensembles. This document also is referenced within CSA Z94.4-11 [33] for selection of CBRN respiratory protection. Table 7-7 outlines the various ensemble configurations. There is extensive guidance within the document outlining the types of events in which each configuration may be used, including contagious outbreak events, and describing how selection should be performed. The document also describes the general limitations of PPE currently in service that does not meet the performance specifications of the standard. The configuration designations include:

- The type of RPD:
  - S or s is SCBA with or without CBRN approval
  - vP, VP, or P are APRs (PAPRs when preceded by PAPR) providing vapor and/or particulate protection to varying degrees; P is also intended for contagious outbreak events
- The type of overall system:
  - 1 is totally encapsulating, high liquid and vapor protection
  - 2 and M provide moderate liquid and vapor protection
  - F is firefighter turnout gear
  - 4 is particulate or contagious outbreak event protection

There are a few significant differences between the Canadian and U.S. standards. First, the Canadian standard contains a comprehensive selection process for the full protective ensemble, based explicitly on the type of event and the user's role and location within the event. Second, it requires demonstration of full system integration, including comprehensive protection performance via SWPF evaluations and user trials, while wearing potentially interfering equipment. Third, respiratory protection requirements for APRs are managed differently. The limitation on APR use to a non-IDLH environment is not present, rather, the requirements are set to be sufficient to provide appropriate levels of protection for broader application. Also, despite the apparent breadth of protection provided by the approach used by NIOSH to demonstrate TIC protection, it has been demonstrated that there may be some significant gaps in the capabilities of existing impregnated carbon systems to remove all toxic industrial gases of potential concern [3] under appropriate test conditions.

TABLE 7-7 Applicability of Ensemble Configurations Specified in CSA Z1610-11

Ensemble Configuration	General Component Types (Clothing, RPD)	Applicability
C1S, C1s	Totally encapsulating NFPA 1991; NIOSH CBRN or NFPA 1981 SCBA	Any, but most relevant to hot-zone response
C2S	Similar to NFPA 1994 class 2; NIOSH CBRN SCBA	Any particulate; chemical outside the hot zone where there is no obvious liquid contamination or dermally active liquids; CF is turnout gear
CFS	Similar to NFPA 1971; NIOSH CBRN SCBA	
C2vP, C2VP, C2PAPR-vP, C2PAPR-VP	Similar to NFPA 1994 class 2; various types of CBRN APR/PAPR with excellent particulate removal and good vapor removal	Any particulate[a]; chemical outside the hot zone with various limitations
CMS	Military-style ensemble; NIOSH CBRN SCBA	Any particulate; chemical outside the hot zone with various limitations
CMvP, CMVP, CM PAPRvP, CM, PAPR-VP	Military-style ensemble; various types of CBRN APR/PAPR with excellent particulate removal and good vapor removal	Any particulate[a]; chemical outside the hot zone with various limitations
C4S or C4s	Particulate and body fluid dermal protection with SCBA	Any particulate, or contagious outbreak
C4P, C4PAPR-P	Particulate and body fluid dermal protection with CBRN APR/PAPR with excellent particulate removal	Any particulate[a], or contagious outbreak

[a]Hot-zone use dependent on sufficient demonstrated SWPF of RPD.
Source: [3].

This observation was based on rapid breakthrough of several toxic test chemicals and/or their toxic decomposition products for a military canister that was not greatly different from an approved NIOSH canister.

A different, more comprehensive list of toxic industrial chemicals is therefore suggested for testing. If an APR is not shown to remove all the chemicals on the list, some alternative means of ensuring that exposure to any particular chemical will not occur must be employed (such as appropriate detectors or limited use below IDLH). The test concentrations are the same for all gases, as it is the hazard environment that determines the concentration, not the IDLH value.

Finally, protection against contagious outbreak events is discussed specifically, with high levels of respiratory aerosol protection required if extremely hazardous

TABLE 7-8 British CBRN Standards: Respiratory

Standard	Title	Comments
BS 8468-1:2006	Respiratory Equipment for Use Against Chemical, Biological, Radiological and Nuclear (CBRN) Agents; Positive Pressure, Self-Contained, Open-Circuit Breathing Apparatus: Specification	
BS 8468-2:2006	Respiratory Protective Devices for Use Against Chemical, Biological, Radiological and Nuclear (CBRN) Agents; Negative Pressure, Air Purifying Devices with Full Face Mask: Specification	
BS 8468-4:2008	Respiratory Protective Devices for Use Against Chemical, Biological, Radiological and Nuclear (CBRN) Agents; Powered Air-Purifying Respirators: Specification	
BS 8468-3.1:2009	Respiratory Protective Devices for Use Against Chemical, Biological, Radiological and Nuclear (CBRN) Agents; Self-Contained Open-Circuit Compressed Air Breathing Apparatus Incorporating a Hood for Escape: Specification	
BS 8468-3.2:2009	Respiratory Protective Devices for Use Against Chemical, Biological, Radiological and Nuclear (CBRN) Agents; Air-Purifying Devices Incorporating a Hood for Escape: Specification	
BS 8468-5:2011	Respiratory Protective Devices for Use Against Chemical, Biological, Radiological and Nuclear (CBRN) Agents; Dual-Mode Apparatus: Specification	Enables the wearer to switch between two modes of use described in any two of BS 8468-1, BS 8468-2, BS 8468-4, BS 8468-6.1, or BS 8468-6.2
BS 8468-6.1:2011	Respiratory Protective Devices for Use Against Chemical, Biological, Radiological and Nuclear (CBRN) Agents; Positive-Pressure Compressed Airline Equipment: Specification	
BS 8468-6.2:2011	Respiratory Protective Devices for Use Against Chemical, Biological, Radiological and Nuclear (CBRN) Agents; Constant Flow Compressed Airline Equipment: Specification	
BS 8468-7 (draft)	Respiratory Protective Devices for Use Against Chemical, Biological, Radiological and Nuclear (CBRN) Agents; Part 7: Closed-Circuit Breathing Apparatus: Specification	

TABLE 7-9 British CBRN Standards: Dermal and Ensemble

Standard	Title	Description of Ensembles
BS 8467:2006	Protective Clothing: Personal Protective Ensembles for Use Against Chemical, Biological, Radiological and Nuclear (CBRN) Agents—Categorization, Performance Requirements, and Test Methods	Category A: gastight suits with independent air supply, conforming to EN 943-2, the standard for emergency team gastight suits, inward leakage 0.01%; RPD[a]: BS 8468-1 SCBA or BS 8468-6.1 or -6.2 airline-supplied RPDs Category B: liquid-tight equipment, e.g., type 3a or type 3b emergency team PPE per BS 8428, inward leakage 0.05%, or permeable CBRN protective system; RPD[a] any of category A RPD, BS 8468-2 PAPR, 8468-4 APR B1, breathable oxygen, high vapor B2, breathable oxygen, high liquid Category C: high mechanical strength, low likelihood of dermal effects, inward leakage 0.1%—urban search and rescue. Garments meeting EN 469:2005—protective clothing for firefighters would meet requirements although no thermal protection is required; RPD of category B Category D: little likelihood of encounter with agent—perimeter activities. Ensemble EN 13034:2005 (not type PB[6]) including a hood, inward leakage 0.1%; RPD of category B
BS 8428:2004	Protective Clothing: Protection Against Liquid Chemicals—Performance Requirements for Chemical Protective Suits with Liquid-Tight Connections Between Different Parts of the Clothing for Emergency Teams (Type 3-ET Equipment)	Type 3a: air supply worn inside; type 3b: air supply worn outside

[a]Includes RPD satisfying BS 8468-5 when equivalent mode is provided.

TABLE 7-10 Categories of PPE Ensembles in BS 8467 CBRN Ensemble Standard

Category	Applicability and Limitations
A	High levels of known or unknown agents; may be oxygen deficient; inhibits high levels of physical activity
B	High levels of known agent; indirect contamination; permits moderate-to-high levels of physical activity; breathable oxygen; B1 for mainly vapor and B2 for mainly liquid
C	Low, non-dermally hazardous airborne levels of agent; permits high levels of physical activity
D	Used where only very low levels of contaminants exist, either because users are operating a long way from the center of the incident, or because users are operating a long time after the incident. The basic chemical suit could consist of a type 6 suit conforming to EN 13034:2005.

organisms such as *Yersinia pestis* (plague) or *Variola* virus (smallpox) could be involved.

Other Relevant North American PPE Standards. There are two relevant standards for eye and face protection, CSA Z94.3-07, *Eye and Face Protectors*, and ANSI/ISEA Z87.1-2010, *American National Standard for Occupational and Educational Personal Eye and Face Protection Devices*. Similarly, Canada and the United States each has current air quality standards for compressed air: CAN/CSA-Z180.1-00 (R2010), *Compressed Breathing Air and Systems*; ANSI CGA G-7.1 (2004), *Commodity Specification for Air*; and NFPA 1989-2008, *Standard on Breathing Air Quality for Emergency Services Respiratory Protection*.

TABLE 7-11 Categories of Persons To Be Protected in BS 8467 CBRN Ensemble Standard

Category	Type	Example	Comments
1	Duty holders: on-site, general, non-CBRN functions	Guard, manager	Short time on scene; evacuation; basic training
2	Initial responders: arrive on-site early in event	Police, ambulance, guard, first aid	Limited time on scene; some CBRN training
3	Professional CBRN responders: arrive on-site later in event	Hazmat or emergency response teams with CBRN training	Lengthy time on scene, including scene mitigation; extensive CBRN training
4	Emergency services: on-site or off-site, later in event	Urban search and rescue, hospital, firefighters	Lengthy time on scene; some CBRN training
5	Victims: throughout event		

TABLE 7-12 EN PPE Standards: Respiratory and Head-borne

Standard	Title	Description of Ensembles
EN 133:2001	Respiratory Protective Devices: Classification	This document explains the distinctions among full, half, and quarter face masks and air purifying, supplied air, etc.
EN 136:1998 (corr. 1:2003)	Respiratory Protective Devices: Full Face Masks—Requirements, Testing, Marking	
EN 137:2006	Respiratory Protective Devices: Self-Contained Open-Circuit Compressed Air Breathing Apparatus with Full Face Mask—Requirements, Testing, Marking	
EN 143:2000	Respiratory Protective Devices: Particle Filters—Requirements, Testing, Marking	Filtration classes (95 $L \cdot min^{-1}$) P1 > 84% P2 > 95% P3 > 99.95%
EN 145:1998	Respiratory Protective Devices: Self-Contained Closed-Circuit Breathing Apparatus Compressed Oxygen or Compressed Oxygen–Nitrogen Type—Requirements, Testing, Marking	
EN 148-1:1999	Respiratory Protective Devices: Threads for Facepieces—Standard Thread Connection	Standard thread connection is used in most CBRN APRs including NIOSH and NATO
EN 149:2001 (+A1:2009)	Respiratory Protective Devices: Filtering Half Masks to Protect Against Particles—Requirements, Testing, Marking	Potentially applicable to contagious outbreak events Filtration, inward leakage classes (95 $L \cdot min^{-1}$) FFP1 > 84% < 22% FFP2 > 95% < 8% FFP3 > 99% < 2%
EN 529:2005	Respiratory Protective Devices: Recommendations for Selection, Use, Care and Maintenance—Guidance Document	
EN 12021:1998	Respiratory Protective Devices: Compressed Air for Breathing Apparatus	

TABLE 7-12 ***(Continued)***

Standard	Title	Description of Ensembles
EN 12083:1998	Respiratory Protective Devices: Filters with Breathing Hoses (Non-mask Mounted Filters); Particle Filters, Gas Filters, and Combined Filters: Requirements, Testing, Marking	
EN 12941:2009	Respiratory Protective Devices: Powered Filtering Devices Incorporating a Helmet or a Hood—Requirements, Testing, Marking	
EN 14458:2004	Personal Eye-Equipment: Faceshields and Visors for Use with Firefighters' and High Performance Industrial Safety Helmets Used by Firefighters, Ambulance and Emergency Services	
EN 14593-1:2005	Respiratory Protective Devices: Compressed Air Line Breathing Apparatus with Demand Valve; Part 1: Apparatus with a Full Face Mask—Requirements, Testing, Marking	

7.3.2 Europe

There are no European EN standards that pertain specifically to CBRN events. There are, however, British CBRN standards that use and add to the EN standards.

Great Britain. British CBRN standards are outlined in Tables 7-8 and 7-9, with further explanatory material in Tables 7-10 and 7-11. The BS CBRN ensemble standard includes four categories of equipment with the applicability described in Table 7-10. People who might require protection in a CBRN scene are described by the categories given in Table 7-11.

European Continent. Tables 7-12 and 7-13 outline some pertinent EN standards. EN 133:2001, *Respiratory Protective Devices: Classification*, provides general information on how RPDs are categorized. In particular, the class 1 to 3 (light duty, general, and special use) full face masks are relevant. Selection guidance for protective clothing in general is provided in CEN/TR 15321:2006, *Guidelines on the Selection, Use, Care and Maintenance of Protective Clothing*, supplemented by CEN/TR 15419:2006, *Protective Clothing: Guidelines for Selection, Use, Care and Maintenance of Chemical Protective Clothing.*

TABLE 7-13 EN PPE Standards: Material, Dermal and Ensemble

Standard	Title	Description of Ensembles
EN 469:2005/AC:2006	Protective Clothing for Firefighters: Performance Requirements for Protective Clothing for Firefighting (Includes Accidental Liquid Splash)	
EN 943-1:2002	Protective Clothing Against Liquid And Gaseous Chemicals, Aerosols and Solid Particles: Performance Requirements for Ventilated and Non-ventilated "Gas-Tight" (Type 1) and "Non-Gas-Tight" (Type 2) Chemical Protective Suits[a]	Encapsulating (with or without overpressure) and nonencapsulating systems—primarily nonemergency use
EN 943-2:2002	Protective Clothing Against Liquid and Gaseous Chemicals, Including Liquid Aerosols and Solid Particles: Performance Requirements for "Gas-Tight" (Type 1) Chemical Protective Suits for Emergency Teams (ET)	Encapsulating systems
EN 1073-1:1998	Protective Clothing Against Radioactive Contamination—Part 1: Requirements and Test Methods for Ventilated Protective Clothing Against Particulate Radioactive Contamination[a]	Overpressure style
EN 1073-2:2002	Protective Clothing Against Radioactive Contamination—Part 2: Requirements and Test Methods for Non-ventilated Protective Clothing Against Particulate Radioactive Contamination[a]	Protective clothing must have already passed the type 5 test (EN ISO 13982-1/2) and subsequently with the test methods specified in EN 14325 (including the test for burst resistance to EN ISO 13938-1) The result is given as a nominal protection factor and is classified as follows: Class/Nominal Protection Factor 3 500 2 50 1 5

EN 1149-5:2008	Protective Clothing: Electrostatic Properties—Part 5: Material Performance and Design Requirements	Material, electrostatic performance design
EN 13034:2005 + A1:2009	Protective Clothing Against Liquid Chemicals: Performance Requirements for Chemical Protective Clothing Offering Limited Protective Performance Against Liquid Chemicals (Type 6 and Type PB [6] Equipment)	
EN 14126:2003	Protective Clothing: Performance Requirements and Test Methods for Protective Clothing Against Infective Agents	
EN 14325:2004	Protective Clothing Against Chemicals: Test Methods and Performance Classification of Chemical Protective Clothing Materials, Seams, Joins and Assemblages	
EN 14605:2005 + A1:2009	Protective Clothing Against Liquid Chemicals: Performance Requirements for Clothing with Liquid-Tight (Type 3) or Spray-Tight (Type 4) Connections, Including Items Providing Protection to Parts of the Body only (Types PB [3] and PB [4])	

[a]"Ventilated" and "non-ventilated," respectively, pertain to whether or not clothing is provided with overpressure/clean air ventilation.

TABLE 7-13 Japanese RPD and Dermal Protection Standards

Standard	Title
JIS T 8150:2006	Guidance for Selection, Use and Maintenance of Respiratory Protective Devices
JIS T 8151:2005	Particulate Respirators
JIS T 8152:2002	Gas Respirators
JIS T 8153:2002	Supplied-Air Respirators
JIS T 8157:2009	Powered Air Purifying Respirator (PAPR)
JIS T 8005:2005	Protective Clothing: General Requirements
JIS T 8006:2005	Clothing for Protection Against Heat and Flame: General Recommendations for Selection, Care and Use of Protective Clothing
JIS T 8115:2010	Protective Clothing for Protection Against Chemicals
JIS T 8116:2005	Protective Gloves for Use Against Chemicals
JIS T 8117:2005	Protective Boots for Use Against Chemicals
JIS T 8122:2007	Protective Clothing for Protection Against Hazardous Biological Agents: Classification and Test Methods
JIS T 8150:2006	Guidance for Selection, Use and Maintenance of Respiratory Protective Devices
JIS Z 4810:2005	Protective Rubber Gloves for Radioactive Contamination
JIS Z 4811:1995	Protective Footwear for Radioactive Contamination

7.3.3 Asia

Japan. The main standards pertinent to, although not directly designed for, CBRN protection are Japanese, and are summarized in Table 7-13.

7.3.4 ISO

ISO has a number of protective clothing standards and is in the process of producing CBRN respirator standards within the overall RPD requirements that are under development; the ISO standards are summarized in Tables 7-14 and 7-15. Footwear selection guidance is provided by ISO/TR 18690:2006, *Guidance for the Selection, Use and Maintenance of Safety, Protective and Occupational Footwear*; clothing guidance is under development in ISO 26061 (draft), on the selection, use, care, and maintenance of protective clothing; and respirator guidance is under development in ISO 16975-1.

At this time it has been proposed by ISO that respirators be classified by various parameters (as relevant for the type) such as:

- Mode of operation
- Class of gases removed and capacity, and/or particulate filtration efficiency, and/or supplied air capacity/air-line use
- Total inward leakage
- Work rate

TABLE 7-14 ISO RPD Standards

Standard	Title	Comments
ISO 8573-1:2010	Compressed Air—Part 1: Contaminants and Purity Classes	Specifies purity grades for air
ISO 16974:2011	Marking and Information Supplied by the Manufacturer	
ISO CD 16975-1	Respiratory Protective Devices—Part 1: Selection, Use and Maintenance	Draft standard
ISO/AWI 17420-1	Respiratory Protective Devices: Performance Requirements; Part 1: Supplied Breathable Gas Devices	In preparation
ISO/AWI 17420-2	Respiratory Protective Devices: Performance Requirements; Part 2: Filtering Devices	In preparation
ISO/CD 17420-3	Respiratory Protective Devices: Performance Requirements; Part 3: Standardized Connector	Draft standard

TABLE 7-15 ISO Dermal Protective Equipment Standards

Standard	Title	Comments
ISO 8194:1987	Radiation Protection: Clothing for Protection Against Radioactive Contamination—Design, Selection, Testing and Use	
ISO 11612:2008	Protective Clothing: Clothing to Protect Against Heat and Flame	Thermal protection
ISO 13982-1:2004/A1:2010	Protective Clothing for Use Against Solid Particulates—Part 1: Performance Requirements for Chemical Protective Clothing Providing Protection to the Full Body Against Airborne Solid Particulates (Type 5 Clothing)	
ISO DIS/14876-1 (draft)	Protective Clothing: Body Armour; Part 1: General Requirements	Draft standard; body armor
ISO 16602:2007	Protective Clothing for Protection Against Chemicals: Classification, Labeling and Performance Requirements	
ISO 26061 (draft)	Protective Clothing: Guidelines on the Selection, Use, Care and Maintenance	Draft standard

The clothing classification is partly systematic.

7.3.5 NATO

NATO has developed standards for the conventional CBRN (cold-war) environment for RPDs [503] and DPE [4], including specialist aircrew requirements for PAPR [504]. Specifications recommended for the asymmetric threat environment, including general DPE [4], protective uniforms, and self-contained breathing apparatus, have been published or are under development. NATO's training standards are covered in STANAG 2150 NBC [505].

REFERENCES

1. Zajtchuk R., Bellamy, R.F. (Eds.) (1997). Textbook of Military Medicine, Part I, Medical Aspects of Chemical and Biological Warfare, Sidell F.R., Takafuji E.T., Franz, D.R. (Eds.). Washington, DC: Office of the Surgeon General, U.S. Department of the Army.
2. Hala J., Navratil J.D. (2003). Radioactivity, Ionizing Radiation, and Nuclear Energy, 2nd ed. Konvoj, Czech Republic.
3. Canadian Standards Association (2011). Standard for Protection of First Responders from Chemical, Biological, Radiological and Nuclear (CBRN) Events. CAN/CGSB/CSA Z1610-2011.
4. North Atlantic Treaty Organization (2011). Operational Requirements, Technical Specifications and Evaluation Criteria for CBRN Protective Clothing, Vol. 1. AEP-38 (Ed. 2). Brussels: Military Agency for Standardization. NATO/PfP unclassified.
5. North Atlantic Treaty Organization (2003). Allied Joint Doctrine for NBC. AJP3.8. Brussels: Military Agency for Standardization. NATO/PfP unclassified.
6. Vincent J.H. (2009). Aerosol exposure: concepts, criteria, standards and applications. J. Phys.: Conf. Ser. 151: 1–9.
7. Hinds W.C. (1999). Aerosol Technology. New York: Wiley.
8. United Nations (2011). Globally Harmonized System of Classification and Labeling of Chemicals (GHS), (4th ed.) ST/SG/AC.10/30/Rev.4.
9. Williams P., Wallace D., (1989). Unit 731: Japan's Secret Biological Warfare in World War II. New York: Free Press.
10. Health Canada, Population and Public Health Branch, Centre for Emergency Preparedness and Response (2004). Laboratory Biosafety Guidelines, 3rd ed.
11. Roy C.J., Milton D.K. (2004). Airborne transmission of communicable infection: the elusive pathway. N. Engl. J. Med. 350: 1701–1803.
12. Tellier R. (2006). Review of aerosol transmission of influenza A virus. Emerging Infecti. Dis. J. 12. Available at http://www.cdc.gov/ncidod/EID/vol12no11/06-0426.htm.
13. Nicas M., Nazaroff W.W., Hubbard A. (2005). Understanding the risk of secondary airborne infection: emission of respirable pathogens. J. Occup. Environ. Hyg. 2: 143–54.
14. Mitsakou C., Helmis C., Housiadas C. (2005). Eulerian modelling of lung deposition with sectional representation of aerosol dynamics. J. Aerosol Sci. 36: 75–94.
15. Centers for Disease Control and Prevention: Emergency Preparedness and Response. http://www.bt.cdc.gov/agent/agentlist-category.asp. Accessed Nov. 2011.
16. Nilsson A. (1970). Airborne transmission of smallpox. B. Med. J. 4: 127.

Personal Protective Equipment for Chemical, Biological, and Radiological Hazards: Design, Evaluation, and Selection, First Edition. Eva F. Gudgin Dickson.

17. Hammarlund E., Lewis M.W., Hansen S.G., Strelow L.I., Nelson J.A., Sexton G.J., Hanifin J.M., Slifka M.K. (2003). Duration of antiviral immunity after smallpox vaccination. Nat. Med. 9: 1131–1137.
18. Zimmerman P.D., Loeb C. (2004). Dirty bombs: the threat revisited. Center for Technology and National Security Policy – Defense Horizons, 38. Storming Media: A093224.
19. Andersson K.G., Mikkelson T., Astrup P., Thykier-Nielsen S., Jaconbsen L.H., Schou-Jensen L., Hoe S.C., Nielsen S.P. (2008). Estimation of health hazards resulting from a radiological terrorist attack in a city. Radiat. Prot. Dosim. 131: 297–307.
20. International Commission on Radiological Protection (2000). Guide for the practical application of the ICRP human respiratory tract model. Ann. ICRP, ICRP Supporting Guidance 3. Oxford, UK: Pergamon Press.
21. Bailey M.R., Ansoborlo E., Guilmette R.A., Paquet F. (2007). Updating the ICRP human respiratory tract model. Radiat. Prot. Dosim. 127: 31–34.
22. Zajtchuk R., Jenkins D.P., Bellamy R.F., Ingram V.M., Walker R.I., Ceveny T.J., (Eds.) (1989). Textbook of Military Medicine, Part I, Warfare, Weaponry, and the Casualty, Vol. 2, Medical Consequences of Nuclear Warfare. Washington DC: Office of the Surgeon General, U.S. Department of the Army.
23. International Atomic Energy Association (1997). Generic Assessment Procedures for Determining Protective Actions During a Reactor Accident. IAEA-TECDOC-955.
24. North Atlantic Treaty Organization (2001). Graduated Levels of NBC Threat and Associated Protection. STANAG 2984 (Ed. 5). Brussels: Military Agency for Standardization. NATO/PfP unclassified.
25. North Atlantic Treaty Organization (1995). Guidelines for Air and Ground Personnel Using Fixed and Transportable Collective Protection Facilities on Land. STANAG 2941. Brussels: Military Agency for Standardization. NATO unclassified.
26. Canadian Department of National Defence, Chief of Force Development (2009). The future security environment 2008-2030 Part 1. Current and Emerging Trends. http://www.cfd-cdf.forces.gc.ca/documents/publications/Signed_Eng_FSE_10Jul09_eng .pdf.
27. North Atlantic Treaty Organization (2003). Nuclear, Biological and Chemical (NBC) Hazard Management Policy for NATO Forces. STANAG 2426 (Ed. 2). Brussels: Military Agency for Standardization. NATO unclassified.
28. North Atlantic Treaty Organization (1998). Nuclear, Biological and Chemical Reconnaissance. STANAG 2112. Brussels: Military Agency for Standardization. NATO unclassified.
29. North Atlantic Treaty Organization (2000). Handbook for the Sampling and Identification of Chemical Warfare Agents. AEP-10 Vol. 1 (Ed. 5). Brussels: Military Agency for Standardization. NATO/PfP unclassified.
30. North Atlantic Treaty Organization (2007). Concept of Operations of Medical Support in Nuclear, Biological and Chemical Environments. AMedP-7(D). Brussels: Military Agency for Standardization. NATO unclassified.
31. National Fire Protection Association (2002). Recommended Practice for Responding to Hazardous Materials Incidents. NFPA 471.
32. National Fire Protection Association (2008). Standard for Competence of Responders to Hazardous Materials/Weapons of Mass Destruction Incidents. NFPA 472.

33. Canadian Standards Association (2011). Selection, Use, and Care of Respirators. CSA Z94.4–11.

34. First Annual Report to the President and the Congress of the Advisory Panel to Assess Domestic Response Capabilities for Terrorism Involving Weapons of Mass Destruction I. Assessing the Threat (1999). http://www.rand.org/nsrd/terrpanel/terror.pdf.

35. Davis L.E., LaTourrette T., Mosher D.E., Davis L.M., Howell D.R. (2003). Individual Preparedness and Response to Chemical, Radiological, Nuclear, and Biological Terrorist Attacks, App. A. Rand Corporation MR1731. http://www.rand.org/pubs/monograph_reports/MR1731/MR1731.pref.pdf.

36. Areal Locations of Hazardous Atmospheres, designed by the U.S. Environmental Protection Agency.

37. Hazard Prediction and Assessment Capability (HPAC), designed by the U.S. Defense Threat Reduction Agency.

38. Chemical/Biological Agent Vapor, Liquid, and Solid Tracking, designed by the U.S. Naval Surface Warfare Center.

39. Linders M.J., van Beest C.A., Brasser P., Geers L.F., van't Hof G., Rumley-van Gurp R.A., Sterkenburg R.P., van Swieten S.C., Zappey H.W., Hin A.R. (2004). Chemical Incident Simulator: A New Approach For Deriving Passive Defense Requirements. U.S. Pentagon Report A726944.

40. Second-Order Closure Integrated PUFF (SCIPUFF) Model, designed by Aeronautics Research Associates, Princeton, NJ.

41. Conjunction of Multizone Infiltration Specialists (COMIS) multizone air flow model. Described in: Allard F., Dorer V.B., Feustel H.E., et al. (1990). Fundamentals of the Multizone Air Flow Model: COMIS. Air Infiltration and Ventilation Centre Technical Note 29, May, 115 pp.

42. Porshnev A.I., Reshetin V.P. (1993). Numerical investigation of aerosols propagation and sedimentation in case of accidents at NPP. Inzh. Fizi. J. 64: 393–370.

43. Sullivan T.J., Ellis J.S., Foster C.S., Foster K.T., Baskett R.L., Nasstrom J.S., Schalk W.W. (1993). Atmospheric release advisory capability: real-time modeling of airborne hazardous materials. Bull. Am. Meteorol. Soc. 74:2343–2361.

44. Baklanov A., Sørensen J.H., Hoe S.C., Amstrup B. (2006). Urban meteorological modelling for nuclear emergency preparedness. J. Environ. Radioact. 85: 154–170.

45. Päsler-Sauer J. (2000). Description of the Atmospheric Dispersion Model ATSTEP. RODOS (WG2)-TN(97)01, progress report of RODOS, July. http://www.rodos.fzk.de.

46. Miller C.W., Hively L.M. (1987). A review of validation studies for the Gaussian plume atmospheric dispersion model. Nucl. Saf. 28: 522–531.

47. Nasstrom J.S., Sugiyama G., Baskett R.L., Larsen S.C., Bradley M.M. (2007). The NARAC modeling and decision support system for radiological and nuclear emergency preparedness and response. Int. J. Emerg. Manage. 4: 524–550.

48. Mikkelsen T., Larsen S.E., Thykier-Nielsen S. (1984). Description of the Risø puff diffusion model. Nucl. Technol. 67: 56–64.

49. Gering F., Weiss W., Wirth E., Stapel R., Jacob P., Müller H., Pröhl G. (2004). Assessment and evaluation of the radiological situation in the late phase of a nuclear accident. Radiat. Prot. Dosim. 109: 25–29.

50. Harper F.T., Musolino S.V., Wente W.B. (2007). Realistic radiological dispersal device hazard boundaries and ramifications for early consequence management decisions. Health Phys. 93: 1–16.
51. Dombroski M.J., Fischbeck P.S. (2006). An integrated physical dispersion and behavioral response model for risk assessment of radiological dispersion device (RDD) events. Risk Anal. 26: 501–514.
52. Leone J.M., Nasstrom J.S., Maddix D.M., Larson D.J., Sugiyama G., Ermak D.L. (2001). Langrangian Operational Dispersion Integrator (LODI) User's Guide, ver. 1.0. Lawrence Livermore National Laboratory UCRL-AM-212798.
53. Reshetin V.P., Regens J.L. (2003). Simulation modeling of anthrax spore dispersion in a bioterrorism incident. Risk Anal., 23: 1135–1145.
54. Reshetin V.P., Regens D. (2004). Evaluation of malignant anthrax spore dispersion in high-rise buildings. J. Eng. Phys. Thermophys. 77: 1155–1166.
55. Kournikakis B., Walker M., Ho J., Duncan S. (2009). Statistical analysis of bacterial spore aerosols created by opening a spore containing "anthrax letter" in an office. J. Aerosol Sci. 40: 514–522.
56. Cantrell W.A., Nutter D. (2008). Factors affecting the spread of a bioterrorist agent throughout a building. Inquiry 9: 23–36.
57. Policastro A.J., Gordon S.P. (1999). The use of technology in preparing subway systems for chemical/biological terrorism. Proc. Commuter Rail/Rapid Transit Conference, Toronto, ON, Canada.
58. Lien F.-S., Ji H., Yee E., Kournikakis B. (2010). Prediction of aerosol hazards arising from the opening of an anthrax-tainted letter in an open office environment using computational fluid dynamics. J. Eng. Sci. Technol. 5: 302–331.
59. Sextro R.G., Lorenzetti D.M., Sohn M.D., Thatcher T.L. (2002). Modeling the spread of anthrax in buildings. Proc. 9th Int. Conf. on Indoor Air Quality and Climate Control 4: 506–511.
60. Isukapalli S.S., Lioy P. J., Georgopoulos, P.G. (2008). Mechanistic modeling of emergency events: assessing the impact of hypothetical releases of anthrax. Risk Anal. 28: 723–740.
61. Knudsen G.R. (1989). Model to predict aerial dispersal of bacteria during environmental release. Appl. Environ. Microbiol. 55: 2641–2647.
62. Nicas M. (2005). Use of a probabilistic infectious dose model for estimating airborne pathogen infection risk. J. Int. Soc. Respir. Prot. 22: 24–37.
63. Hathway A., Noakes C., Sleigh A. (2008). CFD Modeling of a hospital ward: assessing risk from bacteria produced from respiratory and activity sources. Indoor Air 2008, Copenhagen, Denmark.
64. Mazumdar S., Yin Y., Guity A., Marmion P., Gulick B., Chen Q. (2010). Impact of moving objects on contaminant concentration distributions in an inpatient ward with displacement ventilation. HVAC&R Res. 16: 545–563.
65. Holmes N.S., Morawska L. (2006). A review of dispersion modelling and its application to the dispersion of particles: an overview of different dispersion models available. Atmos. Environ. 40: 5902–5928.
66. Transport Canada (2012). Emergency Response Guidebook http://www.tc.gc.ca/media/documents/canutec-eng/ERG2012.pdf.

67. National Fire Protection Association (2008). Standard for Competence of Responders to Hazardous Materials/Weapons of Mass Destruction Incidents. NFPA 472-2008.

68. O'Rourke P.J. (1981). Collective Drop Effects on Vaporizing Liquid Sprays. Technical Report LA-9069-T. Los Alamos, NM: Los Alamos Laboratories.

69. Georgopoulos P.G., Fedele P., Shade P., Lioy P.J., Hodgson M., Longmire A., Sands M., Brown M.A. (2004). Hospital response to chemical terrorism: personal protective equipment, training and operations planning. Am. J. Ind. Med. 46: 432–445.

70. Korlie M.S. (2000). Three-dimensional computer simulation of liquid drop evaporation. Comput. & Math. Appli. 39: 43–52.

71. Eriksson L., Andersson P.L., Johansson E., Tysklind M. (2006). Megavariate analysis of environmental QSAR data: Part I. A basic framework founded on principal component analysis (PCA), partial least squares (PLS), and statistical molecular design (SMD). Mol. Divers. 10: 169–186.

72. Roy C.J., Hale M., Hartings J.M., Pitt L., Duniho S. (2003). Impact of inhalation exposure modality and particle size on the respiratory deposition of ricin in BALB/c mice. Inhal. Toxicol. 15: 619–638.

73. North Atlantic Treaty Organization (2006). Radiological Aerosol Challenge Levels. AC225/LG/7 D(2006) 0003. Brussels. NATO unclassified.

74. Salbu B. (2000). Source-related characteristics of radioactive particles: a review. Radiat. Prot. Dosim. 92: 49–54.

75. Fogh C.L., Byrne M.A., Roed J., Goddard A.J.H. (1997). Size specific indoor aerosol deposition measurements and derived I/O concentrations ratios. Atmos. Environ. 31: 2193–2203.

76. North Atlantic Treaty Organization (1990). Extreme Climatic Conditions and Derived Conditions for Use in Defining Design/Test Criteria for NATO Forces Materiel. STANAG 2895 (Ed. 1). Brussels: Military Agency for Standardization. NATO unclassified.

77. North Atlantic Treaty Organization (1991). Nuclear Survivability Criteria for Armed Forces Material and Installations. STANAG 4145. Brussels: Military Agency for Standardization. NATO/PfP unclassified.

78. Hayes A.W. (1984). Principles and Methods of Toxicology. New York: Raven Press.

79. Ottoboni M.A. (1997). Chronic Toxicity: The Dose Makes the Poison, 3rd ed. New York: International Thomson.

80. Darmer K.I., Kinkead E.R., DiPasquale L.C. (1972). Acute Toxicity in Rats and Mice Resulting from Exposure to HCl Gas and HCl Aerosol for 5 and 30 Minutes. Aerospace Medical Research Laboratory, U.S. Air Force, AMRL-TR-72-21.

81. Bollinger N. (2004). Respirator Selection Logic. Cincinatti, OH: U.S. Department of Health and Human Services Centers for Disease Control and Prevention, NIOSH.

82. Gudgin Dickson E.F. (2008). Estimates of percutaneous toxicity of sulphur mustard vapor suitable for use in protective equipment standards. J. Toxicol. Environ. Health A 71: 1382–1391.

83. Duncan E.J.S., Gudgin Dickson E.F., Fedele P.D., Nelson D. (2005). A model for prediction of probability of developing an adverse physiological symptom in individuals exposed to VX nerve agent, J. Med. CBR. Def. 3. http://www.jmedcbr.org/Issue_0301/Dickson/Dickson_0605.html.

84. North Atlantic Treaty Organization (2003). Assessment of Effects Levels of Classical Chemical Warfare Agents Applied to the Skin to Be Used in the Design of Protective Equipment. AEP-52. Brussels: Military Agency for Standardization. NATO unclassified.

85. Centers for Disease Control and Prevention Airborne Exposure Limits (1988). FR53 no. 50, pp. 8594–8597.

86. Centers for Disease Control and Prevention Airborne Exposure Limits (2003). FR 68 no. 196, pp. 58348–58351.

87. Centers for Disease Control and Prevention Airborne Exposure Limits (2004). FR 69 no. 85, pp. 24164–24168.

88. International Commission on Radiological Protection (1991). 1990 Recommendations of the ICRP. ICRP Publication 60. Ann. ICRP 21(1–3).

89. International Commission on Radiological Protection (1991). Principles for Intervention for Protection of the Public in a Radiological Emergency. ICRP Publication 63. Ann. ICRP 22(4).

90. Institute of Medicine, U.S. National Academy of Science (1997). An Evaluation of Radiation Exposure Guidance for Military Operations, Interim Report. Washington, DC: National Academy Press.

91. North Atlantic Treaty Organization (2003). Commander's Guide on Nuclear Radiation Exposure of Groups (Ed. 6), STANAG 2083. Brussels: Military Agency for Standardization. NATO/PfP unclassified.

92. North Atlantic Treaty Organization (2004). Commander's Guide on Low Level Radiation (LLR) Exposure in Military Operations, STANAG 2473 (Ed. 2). Brussels: Military Agency for Standardization. NATO/PfP unclassified.

93. International Commission on Radiological Protection (1991). ICRP Publication 60: 1990 Recommendations of the International Commission on Radiological Protection—Users' Edition. Ann. ICRP 21(1–3).

94. U.S. Army Medical Research Institute of Infectious Disease (2005). USAMRIID's Medical Management of Biological Casualties, 6th ed. Fort Detrick, MD.

95. Feeley J.C., Patton C.M. (1980). Bacillus. In: Lennette E.H., Balows A.W. (Eds.), Manual of Clinical Microbiology 3rd ed. Washington, DC: American Society for Microbiology, pp. 145–149.

96. Jones R.M., Nicas M., Hubbard A, Sylvester M.D., Reingold A. (2005). The infectious dose of *Francisella tularensis* (tularemia). Appl. Biosaf. 10: 227–239.

97. LaTourette T., Peterson D.J., Bartis J.T., Jackson B.A., Houser A. (2003). Protecting Emergency Responders, Vol. 2, Community Views of Safety and Health Risks and Personal Protection Needs. Santa Monica, CA: Rand Science and Technology Policy Institute.

98. Dickson E. (2009). Minimizing risk to self and others. CBRNE Coroner's Workshop, Burnaby, BC, Canada, Jan. 2009.

99. DiChiara A., Addonizio, M. (2007). Law Enforcement Advanced Protection (LEAP) Requirements Focus Group Report, Army Natick Research Development and Engineering Command. Natick Soldier Center Technical Report Natick/TR-07/021, U.S. Pentagon Report A274274.

100. Creighton T.E, Hibbard B., Doherty S., McManus K. (2008). Massachusetts State Police Special Tactical Operations Team User Focus Group Report: Law Enforcement Advanced

Protection (LEAP) Duty Uniforms, Integrated Head Protection, Chemical/Biological Protection and Human Systems Integration. U.S. Pentagon Report A504284.

101. North Atlantic Treaty Organization (2000). Nuclear, Biological, and Chemical (NBC) Defense Factors in the Design, Testing and Acceptance of Military Equipment. AEP-7 (Ed. 4). Brussels: Military Agency for Standardization.

102. Kowalski W.J., Bahnfleth W.P., Whittam T.S. (1999). Filtration of airborne microorganisms: modeling and prediction. ASHRAE Trans. Res. 105: 4–17.

103. Lin P.J.Y. (2010). Electrospinning of a nanofibrous filtration medium with application in bioprotective clothing. M. Chem. Mat. Eng. thesis, Royal Military College of Canada, Kingston, ON, Canada.

104. Kim C.S., Bao L., Okuyama K., Shimada S., Niinuma H. (2006). Filtration efficiency of a fibrous filter for nanoparticles. J. Nanopart. Res. 8: 215–221.

105. Yang R.T. (1997). Gas Separation by Adsorption Processes. London: Imperial College Press.

106. Ishizaki K., Komarneni S., Nanko M. (1998). Porous Materials, Process Technology and Applications. Boston: Kluwer Academic.

107. British Standards Institution (2007). Personal Protective Equipment: Ergonomic Principles. BS EN 13921: 2007.

108. Adams S.A., Slocum A.C., Keyserling W.M. (1994). A model for protective clothing effects on performance. Int. J. Clothing Sci. Technol. 6: 6–16.

109. International Organization for Standardization (2004). Ergonomics of the thermal environment: determination of metabolic rate. ISO 8996: 2004.

110. International Organization for Standardization (2007). Respiratory Protective Devices: Human Factors; Part 1: Metabolic Rates and Respiratory Flow Rates. ISO 16976-1:2007.

111. Flouris, A.D., Cheung, S.S. (2006). Design and control optimization of microclimate liquid cooling systems underneath protective clothing. Ann. Biomed. Eng. 34: 350–372.

112. Roberge R.J., Kim J.-H., Coca A. (2011). Protective facemask impact on human thermoregulation: an overview. Ann. Occup. Hyg. Online, Sept. 13.

113. Caretti D.M., Gardner P.D., Coyne K.M. (2004). Workplace Breathing Rates: Defining Anticipated Values and Ranges for Respirator Certification Testing. Edgewood Chemical Biological Center, U.S. Army Research, Development and Engineering Command, Report ECBC-TR-316.

114. Clayton M.P., Bancroft B., Rajan B. (2002). A review of assigned protection factors of various types and classes of respiratory protective equipment with reference to their measured breathing resistances. Ann. Occup. Hyg. 46: 537–547.

115. Proctor D.F. (1952). Studies of respiratory air flow in measurement of ventilator function. Dis. Chest 22: 432–446.

116. International Organization for Standardization (2011). Respiratory Protective Devices: Human Factors: Part 3: Physiological Responses and Limitations of Oxygen and Limitations of Carbon Dioxide in the Breathing Environment. ISO/TS 16976-3:2011.

117. International Organization for Standardization (1989). Garment Constructions and Anthropometric Surveys: Body Dimensions. ISO TS 8559: 1989.

118. North Atlantic Treaty Organization (2002). Methodology for Anthropometric Data. STANAG 2177 (Ed. 2). Brussels: Military Agency for Standardization. NATO unclassified.

119. SAE International (2002). Civilian American and European Surface Anthropometry Resource (CAESAR) European Edition. Warrendale, PA: SAE.

120. British Standards Institution (2004). Principles for Selecting and Using Test Persons for Testing Anthropometric Aspects of Industrial Products and Designs. BS ISO 15537: 2004.

121. Committee for European Norms (2008). Safety of Machinery: Human Body Measurements; Part 3: Anthropometric Data. EN 547-3:1996 + A1:2008.

122. SAE International (2002). Civilian American and European Surface Anthropometry Resource (CAESAR), North American Edition. Warrendale, PA: SAE.

123. Young, J.W. (1993). Head and Face Anthropometry of Adult U.S. Civilians. Federal Aeronautics Association, U.S. Department of Transportation, DOT/FAAIANM93/10, DTIC Accession ADA268661.

124. Yang L., Shen H., Wu G. (2007). Racial differences in respirator fit testing: a pilot study of whether American fit panels are representative of Chinese faces. Ann. Occup. Hyg. 51: 415–421.

125. Facefinder (2011). http://www.sizechina.com/facefinder.

126. Kim H., Han D.-H., Roh Y.-M., Kim K., Park Y.-G. (2003). Facial anthropometric dimensions of Koreans and their associations with fit of quarter-mask respirators. Ind. Health 41: 8–18.

127. Chamberland A., Carrier R., Forest F., Hachez, G. (1998). 1997 Anthropometric Survey of the Land Forces. Department of National Defense (Canada) Contractor Report W7711-6-7350.

128. Donelson S.M., Gordon C.C. (1996). 1995 Matched Anthropometric Database of U.S. Marine Corps Personnel: Summary Statistics. Technical Report Natick/TR-06-036 DTIC Accession ADA316646.

129. Gordon C.C., Churchill T., Clauser C.E., Bradtmiller, B., McConville, J.T. (1989). Anthropometric Survey of U.S. Army Personnel: Methods and Summary Statistics 1988. DTIC Accession ADA225094.

130. U.S. military anthropometric data are available for download from the Defense Technical Information Center, http://www.dtic.mil/dticasd/docs-a/anthro_military.html.

131. Other military anthropometric data are available for download from the Defense Technical Information Center, http://www.dtic.mil/dticasd/docs-a/anthro_other_military.html.

132. Stirling M., National Anthropometric Survey of Female Firefighters, CACFOA (Chief Fire Officers Organization, UK), http://www.humanics-es.com/FireFighter Anthropometry.pdf. Accessed Feb. 2012.

133. Data for selected law enforcement and health care personnel are available for download from the Defense Technical Information Center at http://www.dtic.mil/dticasd/docs-a/anthro_civilian.html.

134. Lin Y.-C, Wang M.-J. J., Wang E.M. (2004). The comparisons of anthropometric characteristics among four peoples in East Asia. Appl. Ergon. 35: 173–178.

135. PeopleSize 2008, Open Ergonomics. Contains 289 individual body measurements for U.S., UK, German, and Australian adults and American, Australian, Belgian, British, Chinese, French, German, Japanese, and Swedish people, by age group.

136. Weinberg S.M., Naidoo S., Govier D.P., Martin R.A., Kane A.A., Marazita M.L. (2006). Anthropometric precision and accuracy of digital three-dimensional photogrammetry:

comparing the Genex and 3dMD imaging systems with one another and with direct anthropometry. J. Craniofa. Surg. 17: 477–483.

137. Kau C.H., Richmond S., Zhurov A.I., Knox J., Chestnutt I., Hartles F., Playle R. (2005). Reliability of measuring facial morphology with a 3-dimensional laser scanning system. Am. J. Orthod. Dentofac. Orthop. 128: 424–430.

138. Ghoddousi H., Edler R., Haers P., Wertheim D., Greenhill D. (2007). Comparison of three methods of facial measurement. Int. J. Oral Maxillofac. Surg. 36: 250–258.

139. Hack A.L., Hyatt E.C., Held B.J., Moore T.O., Richards C.P., McConville J.T. (1973). Selection of Respirator Test Panels Representative of U.S. Adult Facial Sizes. Report LA5488. Los Alamos, NM: Los Alamos Scientific Laboratory of the University of California.

140. Hack A.L., McConville J.T. (1978). Respirator protection factors: Part I. Development of an anthropometric test panel. Am. Ind. Hyg. Assoc. J. 39: 970–975.

141. Oestenstad R.K., Elliott L.J., Beasley T.M. (2007). The effect of gender and respirator brand on the association of respirator fit with facial dimensions. J. Occup. Environ. Hyg. 4: 923–930.

142. Riedar K., Oestenstad H., Dillion K., Perkins L.L. (1990). Distribution of faceseal leak sites on a half-mask respirator and their association with facial dimensions. Am. Ind. Hyg. Assoc. J. 51: 285–290.

143. Oestenstad R.K., Perkins L.L. (1992). An assessment of critical anthropometric dimensions for predicting the fit of a half-mask respirator. Am. Ind. Hyg. Assoc. J. 53: 639–644.

144. Zhuang Z., Coffey C.C., Berry Ann, R. (2005). The effect of subject characteristics and respirator features on respirator fit. J. Occup. Environ. Hyg. 2: 641–649.

145. Oestenstad R.K., Elliott L.J., Beasley T.M. (2007). The effect of gender and respirator brand on the association of respirator fit with facial dimensions. J. Occup. Environ. Hyg. 4: 923–930.

146. Zhuang Z., Guan J., Hsiao H., Bradtmiller B. (2004). Evaluating the representativeness of the LANL respirator fit test panels for the current U.S. civilian workers. J. Int. Soc. Respir. Prot. 21: 83–93.

147. Zhuang Z., Bradtmiller B., Shaffer R.E. (2007). New respirator fit test panels representing the current U.S. civilian work force. J. Occup. Environ. Hyg. 4: 647–659.

148. Zhuang Z., Bradtmiller B. (2005). Head and face anthropometric survey of U.S. respirator users. J. Occup. Environ. Hyg. 2: 567–576.

149. International Organization for Standardization (2010). Respiratory Protective Devices: Human Factors; Part 2: anthropometrics. ISO TS 16976-2.

150. Canadian Standards Association (2002). Specifications for Facially Featured Headforms. CSA Z262.6-02.

151. American Society for Testing and Materials (2009). Standard Specification for Headforms. ASTM F2220-02.

152. British Standards Institution (2006). Headforms for Use in the Testing of Protective Helmets. BS EN 960: 2006.

153. American Society for Testing and Materials (2004). Standard Practices for Qualitatively Evaluating the Comfort, Fit, Function, and Integrity of Chemical-Protective Suit Ensembles. ASTM F1154-99a.

154. Scanlan S., Roberts W., McCallum R., Robinson D. (2004). A Dexterity and Tactility Evaluation of the Australian Nuclear Biological Chemical (NBC) Glove. Defense Science and Technology Organization (Australia) DSTO-TN-0576.

155. Greiner T.M. (1991). Hand Anthropometry of U.S. Army Personnel. Technical Report Natick/TR-92/011. U.S. Army.

156. Hidson D. (1991). Development of a Standard Anthropometric Dimension Set for Use of Computer-Aided Glove Design. Technical Note 91–22. Defense Research Establishment, Ottawa, ON, Canada.

157. Garrett J.W. (1971). The adult human hand: some anthropometric and biomechanical considerations. Human Factors: J. Hum. Factors Ergon. Soc. 13: 117–131.

158. Wunderlich R.E.; Cavanagh P.R. (2001). Gender differences in adult foot shape: implications for shoe design. Medi. Sci. Sports Exercise 33: 605–611.

159. British Standards Institution (2003). Respiratory Protective Devises: Full Face Masks—Requirements, Testing, Marking. BS EN 136: 1998.

160. Johnson A.T., Weiss R.A., Grove C. (1992). Respirator performance rating tables for mask design. Am. Ind. Hyg. Assoc. J. 53: 193–202.

161. International Organization for Standardization (in preparation). Respiratory Protective Devices: Human Factors; Part 7: Hearing and Speech. ISO TS 16976-7.

162. Plummer R., Stobbe T., Ronk R., Myers W., Kim H., Jaraiedi M. (1985). Manual dexterity evaluation of gloves used in handling hazardous materials. Proc. Hum. Factors Soc. 29: 819–823.

163. Dianat I., Haslegrave C., Stedmon A. (2010). Short and longer duration effects of protective gloves on hand performance capabilities and subjective assessments in a screw-driving task. Ergonomics 53: 1468–1483.

164. Barile F.A. (2008). Principles of Toxicity Testing. Boca Raton, FL: CRC Press.

165. North Atlantic Treaty Organization (2009). Physiological Requirements for Aircrew CBRN Defence Assemblies Used in Flight. STANAG 3943 (Ed. 3). Brussels: Military Agency for Standardization. NATO unclassified.

166. North Atlantic Treaty Organization (1990). Mechanical Environmental Conditions to which Material Intended for Use by NATO Forces Could be Exposed. STANAG 2914. Brussels: Military Agency for Standardization. NATO unclassified.

167. North Atlantic Treaty Organization (2001). Mechanical Environmental Description. AECP-1 (Ed. 2). Brussels: Military Agency for Standardization. NATO unclassified.

168. North Atlantic Treaty Organization (1999). NATO Levels of Packaging. STANAG 4280 PPS (Ed. 2). Brussels: Military Agency for Standardization. NATO unclassified.

169. Zhou W., Reddy N., Yang Y. (2005). Overview of protective clothing. In: Scott R.A. (Ed.), Textiles for Protection, pp. 3–30. Boca Raton, FL: CRC Press.

170. Altman M., Enache S., Magyar M. (2007). Barrier laminates and articles made therefrom. U.S. Patent App. 20070021023.

171. Blake H., Dacey P. (2007). Fire resistant barrier laminates and articles made therefrom. U.S. Patent Appl. 20070037020.

172. Horrocks A.R., Price D. (2001). Fire Retardant Materials. Cambridge, UK: Woodhead Publishing.

173. Hearle J.W.S. (2005). Fibres and fabrics for protective textiles. In: Scott R.A. (Ed.), Textiles for Protection, pp. 118–150. Boca Raton, FL: CRC Press.

174. Forsberg K., Keith L.H. (1999). Chemical Protective Clothing Perfomance Index, 2nd ed. Toronto: Wiley.

175. Ivanov V.S. (1992). Radiation Chemistry of Polymers (New Concepts in Polymer Science Series). Utrecht: VSP.

176. North Atlantic Treaty Organization (2001). Specification for Paints and Paint Systems, Resistant to Chemical Agents and Decontaminants, for the Protection of Land Military Equipment. STANAG 4360 (Ed. 1). Brussels: Military Agency for Standardization.

177. Thomas, S., Zaikov, G. (Eds.) (2008). Polymer Nanocomposite Research Advances. New York: Nova Science Publishers.

178. Krishnamoorti R., Vaia R.A., Giannelis E.P. (1996). Structure and dynamics of polymer-layered silicate nanocomposites. Chem. Mater. 8: 1728–1734.

179. Garces J.M., Moll D.J., Bicerano J., Fibiger R., McLeod D.G. (2000). Polymeric nanocomposites for automotive applications. Adv. Mater. 12: 1835–1839.

180. Yano K., Usuki A., Okada A. (1997). Synthesis and properties of polyimide–clay hybrid films. J. Polym. Sci. A: Polym. Chem. 35: 2289–2294.

181. Giannelis E.P. (1996). Polymer layered silicate nanocomposites. Adv. Mater. 8: 29–35.

182. Xu R.J., Manias E., Snyder A.J., Runt J. (2001). New biomedical poly(urethane urea)–layered silicate nanocomposites. Macromolecules 34: 337–339.

183. LeBaron P.C., Wang Z., Pinnavaia T.J. (1999). Polymer-layered clay composites. Appl. Clay Sci. 15: 11–29.

184. Gusev A.A., Lusti H.R. (2001). Rational design of nanocomposites for barrier applications. Adv. Mater. 13: 1641–1643.

185. Duncan S., Pedersen D. (2005). Next generation chemical biological protective materials for the Canadian military. 1st Int. Conf. on Individual Protective Equipment, Bath, UK.

186. Grove C. (2005). Composite materials and filtration media. 1st Int. Conf. on Individual Protective Equipment, Bath, UK.

187. Wilusz E., Zukas W.X., Fekete Z.A., Karasz F.E. (2004). Ion beam modification of permselective membranes for chemical biological protective clothing. 8th Int. Symp. on Protection Against Chemical and Biological Warfare Agents, Gothenburg, Sweden, June.

188. Potluri P., Needham P. (2005). Technical textiles for protection. In: Scott R.A. (Ed.), Textiles for Protection, pp. 151–175. Boca Raton, FL: CRC Press.

189. Ashley S. (2003). X-ray proofing. Sci. Am., May.

190. Corcoran E.C., Forrest W., Horton R., Kelly D.G., Mattson K., McDonald C., Nielsen K.S., Topping K., Weir R.D., Yonken A. (2011). A performance study of radio opaque personal protective fabrics for the reduction of transmittance of gamma-rays and neutrons. J. Radioanal. Nucl. Chem. 291: 251–256.

191. Hsieh A.J., Yi J., Pate B.D., Boyce M.C. (2006). Unraveling Dynamic Mechanical Deformation in Segmented Polyurethanes: from High Strain-Rate Hardening to Complete Fold Recovery. U.S. Pentagon Report A837184.

192. Mukhopadhyay A., Midha V.M. (2008). A review on designing the waterproof breathable fabrics: Part I. Fundamental principles and designing aspects of breathable fabrics. J. Ind. Text. 37: 225–262.

193. Lee, S., Obendorf S. K. (2001). A statistical model to predict pesticide penetration through nonwoven chemical protective fabrics. Text. Res. J. 71: 1000–1009.

194. Tan K., Obendorf S.K. (2006). Surface modification of microporous polyurethane membrane with poly(ethylene glycol) to develop a novel membrane. J. Membr. Sci. 274: 150–158.

195. Lomax G.R. (1985). The design of waterproof, water vapor–permeable fabrics. J. Ind. Text. 15: 40–66.

196. Gore R.W., Allen S.B. (1980). Waterproof laminate. U.S. Patent 4,194,041.

197. Farnworth B., Lotens W.A., Wittgen P.P.M.M. (1990). Variation of water vapor resistance of microporous and hydrophilic films with relative humidity. Text. Res. J. 60: 50–53.

198. Jain M.K., Wu H.Sh. (2005). Waterproof and high moisture vapor permeable fabric laminate. U.S. Patent Appli. 20050014432.

199. Beadle B, Garrett G., Clarke A. (2006). Polymer material. World Patent WO/2006/035236.

200. Willis C., Brewer S., Stone C., Dennis M., Beadle B. (2004). Selectively permeable membranes based on interpolymer complexes, 8th Int. Symp. on Protection Against Chemical and Biological Warfare Agents, Gothenburg, Sweden, June.

201. Aquanarijn Micro Filtration BV. http://www.microfiltration.nl/polym_microsieves.php.

202. Tong H.D., Jansen H.V., Gadgil V.J., Bostan C.G., Berenschot E., van Rijn C.J.M., Elwenspoek M. (2004). Silicon nitride nanosieve membrane. Nano Lett., 4: 283–287.

203. Jahn S.F., Engisch L., Baumann R.R., Ebert S., Goedel W.A. (2009). Polymer microsieves manufactured by inkjet technology. Langmuir 25: 606–610.

204. Gironès M., Akbarsyah I.J., Nijdam W., van Rijn C.J.M., Jansen H.V., Lammertink R.G.H., Wessling M. (2006). Polymeric microsieves produced by phase separation micromolding. J. Membr. Sci. 283: 411–424.

205. Bikel M., Çulfaz P.Z., Bolhuis-Versteeg L.A.M., Garduño Pérez J., Lammertink R.G.H., Wessling M. (2010). Polymeric microsieves via phase separation microfabrication: process and design optimization. J. Membr. Sci. 347: 93–100.

206. Prenen A.M., van der Werf J.C.A., Bastiaansen C.W.M., Broer D.J. (2009). Monodisperse, polymeric nano- and microsieves produced with interference holography. Adv. Mater. 21: 1751–1755.

207. Van Rijn C.J.M. (2004). Nano and Micro Engineered Membrane Technology. Amsterdam: Elsevier.

208. Angadjivand S.A., Jones M.E., Meyer D.E. (2000). Electret filter media. U.S. Patent 6,119,691.

209. Rengasamy S., Zuang Z., Roberge R., Shaffer R.E. (2010). Particulate respiratory protection: overview, emerging issues and research needs. In: Argosyan V.E. (Ed.), Protective Devises: Types, Uses and Safety, pp. 131–160. New York: Nova Science Publishers.

210. Jasper W.J., Mohan A., Hinestroza J., Barker R. (2007). Degradation processes in corona-charged electret filter-media with exposure to ethyl benzene. J. Eng. Fibers Fabrics 2: 19–24.

211. Rengasamy S., Eimer B., Shaffer R. (2009). Comparison of nanoparticle filtration performance of NIOSH-approved and CE marked filtering facepiece respirators. Ann. Occup. Hyg. 53: 117–128.

212. Kallio G.A., Dietz P.W., Gutfinger C. (1981). Filtration efficiencies in electrostatically augmented granular beds. Environ. Int. 6: 415–422.

213. Graham K., Ouyang M., Raether T., Grafe T., McDonald B., Knauf P. (2002). Polymeric nanofibers in air filtration applications. Proc. 15th Annual Technical Conference and Expo of the American Filtration and Separations Society, Galveston, TX.
214. Gibson P., Schreduer-Gibson H., Rivin D. (2001). Transport properties of porous membranes based on electrospun nanofibers. Colloids Surf. A: Physicochem. Eng. Aspects 187–188: 469–481.
215. Wang J., Kim S.C., Pui D.Y.H., (2008). Investigation of the figure of merit for filters with a single nanofiber layer on a substrate. J. Aerosol Sci. 39: 323–334.
216. Huang Z.-M., Zhang Y.-Z., Kotaki M., Ramakrishna S. (2003). A review on polymer nanofibers by electrospinning and their applications in nanocomposites. Compos. Sci. Technol. 63: 2223–2253.
217. Subbiah T., Bhat G.S., Tock R.W., Parameswaran S., Ramkumar S.S. (2005). Electrospinning of nanofibers. J. Appl. Polym. Sci. 96: 557–569.
218. Burger C., Hsiao B.S., Chu B. (2006). Nanofibrous materials and their applications. Annu. Rev. Mater. Res. 36: 333–368.
219. Grafe T.H., Graham K.M. (2003). Nanofiber webs from electrospinning. 5th Int. Conf. on Nonwovens in Filtration, Stuttgart, Germany.
220. Commercial nanofibre centrifuge technology. Smart Texti. and Nanotechnol. June 2011.
221. Acatay K., Simsek E., Yang C.O., Menceloglu Y.Z. (2004). Tunable, superhydrophobically stable polymeric surfaces by electrospinning. Angew. Chemi. Int. Ed. 43: 5210–5213.
222. Gleason K.K., Rutledge G.C., Gupta M., Ma M., Mao Y. (2007). Superhydrophobic fibers produced by electrospinning and chemical vapor deposition. U.S. Patent Appl. 20070237947.
223. Li H.-W., Wu C.-Y., Tepper F., Lee J.-H., Lee C.N. (2009). Removal and retention of viral aerosols by a novel alumina nanofiber filter. J. Aerosol Sci. 40: 65–71.
224. March H. (Ed.) (2001). Activated Carbon Compendium. Kidlington, UK: Elsevier Science.
225. Gudgin Dickson E.F., Black R., Cordoneanu A., Evans M., Hordry N., Iutzi R., Cunningham N. (2007). Novel carbon-based materials for respiratory protection. Proc. IPE 2007, Melbourne, Australia.
226. Liang C., Dai S., Guiochon G. (2003). A graphitized-carbon monolithic column. Anal. Chem. 75: 4904–4912.
227. Taguchi A., Smått J.-H., Lindén M. (2003). Carbon monoliths possessing a hierarchical, fully interconnected porosity. Adv. Mater. 15: 1209–1211.
228. Gadkaree K.P. (1998). Carbon honeycomb structures for adsorption applications. Carbon 36: 981–989.
229. Cunningham N., Evans M., Bang S. (2008). Production of carbonaceous porous bodies for use in filtration systems, Canadian Patent Appl. 2,640,893 2008/10/12, U.S. Patent Appl. 12/285,399 2008/10/03.
230. Lozano-Castelló D., Cazorla-Amorós D., Linares-Solano A., Quinn D.F. (2002). Activated carbon monoliths for methane storage: influence of binder. Carbon 40: 2817–2825.
231. Liang S.H.C., Harrison B.H., Pagotto J. (1995). Organic amine-impregnated activated carbon. U.S. Patent 4,531,953.

232. Moermans B., Beuckelaer W. de, Vankelecom I.F.J., Ravishankar R., Martens, J.A., Jacobs P.A. (2000). Incorporation of nano-sized zeolites in membranes. Chem. Commun. 2000: 2467–2468.

233. Edwards J.V., Buschle-Diller G., Goheen S. (Eds.) (2006). Modifed Fibers with Medical and Specialty Applications. Dordrecht, The Netherlands: Springer.

234. Giletto A., White W., Cisar A.J., Hitchens G.D., Fyffe J. (2003). Reactive decontamination formulation. U.S. Patent 6,569,353.

235. Bannard R.A.B., Casselman A.A., Purdon J.G., Mendoza C.E. (1991). Metal oximate/polyethylene glycols chemicals decontaminant system. U.S. Patent 5,071,877.

236. Cao L. (2005). Carrier-Bound Immobilized Enzymes: Principles, Application and Design. Weinheim, Germany: Wiley-VCH.

237. Russell A.J., Drevon G.F. (2003). Irreversible immobilization of diisopropylfluorophosphatase into polyurethane coatings. World Patent WO/2003/066851.

238. Rajan K.S. (1991). Enzymatic Detoxification of Chemical Warfare Agents: Immobilization of the Enzyme for Material Surfaces. U.S. Office of Naval Research DTIC Accession ADA231056.

239. Hill C.L., Delannoy L., Duncan D.C., Weinstock I.A., Renneke R.F., Reiner R.S., Atalla R.H., Han J.W., Hillesheim D.A., Cao R., Anderson T.M., Okun N.M., Musaev D.G., Geletii Y.V. (2007). Complex catalysts from self-repairing ensembles to highly reactive air-based oxidation systems. C. R. Chim. 10: 305–312.

240. Gall R.D., Hill C.L., Walker J.E. (1996). Selective oxidation of thioether mustard (HD) analogs by *tert*-butylhydroperoxide catalyzed by $H_5PV_2Mo_{10}O_{40}$ supported on porous carbon materials. J. Catal. 159: 473–478.

241. Grandcolas M., Louvet A., Keller N., Keller V. (2008). Layer-by-layer deposited titanate-based nanotubes for solar photocatalytic removal of chemical warfare agents from textiles. Angew. Chem. Int. Ed. 48: 161–164.

242. Prasad G.K., Mahato T.H., Singh B., Ganesan K., Srivastava A.R., Kaushik M.P., Vijayraghavan R. (2008). Decontamination of sulfur mustard and sarin on titania nanotubes. AIChE J. 54: 2957–2963.

243. Grandcolas M., Sinault L., Mosset F., Louvet A., Keller N., Keller V. (2011). Self-decontaminating layer-by-layer functionalized textiles based on WO_3-modified titanate nanotubes: application to the solar photocatalytic removal of chemical warfare agents. Appl. Catal. A: General 391: 455–467.

244. Allen, S.D. (2006). Light-mediated air purification system and method. U.S. Patent Appl. 20060182670.

245. Axtell H., Pease T. (2004). The incorporation of reactive nanoparticles into textiles for protection against chemical and biological warfare agents. 8th Int. Symp. on Protection Against Chemical and Biological Warfare Agents, Gothenburg, Sweden.

246. Maillard J.Y. (2002). Bacterial target sites for biocide action. J. Appl. Microbiol. Symp. Suppl. 92: 16S–27S.

247. TransTex Technologies, Inc. http://www.silverclear.ca/html/product.html.

248. Sambhy V., MacBride M.M., Peterson B.R., Sen A. (2006). Silver bromide nanoparticle/polymer composites: dual action tunable antimicrobial materials. JACS Articles (Web 07-07-2006).

249. Thomas V., Bajpai M., Bajpai S.K. (2011). In situ formation of silver nanoparticles within chitosan-attached cotton fabric for antibacterial property. J. Ind. Text. 40: 229–245.

250. Borkow G., Gabbay, J. (2004). Putting copper into action: copper impregnated products with potent biocidal activities. FASEB J. 18: 1728–1730.
251. Borkow G., Gabbay J. (2005). Copper as a biocidal tool. Curr. Medi. Chem. 12: 2163–2175.
252. Borkow G., Gabbay J. (2009). Copper, an ancient remedy returning to fight microbial, fungal and viral infections. Curr. Chem. Biol. 3: 272–278.
253. Borkow G., Gabbay J. (2006). Endowing textiles with permanent potent biocidal properties by impregnating them with copper oxide. J. Text. Apparel Technol. Manage. 5: 1–3.
254. Obendorf S.K. (2010). Improving personal protection through novel materials. AATCC Rev. July–Aug., pp. 44–50.
255. Lee S. (2009). Multifunctionality of layered fabric systems based on electrospun polyurethane/zinc oxide nanocomposite fibers. J. Appl. Polym. Sci. 114: 3652–3658.
256. Kenawy E.-R., Abdel-Hay F.I., El-Magd A.A., Mahmoud Y. (2006). Biologically active polymers: VII. Synthesis and antimicrobial activity of some crosslinked copolymers with quaternary ammonium and phosphonium groups. React. Funct. Polym. 66: 419–429.
257. Tan K., Obendorf S.K. (2007). Fabrication and evaluation of electrospun nanofibrous antimicrobial nylon 6 membranes. J. Membr. Sci. 305: 287–298.
258. Acatay K., Simsek E., Akel M., Menceloglu Y.Z. (2004). Electrospinning of low surface energy quaternary ammonium salt containing polymers and their antibacterial activity. In: Guceri S., Gogotsi Y.G., Kuznetsov V. (Eds.), Nanoengineered Nanofibrous Materials, NATO Science Series II, Vol.169, pp. 97–106. Dordrecht, The Netherlands: Kluwer Academic.
259. Chen C.Z., Cooper S.L. (2002). Interactions between dendrimer biocides and bacterial membranes. Biomaterials 23: 3359–3368.
260. Harney M.B., Pant R.R., Fulmer P.A., Wynne J.H. (2009). Surface self-concentrating amphiphilic quaternary ammonium biocides as coating additives. ACS Appl. Mater. Interf. 1: 39–41.
261. Ohayon D., Lee S., Wander J., Wu C.Y., Messier P.J. (2005). Microbiological filtering performance of NBC individual protection canisters. 1st Int. Conf. on Individual Protective Equipment, Bath, UK.
262. Lee J.H., Wu C.-Y., Wysocki K.M., Farrah S., Wander J. (2008). Efficacy of iodine-treated biocidal filter media against bacterial spore aerosols. J. Appl. Microbiol. 105: 1318–1326.
263. Ratnesar-Shumate S., Wu C.-Y., Wander J., Lundgren D., Farrah S., Lee J.-H., Wanakule P., Blackburn M., Lan B.-F. (2008). Evaluation of physical capture efficiency and disinfection capability of an iodinated biocidal filter medium. Aerosol Air Qual. Res. 8: 1–18.
264. Rengasamy S., Fisher E., Shaffer R.E. (2010). Evaluation of the survivability of MS2 viral aerosols deposited on filtering face piece respirator samples incorporating antimicrobial technologies. Am. J. Infect. Control 38: 8–17.
265. Thorn R.M.S., Austin A.J., Greenman J., Wilkins J.P.G., Davis P.J. (2009). In vitro comparison of antimicrobial activity of iodine and silver dressings against biofilms. J. Wound Care 18: 343–346.

266. Williams J.F., Suess J.C., Cooper M.M., Santiago J.I., Chen T.-Y., Mackenzie C.D., Fleiger C. (2006). Antimicrobial functionality of healthcare textiles: current needs, options, and characterization of *N* halamine–based finishes. Res. J. Text. Apparel 10: 1–11.

267. Liang J., Chen Y., Barnes K., Wu R., Worley S.D. (2006). *N*-Halamine/quat siloxane copolymers for use in biocidal coatings. Biomaterials 27: 2495–2501.

268. Tuinman, I. (2006). Electrospinning of antimicrobial polymers into nanofibrous protective layers. Int. Nonwovens Tech. Conf. 2006, Houston, TX.

269. Eby D.M., Farrington K.E., Johnson G.R. (2008). Combination Antimicrobial Nanocomposite Materials for Neutralization of Biological Threat Agents. U.S. Pentagon Report A131784.

270. O'Hanlon S.J., Enright M.C. (2009). A novel bactericidal fabric coating with potent in vitro activity against meticillin-resistant *Staphylococcus aureus* (MRSA). Int. J. Antimicrob. Agents 33: 427–431.

271. Van Langenhove L., Puers R., Matthys D. (2005). Intelligent textiles for protection. In Scott, R.A. (Ed.), Textiles for Protection, Chap. 7. Boca Raton, FL: CRC Press.

272. Farid M.M., Khudhair A.M., Razack S.A.K., Al-Hallaj S. (2004). A review on phase change energy storage: materials and applications. Energy Conversion Manage. 45: 1597–1615.

273. Shim H., McCullough E.A., Jones B.W. (2001). Using phase change materials in clothing. Text. Res. J. 71: 495–502.

274. McKee K.W., Tack, D.W. (2007). Active Camouflage for Infantry Headwear Applications. CR-2007-023. Toronto, ON, Canada: Defense Research and Development Canada.

275. Porter D., Metcalfe E., Thomas M.J.K. (2000). Nanocomposite fire retardants: a review. Fire Mater. 24: 45–52.

276. Breuer O., Sundararaj U. (2004). Big returns from small fibers: a review of polymer/carbon nanotube composites. Polym. Compos. 25: 630–645.

277. Aergerter M.A., Almeida R., Soutar A., Tadanaga K., Yang H., Watanabe T. (2008). Coatings made by sol–gel and chemical nanotechnology. J. Sol-Gel Sci. Technol. 47: 203–236.

278. Horrocks A.R., Kandola B.K., Davies P.J., Zhang S., Padbury S.A. (2005). Polym. Degrad. Stabil. 88: 3–12.

279. Grand A.F., Wilkie C.A (2000). Fire Retardency of Polymeric Materials. New York: Marcel Dekker.

280. Scott R.A. (2000). Textiles in defence. In: Horricks A.R., Anand S.C. (Eds.), Handbook of Technical Textiles. Cambridge, UK: Woodhead.

281. Sitti M., Fearing R.S. (2003). Synthetic gecko foot-hair micro/nano-structures as dry adhesives. J. Adhes. Sci. Technol. 17: 1055–1073.

282. Kururay C. (2010). New elastomer possesses transparency, softness and adhesive properties. http://www.kuraray.co.jp/en/release/2010/100330.html. Accessed Oct. 17, 2011.

283. Patel P. (2011). MIT Technology Review: Super Velcro. http://www.technologyreview.com/computing/24539/?a=f. Accessed Jan. 3, 2011.

284. Duerig T.W., Pelton A.R. (1996). The use of superelasticity in medicine. Fachz. Handel Wirtsch. Tech. Wissensch: Metall. Sonderdruck Heft 9/96: 569–574.

285. Lee H.J., Michielsen S. (2007). Preparation of a superhydrophobic rough surface. J. Polym. Sci. B: Polym. Phys. 45: 253–261.

286. Brewer S.A., Willis C.R. (2008). Structure and oil repellency: textiles with liquid repellency to hexane. Appl. Surf. Sci. 254: 6450–6454.

287. Xue C.-H., Jia S.-T., Zhang J., Ma J.-Z. (2010). Large-area fabrication of superhydropho bic surfaces for practical applications: an overview. Sci. Technol. Adv. Mater. 11: 1–15.

288. Ma M., Hill R.M., Lowery J.L., Fridrikh S.V., Rutledge G.C. (2005). Electrospun poly(styrene-*block*-dimethylsiloxane) block copolymer fibers exhibiting superhydrophobicity. Langmuir 21: 5549–5554.

289. Lombardi J.L., Michielsen S., Connors R., Churchward G. (2007). Lotus LADM Based Self-Decontaminating Surfaces. U.S. Pentagon Report A620174.

290. Zhu L., Xiu Y., Xu J., Tamirisa P.A., Hess D.W., Wong C.-P. (2005). Superhydrophobicity on two-tier rough surfaces fabricated by controlled growth of aligned carbon nanotube arrays coated with fluorocarbon. Langmuir 21: 11208–11212.

291. Manca M., Cannavale A., de Marco L., Arico A.S., Cingolani R., Gigli G. (2009). Durable superhydrophobic and antireflective surfaces by trimethylsilanized silica nanoparticles-based sol–gel processing. Langmuir 25: 6357–6362.

292. Lopata E.S., Nakanishi J.S. (1996). Process for plasma-enhanced chemical vapor deposition of anti-fog and anti-scratch coatings onto various substrates. U.S. Patent 5,487,920.

293. Howarter, J.A., Youngblood J.P. (2008). Self-cleaning and next generation anti-fog surfaces and coatings. Macromol. Rapid Commun. 29: 455–466.

294. Wu L.Y.L., Soutar A.M., Zeng X.T. (2005). Increasing hydrophobicity of sol–gel hard coatings by chemical and morphological modifications. Surf. Coat. Technol. 198: 420–424.

295. Patel P.J., Hsieh A.J., Gilde G.A. (2006). Improved Low-Cost Multi-Hit Transparent Armor. U.S. Pentagon Report A470184.

296. Robinette J., O'Brien D.J., Heflin J.R., Ridley J. (2007). Nanocomposite Interphases for Improved Transparent Polymer Composite Materials. U.S. Pentagon Report A416684.

297. Tan N.C.B., Napadensky E., Sloan J.M., Ma S., Cooper S.L. (2000). Barrier Urethanes Based on Polyethylene Soft Segments. U.S. Pentagon Report A714283.

298. Mitcheson P.D., Yeatman E.M., Rao G.K., Holmes A.S., Green T.C. (2008). Energy harvesting from human and machine motion for wireless electronic devices. Proc. IEEE 96: 1457–1486.

299. Devaux E., Koncar V., Kim B., Campagne C., Roux C., Rochery M., Saihi D. (2007). Processing and characterization of conductive yarns by coating or bulk treatment for smart textile applications. Trans. Inst. Meas. Control 29: 355–376.

300. Hodgson M. (2008). From paper hats to protective devices: the design process for a very compact CBRN hood. 14th Meeting of the International for Society for Respiratory Protection, Dublin, Ireland.

301. Ducharme M. (2006).Thermal insulation of the LASA ventilated-NBC suit during immersion in cold water. Proc. SAFE Symp. 2006, pp. 292–300.

302. Ducharme M. (2006). Heat tolerance of aircrew wearing charcoal NBC coveralls versus the LASA ventilated NBC suit. Proc. SAFE Symp. 2006, pp. 301–310.

303. Bodurtha P., Gudgin Dickson E.F., Daynard D., McPherson D., Bouillon C., Harrison B., Winship J. (2008). A full system assessment of the chemical/biological protection provided by the advanced NBC protective (ANBCP) Jet and HT systems. SAFE Europe, Geneva, Switzerland.

304. Tack D.W., Colbert H.J. (2007). Soldier Integrated Headwear System: Conceptual Design Phase Summary Report. CR-2007-031. Toronto, ON, Canada: Defense Research and Development Canada.

305. Tack D.W., Vilhena P. (2007). Operational User Requirements and Priorities for a Soldier's Integrated Headwear System. CR-2007-026 Toronto, ON, Canada: Defense Research and Development Canada.

306. Tack D.W. (2006). Soldier Integrated Headwear System: System Design Process. CR 2006-301 Toronto, ON, Canada: Defense Research and Development Canada.

307. Clark R.P., de Calcina-Goff M.L. (2009). Some aspects of the airborne transmission of infection. J. R. Soc. Interf. 6: S767–S782.

308. Gardner P. (2008). Investigation of novel respirator seal concepts. 14th Meeting of the International Society for Respiratory Protection, Dublin, Ireland.

309. Duncan S., Tremblay-Lutter J., Grant T., Gudgin Dickson E. (2001). System protection performance of a next-to-skin chemical vapour protective suit. 7th Int. Symp. on Protection Against Chemical and Biological Warfare Agents, Stockholm, Sweden.

310. Duncan E.J.S., Tremblay-Lutter J.F., Gudgin Dickson E.F., Lacroix B., Grant T.L. (2006). Thin, stretchable chemical protective garment worn next-to-skin. U.S. Patent 7,062,788 B2.

311. Duncan S., Gudgin Dickson E. (2001). Degradation in whole-body vapour protection performance of air-permeable CB protective ensembles with increasing wind speed. 7th Int. Symp. on Protection Against Chemical and Biological Warfare Agents, Stockholm, Sweden.

312. Croll L.M., Smith S., Harrison B., Narayan S., Evans M., Reynolds J. (2006). Unexpected results obtained during hydrogen cyanide and cyanogen chloride testing of military carbons. 13th Meeting of the International Society for Respiratory Protection, Toronto, ON, Canada.

313. Lundgren D.A., Whitby K.T. (1965). Effect of particle electrostatic charge on filtration by fibrous filters. Ind. Eng. Chem. Process Des. Dev. 4: 345–349.

314. Backer S., Tanenhaus J. (1951). The relationship between the structural geometry of a textile fabric and its physical properties: Part III. Textile geometry and abrasion-resistance. Text. Res. J. 21: 635–654.

315. Tremblay-Lutter J., Caron P., Duncan S., Claesson O., van Houwelingen T., Stromqvist M., Gudgin Dickson E. (2004). CB*plus* combat operations uniform innovation: innovation in protection and benefits of international collaboration. 8th Int. Symp. on Protection Against Chemical and Biological Warfare Agents, Gothenburg, Sweden.

316. Cadarette B.S., Cheuvront S.N Kolka M.A., Stephenson L.A., Montain S.J., Sawka, M.N. (2006). Intermittent microclimate cooling during exercise heat stress in U.S. Army chemical protective clothing. Ergonomics 49: 209–219.

317. Yarborough P., Nelson C.N. (Eds.) (2005). Performance of Protective Clothing: Global Needs and Emerging Markets, Vol. 8. STP 1462. West Conshohocken, PA: American Society for Testing and Materials International.

318. Farnworth B., Dacey P. (2006). Gas distribution garment. U.S. Patent Application 2006/0026743A1.

319. Farnworth B., Dacey P. (2006). Gas distribution garment having a spacer element. U.S. Patent Appl. 2006/0174392.

320. Caretti D.M., Gardner P.D. (2003). Quantifying the heat stress attributable to respirator wear. J. Int. Soc. Respir. Prot. 20: 110–124.

321. Levine L., Quigley M.D., Cadarette B.S., Sawka M.N., Pandolf K.B. (1990). Physiologic strain associated with wearing toxic-environment protective systems during exercise in the heat. In: Das B. (Ed.), Advances in Industrial Ergonomics and Safety, Vol. II, pp. 897–904. London: Taylor & Francis.

322. Szlyk P.C., Sils I.V., Francesconi R.P., Hubbard R.W. (1989). Patterns of Human Drinking: Effects of Exercise, Water Temperature and Food Consumption. Report ADA206031. Natick, MA: Army Research Institute for Environmental Medicine.

323. McLellan T.M. (2008). Chemical–biological protective clothing: effects of design and initial state on physiological strain. Aviat. Space Environ. Med. 79: 500–508.

324. Duncan S., McLellan T., Gudgin Dickson E. (2011). Improving Comfort in Military Protective Clothing. In: Song, G. (Ed.), Improving Comfort in Clothing, Woodhead Publishing Series in Textiles No. 106, pp. 320–369. Cambridge UK: Woodhead.

325. Huang J.T., Huang V.J. (2007). Evaluation of the efficiency of medical masks and the creation of new medical masks. J. Int. Med. Res. 35: 213–223.

326. Lafferty E.A., McKay R.T. (2006). Physiologic effects and measurement of carbon dioxide and oxygen levels during qualitative respirator fit testing. J. Chem. Health Saf. 13: 22–28.

327. International Organization for Standardization (2002). Ergonomics of Human-System Interaction: Usability Methods Supporting Human-Centred Design. ISO/TR 16982: 2002.

328. International Organization for Standardization (2010). Ergonomics of Human-System Interaction: Specification for the Process Assessment of Human-System Issues. ISO/TS 18152:2010.

329. International Organization for Standardization (in preparation). Respiratory Protective Devices: Human Factors; Part 8: Ergonomic Factors. ISO/TS 16976-8 (draft).

330. Krausman A.S., Nussbaum M.A. (2007). Effects of wearing chemical protective clothing on text entry when using wearable input devices. Int. J. Ind. Ergonom. 37: 525–530.

331. Avon Protection, 50 Series Protective Mask. http://www.avon-protection.com/Downloads/protection-us-datasheets/50-series_low.pdf. Accessed Jan. 2, 2011.

332. North Atlantic Treaty Organization (1984). Aircrew Nuclear Flash Blindness Protection. STANAG 3830 (Ed. 1). Brussels: Military Agency for Standardization. NATO unclassified.

333. Caretti D.M., Strickler L.C. (2003). Speech intelligibility during respirator wear: influences of respirator speech diaphragm size and background noise. Am. Ind. Hyg. Assoc. J. 64: 846–850.

334. Johnson A.T., Scott W.H., Laustead C.G., Coyne K.M., Sahota M.S., Johnson M.M., Yeni-Komshian G., Caretti D.M. (2000). Communications using a telephone while wearing a respirator. Am. Ind. Hyg. Assoc. J. 61: 254–267.

335. McMahon R.W., Stemann K.M. (1991). Speech Intelligibility with an Amplified Voicemitter Designed for the M40 Series Protective Mask. USAHEL Technical Note 3-01. Aberdeen Proving Ground, MD: U.S. Army Human Engineering Laboratory.

336. North Atlantic Treaty Organization (1999). Personnel Identification while in NBC Individual Protective Equipment (IPE). STANAG 2429. Brussels: Military Agency for Standardization. NATP/PfP unclassified.

337. Gudgin Dickson E.F., Bodurtha P., Fedele P. (2005). Predicting protection performance behaviour of protective clothing against airborne contaminants: modeling and measurement. 1st Int. Conf. on Individual Protective Equipment, Bath, UK.

338. Lai M. (2007). Usability assessment and redesign of BoSS-21 user interface. B.A.Sc. thesis, University of Toronto, Toronto, ON, Canada.

339. North Atlantic Treaty Organization (2000). Nuclear, Biological, and Chemical (NBC) Defense Factors in the Design, Testing and Acceptance of Military Equipment. STANAG 4521 (Ed. 1). Brussels: Military Agency for Standardization. NATO unclassified.

340. Black S., Kapsall V., Bougourd J., Geesin F. (2005). Fashion and function: factors affecting the design and use of protective clothing. In: Scott R.A. (Ed.), Textiles for Protection, pp. 60–89. Boca Raton, FL: CRC Press.

341. Geer L., Brasser P., van Gulijk C., van der Gijp S. (2005). The power of CFD in modeling of protective equipment. 1st Int. Conf. on Individual Protective Equipment, Bath, UK.

342. Hill R.W., Barry J.J. (2004). New developments in the assessment of protective fabrics using computational models. Int. Nonwovens J., Winter, pp. 22–30.

343. Barry J.J., Hill R. W., Brasser P., Sobera M., Kleijn C., Gibson P. (2003). Computational fluid dynamics modeling of fabric systems for intelligent garment design. MRS Bull. 28: 568–573.

344. Brasser P., van Houwelingen T. (2008). Theoretical and experimental study of vapor deposition onto a dressed body part. AIChE J. 54: 844–849.

345. Sobera M.P., Kleijn C.R., Brasser P., van den Akker H.E.A. (2004). A multi-scale numerical study of the flow, heat, and mass transfer in protective clothing. Lect. Notes Comput. Sci. 3039: 637–644.

346. Lodewyckx P., Wood G.O., Ryu S.K. (2004). The Wheeler–Jonas equation: a versatile tool for the prediction of carbon bed breakthrough times. Carbon 42: 1351–1355.

347. Liu Z.G., Wang P.K. (1997). Pressure drop and interception efficiency of multifiber filters. Aerosol Sci. Technol. 26: 313–325.

348. Liu X.L.G., Wang P.K. (1996). Numerical investigation of viscous flow fields around multiber filters. Aerosol Sci. Technol. 25: 375–391.

349. Shapiro M. (1996). An analytical model for aerosol filtration by nonuniform filter media. J. Aerosol Sci. 27: 263–280.

350. Maschio C., de Arruda A.C.F. (2001). Modeling of the efficiency of fibrous filters through numerical simulation and x-ray tomography. Adv. Powder Technol. 23: 311–329.

351. Dhaniyala S., Liu B.Y.H. (2001). Theoretical modeling of filtration by nonuniform fibrous filters. Aerosol Sci. Technol. 34: 170–178.

352. Quintard M., Whitaker S. (1995). Aerosol filtration: an analysis using the method of volume averaging. J. Aerosol Sci. 26: 1227–1255.

353. Przekop R., Moskal A., Gradon L. (2003). Lattice–Boltzmann approach for description of the structure of deposited particulate matter in fibrous filters. J. Aerosol Sci. 34: 133–147.

354. Zhong W., Pan N. (2007). Aerosol filtration by fibrous filters: a statistical mechanics approach. Text. Res. J. 77: 284–289.

355. Wang Q., Maze B., Tafreshi H.V., Pourdeyhimi B. (2006). Approaches for predicting collection efficiency of fibrous filters. J. Text. Apparel Technol. Manage. 5: 1–7.

356. Fedele P.D. (1992). Model of aerosol protection offered by permeable protective garments. In: McBriaty J.P., Henry N.W. (Eds.), Performance of Protective Clothing, Vol. 4. ASTM STP 1133. Philadelphia: American Society for Testing and Materials.

357. Extrand C.W. (2004). Criteria for ultralyophobic surfaces. Langmuir 20: 5013–5018.

358. Extrand C.W. (2006). Designing for optimum liquid repllency. Langmuir 22: 1711–1714.

359. Semple S., Brouwer D.H., Dick F., Cherrie J.W. (2001). A dermal model for spray painters: Part II. Estimating the deposition and uptake of solvents. Ann. Occup. Hyg. 45: 25–33.

360. Van Houwelingen T., Gudgin Dickson E., Tremblay-Lutter J. (2004). Human–mannequin comparisons in testing NBC protective clothing. 8th Int. Symp. on Protection Against Chemical and Biological Warfare Agents, Gothenburg, Sweden.

361. Matsumura Y., Suzuki K., Ogowa A., Iijima N. (2008). Robotic torso for inward leakage test of PAPR. 14th Meeting of the International Society for Respiratory Protection, Dublin, Ireland.

362. Fiala D., Psikuta A., Jendritzky G., Paulke S., Nelson D.A., van Marken Lichtenbelt W.D., Frijns A.J.H. (2010). Physiological modeling for technical, clinical and research applications. Front. Biosci. S2: 939–968.

363. The Technical Cooperation Program Human Resources and Performance Group (2004). Human Performance Modeling in Military Simulation: Current State of the Art and the Way Ahead (2002 TTCP HUM Group Meeting). U.S. Pentagon Report A144924.

364. Anthony T.R., Flynn M.R. (2006). CFD model for a 3-D inhaling mannequin: verification and validation. Ann. Occup. Hyg. 50: 157–173.

365. Yang T., Cropper P.C., Cook M.J., Yousaf R., Fiala D. (2007). A new simulation system to predict human–environment thermal interactions in naturally ventilated buildings. Proc. Building Simulation (International Conference, Beijing), pp. 751–756.

366. Mao A. (2009). Computer-aided clothing thermal functional engineering design. Ph.D. dissertation, Polytechnic University, Hong Kong.

367. Lin G., Meng S., Wang R., Luo X., Li Y. (2011). The heat and moisture transfer balance theory of garment simulation. J. Comput. Appl. Math. 236: 980–987.

368. International Organization for Standardization (1993). Textiles: Physiological Effects—Measurement of Thermal and Water-Vapour Resistance Under Steady-State Conditions (Sweating Guarded-Hotplate Test). ISO 11092:1993.

369. Blanchard L., Santee W. (2008). Comparison of USARIEM Heat Strain Decision Aid to Mobile Decision Aid and Standard Army Guidelines for Warm Weather Training. USARIEM Technical Report T08-07.

370. Endrusick T.L., Stroschein L.A., Gonzalez R.R., US Army Research Institute of Environmental Medicine, Biophysics and Biomedical Modeling Division. http://www.usariem.army.mil/pages/ThermalManikinHistory.htm. Accessed May 7, 2011.

371. Individual Protective Ensemble Mannequin system. http://www.mtnw-usa.com/measurement-expertise/blog/tag/thermal-testing/. Accessed July 2010.

372. Psikuta A. (2009). Development of an "artificial human" for clothing research. Ph.D. dissertation, De Montfort University, Switzerland.

373. Psikuta A., Richards M., Fiala D. (2008). Single-sector thermophysiological human simulator. Physiol. Meas. 20: 181–192.

374. Fontana M., Rizzi C., Cugini U. (2004). Physics-based modelling and simulation of functional cloth for virtual prototyping applications. SM '04 Proc. 9th ACM Symp. on Solid Modeling and Applications. Aire-la-Ville, Switzerland: Eurographics Association.

375. Butler K.M. (2008). A computational method for evaluating fit. 14th Meeting of the International Society for Respiratory Protection, Dublin, Ireland.

376. Duncan E.J.S., Kournikakis B., Ho J., Hill I. (2009). Pulmonary deposition of aerosolized *Bacillus atrophaeus* in a swine model due to exposure from a simulated anthrax letter incident. Inhal. Toxicol. 21: 141–152.

377. Hatanaka T., Inuma M,. Sugibayashi K., Morimoto Y. (1990). Prediction of skin permeability of drugs: I. Comparison with artificial membrane. Chem. Pharm. Bull. (Tokyo) 38: 3452–3859.

378. Swift D.L., Proctor D.F. (1988). A dosimetric model for particles in the respiratory tract above the trachea. Ann. Occup. Hyg. 32(Suppl. 1): 1035–1044.

379. Bailey M.R. (1994). The new ICRP model for the respiratory tract. Radiat. Prot. Dosim. 53: 107–114.

380. Brown J.S. (2005). Particle inhalability at low wind speeds. Inhal. Toxicol. 17: 831–837.

381. Moody R.P., Akram M., Dickson E., Chu I. (2007). In vitro dermal absorption of methyl salicylate, ethyl parathion, and malathion: first responder safety. J. Toxicol. Environ. Health A 70: 985–999.

382. Yannas I.V., Burke J.F. (1980). Design of an artificial skin: I. Basic design principles. J. Biomed. Mater. Res. 14: 65–81.

383. Stone W.L., Paromov V., Yang H. (2007). Topical Application of Liposomal Antioxidants for Protection Against CEES Induced Skin Damage. U.S. Pentagon Report A418074.

384. Danon A., Ben-Shimon S., Ben-Zvi Z. (1986). Effect of exercise and heat exposure on percutaneous absorption of methyl salicylate. Eur. J. Clin. Pharmacol. 31: 49–52.

385. Duncan E.J.S., Brown A., Lundy P., Sawyer T.W., Hamilton M., Hill I., Conley J.D. (2002). Site-specific percutaneous absorption of methyl salicylate and VX in domestic swine. J. Appl. Toxicol. 22: 141–148.

386. Flynn M.R., Koto Y., Fent K., Nylander-French, L.A. (2006). Modeling dermal exposure: an illustration for spray painting applications. J. Occup. Environ. Hyg. 3: 475–480.

387. Byrne M.A., Bell K.F., Goddard A.J.H., Lange C., Roed J. (1995). Aerosol deposition velocity measurements to human body surfaces. J. Aerosol Sci. 26: 1315–1315(1).

388. Andersson K.G., Fogh C.L., Byrne M.A., Roed J., Goddard A.J.H., Hotchkiss S.A. (2002). Radiation dose implications of airborne contaminant deposition to humans. Health Phys. 82: 226–232.

389. Fogh C.L., Andersson K.G. (2000). Modelling of skin exposure from distributed doses. Ann. Occup. Hyg. 44: 520–532.

390. Byrne M.A. (2009). Contamination of humans: in the respiratory tract and on body surfaces. In: Andersson K.G. (Ed.), Airborne Radioactivity in Contaminated Areas: Radioactivity in the Environment, Vol. 15, pp. 77–103. Amsterdam: Elsevier.

391. Schneider T., Bohgard M., Gudmundsson A. (1994). A semiempirical model for particle deposition onto facial skin and eyes: role of air currents and electric fields. J. Aerosol Sci. 25: 583–593.

392. American Society for Testing and Materials (2011).Terminology Relating to Protective Clothing. ASTM F1494-03.

393. International Organization for Standardization (2004). Protective Clothing: Vocabulary. ISO/TR 11610:2004.

394. Japanese Institute of Standards (2006). Glossary of Terms for Respiratory Protective Devices. JIS T 8001: 2006.

395. International Organization for Standardization (2010). Respiratory Protective Devices: Terms, Definitions, Graphical Symbols and Units of Measurement. ISO 16972:2010.

396. British Standards Institution (2005). Personal Eye-Protection. Vocabulary. BS EN 165: 2005.

397. North Atlantic Treaty Organization (2003). Glossary of NBC Terms and Definitions. AAP-21(A). Brussels: Military Agency for Standardization. NATO unclassified.

398. North Atlantic Treaty Organization (1981). NBC Protective Mask and Filter Canister Screw Threads. STANAG 4155. Brussels: Military Agency for Standardization. NATO unclassified.

399. International Organization for Standardization (draft in preparation). Respiratory Protective Devices: Performance Requirements; Part 3: Thread Connection. ISO/FDIS 17420-3.

400. Gamage B., Moore D., Copes R., Yassi A., Bryce E. (2005). Protecting health care workers from SARS and other respiratory pathogens: a review of the infection control literature. Am. J. Infect. Control 33: 114–121.

401. Loeb M., Dafoe N., Mahony J., John M., Sarabia A., Glavin V., Webby R., Smieja M., Earn D.J.D., Chong S., Webb A., Walter S.D. (2009). Surgical mask vs. N95 respirator for preventing influenza among health care workers: a randomized trial. J. Am. Med. Assoc. 302: 1865–1871.

402. Lee S.-A., Grinshpun, S.A., Reponen T. (2008). Respiratory performance offered by N95 respirators and surgical masks: human subject evaluation with NaCl aerosol representing bacterial and viral particle size range. Ann. Occup. Hyg. 52: 177–185.

403. Duling M.G., Lawrence R.B., Slaven J.E., Coffey C.C. (2007). Simulated workplace protection factors for half-facepiece respiratory protective devices. J. Occup. Environ. Hyg. 4: 420–431.

404. Oberg T., Brosseau L. (2008). Surgical mask filter and fit performance. Am. J. Infect. Control 36: 276–282.

405. Alvey J. (2008). Breathing apparatus. U.S. Patent 7,380,551 B2.

406. Richardson A.W., Hofacre K.C., Gardner P.D. (2008). Technology Survey for Enhancement of Chemical, Biological, Radiological, and Nuclear Respiratory Protection. U.S. Army Edgewood Chemical Biological Center Report ECBC-CR-094.

407. International Organization for Standardization (1998). Protective Clothing: General Requirements. ISO 13688:1998.

408. Committee for European Norms (2003). Protective Clothing: General Requirements. EN 340:2003.

409. National Fire Protection Association (2005). Standard on Vapor-Protective Ensembles for Hazardous Materials Emergencies. NFPA 1991–05.

410. National Fire Protection Association (2012). Standard on Protective Ensembles for Chemical/Biological Terrorism Incidents. NFPA 1994–12.

411. Marshall S.M., Fedele P.D., Lake W.A. (2001). Guidelines for Incident Commander's use of Firefighter Protecive Ensemble (FFPE) with Self-Contained Breathing Apparatus

(SCBA) for Rescue Operations During a Terrorist Chemical Agent Incident. SBCCOM final report, Aug. 1999. Report A237783.

412. National Fire Protection Association (2007). Standard on Protective Ensembles for Structural Firefighting and Proximity Firefighting. NFPA 1971–07.

413. Dickson E.F., Tremblay J. (2010). Double plus good. CBRNE World, Winter, pp. 52–54.

414. International Organization for Standardization (2004). Protective Clothing for Use Against Solid Particulates; Part 1: Performance Requirements for Chemical Protective Clothing Providing Protection to the Full Body Against Airborne Solid Particulates (Type 5 Clothing). ISO 13982-1:2004.

415. U.S. Department of Justice (2010). CBRN Protective Ensemble Standard for Law Enformcement. NIJ Standard 0116.00.

416. Protect and detect: enhancing CBRN readiness. (2011). Asian Military Rev. 19 (online).

417. Beardall T., Bodurtha P., Daynard D., Gosselin A., Dickson E., Harrison B. (2008). Approaches to field expedient determination of protection for first responders. 14th Meeting of the International Society for Respiratory Protection Dublin, Ireland.

418. Gudgin Dickson E.F., Bodurtha P., Beardall T., Harrison B., Daynard D. (2011). Respiratory Protection Program for CBRN Protection: Summary Report. Royal Military College of Canada Report RMCC TR CPT-1101. Kingston ON, Canada: Royal Military College of Canada.

419. Favas G. (2005). End of Service Life Indicator (ESLI) for Respirator Cartridges—Part I: Literature Review. Defense Science and Technology Organization, Australia.

420. Friday D., Shrewsbury M., Peterson G.W. (2008). Residual Life Indicator for Physical Adsorption Capacity of NBC Filters—Part 1: Acetone Vapor Pulses and the Effect of Moisture Content on Retention Characteristics. U.S. Pentagon Report A806384.

421. Watson E., Caraher T.W., Bennet M.R., Roehl J.E., Greenfield M.J., Ericsson A.J. (2004). Residual life indicator. U.S. Patent 6,701,864.

422. I-bodi. http://www.i-bodi.com/Products.aspx. Accessed June 2011.

423. Gudgin Dickson E., Tremblay-Lutter J., Prior L., Brown J., Gibson-Harris J. 2010. Human–mannequin comparisons in testing NBC protective clothing: a next generation mannequin system. 10th Int. Symp., on Protection Against Chemical and Biological Warfare Agents, Stockholm Sweden.

424. American Society for Testing and Materials (2006). Standard Guide for Selection of Chemicals to Evaluate Protective Clothing Materials. ASTM F1001-99a.

425. National Institute of Occupational Safety and Health (2003). Full facepiece air purifying respirators (APR) to protect emergency response workers against CBRN agents (online).

426. International Organization for Standardization (2001). Determination of Resistance of Protective Clothing Materials to Permeation by Liquids and Gases. ISO 6529: 2001.

427. American Society for Testing and Materials (2003). Standard Classification System for Chemicals According to Functional Groups. ASTM F1186-03.

428. Wu J., Hammarstrom L.-G., Claesson O., Fangmark I., (2003). Modeling the influence of physico-chemical properties of volatile organic compounds on activated carbon adsorption capacity. Carbon 4: 1322–1325.

429. Coon P.A., Famini G.R., White W.E, Thornton R.C. (1988). Data Base Users' Guide for the Chemical Agent Simulant Data Center, Special Publication. CRDEC-SP-88032. Aberdeen Proving Ground, MD: Chemical Research Development and Engineering Center.

430. Seitzinger A.T., Grasso P.S., White W.E., Stuempfle A.K., Birenzvige A. (1990). Use of Methyl Salicylate as a Trialing Chemical Agent Simulant, U.S. Department of Defense Report A373322.

431. American Society for Testing and Materials (2001). Standard Test Method for Evaluating the Bacterial Filtration Efficiency (BFE) of Medical Face Mask Materials, Using a Biological Aerosol of *Staphylococcus aureus*. ASTM F2101-01.

432. Greenberg D.L., Busch J.D., Keim P., Wagner D.M. (2010). Identifying experimental surrogates for *anthracis* spores: a review. Invest. Genet. 1: 1–12.

433. Carrera, M., Zandomeni, R., Fitzgibbon, J., Sagripanti, J.-L. (2007). Difference between the spore sizes of *Bacillus anthracis* and other *Bacillus* species. J. Appl. Microbiol. 102: 303–312.

434. American Society for Testing and Materials (2003). Standard Test Method for Resistance of Materials Used in Protective Clothing to Penetration by Blood-Borne Pathogens Using *Phi-X174* Bacteriophage as a Test System. ASTM F 1671-03.

435. International Organization for Standardization (1997). Road Vehicles: Test Dust for Filter Evaluation; Part 1: Arizona Test Dust. ISO 12103-1:1997.

436. Martin C.W., Golden J. (2009). Simulant of radiological contamination. U.S. Patent Appl. 20090057622.

437. Fedele P., Georgopoulos P., Shade P., Lioy P., Hodgson M., Brown M.A. (2004). In-Hospital Response to External Chemical Emergencies: Personal Protective Equipment, Training, Site Operations Planning, and Medical Programs: Technical Report—Final Draft. Edgewood, MD: US Army Edgewood Chemical Biological Center.

438. Ho J., Duncan S. (2005). Estimating aerosol hazards from an anthrax letter. Aerosol Sci. 36: 701–719.

439. Reshetin V.P., Regens J.L. (2003). Simulation modeling of anthrax spore dispersion in a bioterrorism incident. Risk Anal. 23: 1135–1145.

440. Reshetin V.P., Regens J.L. (2004). Estimating *Bacillus anthracis* spore dispersion in a high-rise building. J. Eng. Phys. Thermophys. 77: 1–10.

441. Bourget S., Ohayon D., Tanelli J., Gendron A.M., Messier P.J. (2004). Microbiocidal filtering media for individual protection: facemasks and canisters, 8th Int. Symp. on Protection Against Chemical and Biological Warfare Agents, Gothenburg, Sweden.

442. Duncan S., Ho J. (2008). Estimation of viable spores in *Bacillus atrophaeus* (BG) particles of 1 to 9 μm size range. CLEAN—Soil, Air, Water 36: 584–592.

443. Heimbuch B.K., Kinney K.R., Nichols R.K., Wander J.D. (2008). The Dry Aerosol Deposition Device (DADD): An Instrument for Depositing Microbial Aerosols onto Surfaces. Air Force Research Laboratories AFRL RX-TY-TR-2008–4502.

444. Woo M.-H., Hsu Y.-M., Wu, C.-Y., Heimbuch, B., Wander, J. (2010). Method for contamination of filtering facepiece respirators by deposition of MS2 viral aerosols. J. Aerosol Sci. 41: 944–952.

445. International Atomic Energy Agency. http://www.iaea.org/Publications/Factsheets/English/ines.pdf. Accessed Dec. 31, 2011

446. Gao P., King W.P., Shaffer, R. (2007). Review of chamber design requirements for testing of personal protective clothing ensembles. J. Occup. Environ. Hyg. 4: 562–571.

447. Stull J.O., Stull G.G. (2010). Risk-Based Protective Clothing Material Permeation Criteria, Contract W91CRB-07-C-0006. Project Final Report for U.S. Technical Support Working Group.

448. Kowalski W.J., Bahnfleth W.P., Whittam T.S. (1999). Filtration of airborne microorganisms: modeling and prediction. ASHRAE Trans. Res. 4273: 4–17.

449. American Society for Testing and Materials (2011). Standard Test Method for Man-in-Simulant Test (MIST) for Protective Ensembles. ASTM F2588-11.

450. Lundgren L., Skare L., Lidén C. (2006). Measuring dust on skin with a small vacuuming sampler: a comparison with other sampling techniques. Ann. Occup. Hyg. 50: 95–103.

451. Cherrie J.W., Brouwer D.H., Roff M., Vermeulen R., Kromhour H. (2000). Use of qualitative and quantitative fluorescence techniques to assess dermal exposure. Ann. Occup. Hyg. 44: 519–522.

452. American Society for Testing and Materials (2007). Standard Test Method for Resistance of Protective Clothing Materials to Permeation by Liquids or Gases Under Conditions of Continuous Contact. ASTM F739-07.

453. Baron PA, Willeke K (Eds.) (2005). Aerosol Measurement: Principles, Techniques, and Applications 2nd ed. Hoboken, NJ: Wiley.

454. Hesketh H.E. (1977). Fine Particles in Gaseous Media. Ann Arbor, MI: Ann Arbor Science.

455. Volckens J., Peters T.M. (2005). Counting and particle transmission efficiency of the aerodynamic particle sizer. J. Aerosol Sci. 36: 1400–1408.

456. May K.R. (1973). The Collison nebulizer: description, performance and application. J. Aerosol Sci. 4: 235–238.

457. Mainelis G., Berry D., An H.R., Yao M., DeVoe K., Fennell D.E., Jaeger R. (2005). Design and performance of a single-pass bubbling bioaerosol generator. Atmos. Environ. 39: 3521–3533.

458. Andersen A.A. (1958). A new sampler for the collection sizing and enumeration of viable airborne particles. J. Bacteriol. 76: 471–484.

459. Ioannou C.J., Hanlon G.W., Denyer S.P. (2007). Action of disinfectant quaternary ammonium comnpounds against *Staphylococcus aureus*. Antimicro. Agents Chemother. 51: 296–306.

460. Richardson G.S., Winskill J., McLeod A. (2004). The development of an improved system to assess respirator performance on the battlefield, 8th Int. Symp. on Protection Against Chemical and Biological Warfare Agents, Gothenburg, Sweden.

461. Christiansen A.F., Richardson G.S., Steeenweg L., van der Gijp S., Nijboer Br. (2005). Test of military respirators in the field. 1st Int. Conf. on Individual Protective Equipment, Bath, UK.

462. Duncan E.J.S., Gudgin Dickson E.F. (2003). A new whole-body vapor exposure chamber for protection performance research on chemical protective ensembles. Am. Ind. Hyg. Assoc. J. 64: 212–221.

463. Dickson E., Duncan S., Chu I., Birkis U. (2003). Risks to first responders at a toxic chemical terrorist incident: methods for quantifying cutaneous exposure and assessing risk. Proc. CBMTS 2nd World Congress on CBRN Terrorism, Dubrovnik, Croatia.

464. Gudgin Dickson E., Fedele P., Marshall S. (2004). Quantifying cutaneous exposure to first responders responding to a liquid chemical agent incident: methods and results. 8th Int. Symp. on Protection Against Chemical and Biological Warfare Agents, Gothenburg, Sweden.

465. U.S. Code of Federal Register Title 42—Public Health, Chapter I—Public Health Service, Department of Health and Human Services, Subchapter G—Occupational Safety and Health Research and Related Activities, Part 84—Approval of Respiratory Protective Devices.

466. U.S. Department of Defense (1995). Filter Units, Protective Clothing, Gas-Mask Components and Related Products: Performance Test Methods (Version 01), Method 102.9.1.

467. British Standards Institution (2009). High Efficiency Air Filters (EPA, HEPA and ULPA): Classification, Performance Testing, Marking. BS EN 1822-1:2009.

468. British Standards Institution (2009). High Efficiency Air Filters (EPA, HEPA and ULPA): Testing Flat Sheet Filter Media. BS EN 1822-3:2009.

469. British Standards Institution (2009). High Efficiency Air Filters (EPA, HEPA and ULPA): Aerosol Production, Measuring Equipment, Particle Counting Statistics. BS EN 1822-2:2009.

470. British Standards Institution (2009). High Efficiency Air Filters (EPA, HEPA and ULPA): Determining Leakage of Filter Elements (Scan Method). BS EN 1822-4:2009.

471. British Standards Institution (2009). High Efficiency Air Filters (EPA, HEPA and ULPA): Determining the Efficiency of Filter Elements. BS EN 1822-5:2009.

472. Hill J.S., Hanley J.T., Elion J.M., Owen M.K., Clayton A.C. (2006). An Evaluation of Re-aerosolization from Contaminated Garments During Doffing. Contract W911S6-05-D-0002.

473. Donaldson K., Li X.Y., MacNee W. (1998). Ultrafine (nanometer) particle mediated lung injury. J. Aerosol Sci. 7: 533–560.

474. Waters K.M., Masiello L.M., Zangar R.C., Tarasevich B.J., Karin N.J., Quesenberry R.D., Bandyopadhyay S., Teeguarden J.G., Pounds J.G., Thrall B.D. (2009). Macrophage responses to silica nanoparticles are highly conserved across particle sizes. Toxicol. Sci. 107: 553–569.

475. Touch-Test Sensory Evaluators. http://www.touchtest.com.

476. Lafayette Instrument (2002). Purdue Pegboard Test: user instructions. http://www.si-instruments.com/files/catalog/Download%20Centre/Lafayette/Operating%20Manuals/lafayette__32020_purdue_pegboard_test_manual.pdf.

477. American Society for Testing and Materials (2010). Standard Test Method for Evaluation of Glove Effects on Wearer Hand Dexterity Using a Modified Pegboard Test. ASTM F2010/F2010M-10.

478. American Society for Testing and Materials (1996). Standard Method for Static Coefficient of Shoe Sole and Heel Materials as Measured by the James Machine. ASTM F489-96.

479. Kawabata S., Niwa M., Wang, F. (1994). Objective hand measurement of nonwoven fabrics: Part I. Development of the equations. Text. Res. J. 64: 597–610.

480. Cardello A.V., Winterhalter C., Schutz H.G. (2003). Predicting the handle and comfort of military clothing fabrics from sensory and instrumental data: development and application of new psychophysical methods. Text. Res. J. 73: 221–237.

481. Pan N. (2007). Quantification and evaluation of human tactile sense towards fabrics. Int. J. Des. Nat. 1: 48–60.

482. Marzulli F.N., Maibach H.I. (1973). Antimicrobials: experimental contact sensitization in man. J. Soc. Cosmet. Chem. 24: 399–421.

483. Oeko-Tex Standard 100. http://www.oekotex.com/OekoTex100_PUBLIC/content5.asp?area=hauptmenue&site=oekotexstandard100 &cls=02. Accessed Jan. 28, 2012.

484. American Association of Textile Chemists and Colorists (2010). Absorbency of Textiles. AATCC 79-2010.

485. Pinkus A.R., Task H.L., Hausmann M.A., Barbato M.H., Goodyear C.D. (2004). Dynamic Visual Acuity Assessment Through Visors. U.S. Pentagon Report A521824.

486. Pickering C. (2008). Fogging and respiratory simulation system. 14th Meeting of the International Society for Respiratory Protection, Dublin, Ireland.

487. House A.S., Williams C.E., Hecker M.H.L., Kryter K.D. (1965). Articulation-testing methods: consonantal differentiation with a closed-response set. J. Acoust. Soc. Am. 37: 158–166.

488. Coyne K.M., Barker D.J., Eshbaugh J.P. (2008). Automated Speech Intelligibility System for Head-borne Personal Protective Equipment: Proof of concept. U.S. Pentagon Report A100084.

489. Coyne K.M., Johnson A.T., Yeni-Komshian G.H., Dooly C.R. (1998). Respirator performance ratings for speech intelligibility. Am. Ind. Hyg. Assoc. J. 59: 257–360.

490. American National Standards Institute (2007). Methods for the Calculation of the Speech Intelligibility Index. ANSI S3.5-1997 (R2007).

491. Acoustical Society of America. http://www.sii.to/html/programs.html.

492. International Organization for Standardization (2004). Ergonomics of the Thermal Environment: Risk Assessment Strategy for the Prevention of Stress or Discomfort in Thermal Working Conditions. ISO 16265:2004.

493. International Organization for Standardization (2008). Ergonomics of the Thermal Environment: Cold Workplaces—Risk Assessment and Management. ISO 15743: 2008.

494. Holmér I. (2004). Thermal manikin history and applications. Eur. J. Appl. Physiol. 92: 614–618.

495. Richards M.G.M., Mattle N.G. (2001). A sweating agile thermal manikin (SAM) developed to test complete clothing systems under normal and extreme conditions. RTO HFM Symposium "Blowing Hot and Cold: Protecting Against Climatic Extremes." RTO-MP-076.

496. Warmé-Janville B., Pélicand J.-Y. (2001). Assessment of clothing permeation using an instrumented heated and sweating manikin. RTO HFM Symposium "Blowing Hot and Cold: Protecting Against Climatic Extremes." RTO-MP-076.

497. Fan J., Qian X. (2004). New functions and applications of Walter, the sweating fabric manikin. Eur. J. Appl. Physiol. 92: 641–644.

498. Wang F., Gao C., Kuklane K., Holmér I. (2011). Determination of clothing evaporative resistance on a sweating thermal manikin in an isothermal condition: Heat loss method or mass loss method? Ann. Occup. Hyg. 55: 775–783.

499. Forsberg K., Mansdorf S.Z. (2007). Quick Selection Guide to Chemical Protective Clothing. Hoboken, NJ: Wiley.

500. Anna D.H. (Ed.) (2003). Chemical Protective Clothing. Fairfax VA: Americal Industrial Hygiene Association.

501. American National Standards Institute (1992). Respiratory Protection Standard. ANSI Z88.2-92.

502. British Standards Institution (2005). Respiratory Protective Devices: Recommendations for Selection, Use, Care and Maintenance—Guidance Document. BS EN 529: 2005.

503. North Atlantic Treaty Organization AC/225 (Panel VII) (1995). NBC Respirator Triptych D/103. Brussels: Military Agency for Standardization. NATO confidential.

504. North Atlantic Treaty Organization (1991). The Measurement of Protection Provided to the Respiratory Tract and Eyes by Aircrew Equipment Assemblies Against NBC Agents in Particulate, Aerosol and Vapour Form. STANAG 3864. Brussels: Military Agency for Standardization. NATO unclassified.

505. North Atlantic Treaty Organization (2002). NATO Standards of Proficiency for NBC Defense. STANAG 2150 NBC (Ed. 7). Brussels: Military Agency for Standardization. NATO/PfP unclassified.

INDEX

Personal Protective Equipment for Chemical, Biological, and Radiological Hazards: Design, Evaluation, and Selection, First Edition. Eva F. Gudgin Dickson.

Printed and bound by CPI Group (UK) Ltd, Croydon, CR0 4YY
16/09/2026
14963459-0005